Building Automation
System Integration with Open Protocols

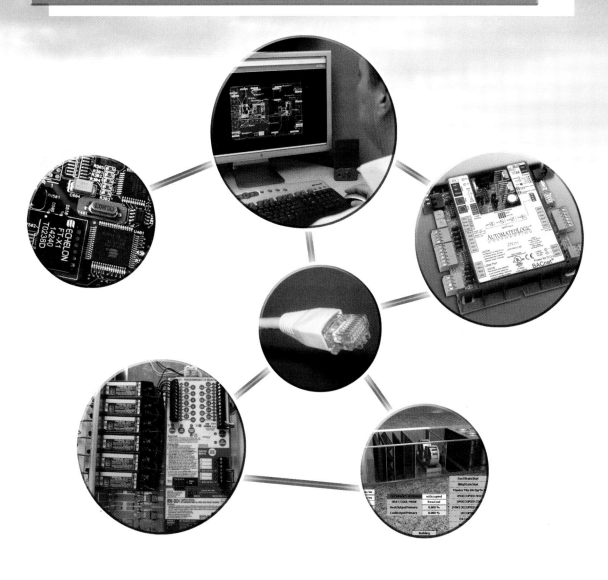

AMERICAN TECHNICAL PUBLISHERS, INC.
ORLAND PARK, ILLINOIS 60467-5756

Building Automation: System Integration with Open Protocols contains procedures commonly practiced in industry and the trade. Specific procedures vary with each task and must be performed by a qualified person. For maximum safety, always refer to specific manufacturer recommendations, insurance regulations, specific job site and plant procedures, applicable federal, state, and local regulations, and any authority having jurisdiction. The material contained is intended to be an educational resource for the user. Neither American Technical Publishers, Inc. nor the National Joint Apprenticeship and Training Committee for the Electrical Industry is liable for any claims, losses, or damages, including property damage or personal injury, incurred by reliance on this information.

American Technical Publishers, Inc., Editorial Staff

Editor in Chief:
    Jonathan F. Gosse
Vice President—Production:
    Peter A. Zurlis
Art Manager:
    James M. Clarke
Technical Editor:
    Julie M. Welch
Copy Editor:
    Diane J. Weidner
Cover Design:
    Samuel T. Tucker

Illustration/Layout:
    Jennifer M. Hines
    Samuel T. Tucker
    Mark S. Maxwell
    Thomas E. Zabinski
Multimedia Coordinator:
    Carl R. Hansen
CD-ROM Development:
    Robert E. Stickley
    Nicole S. Polak
    Daniel Kundrat
    Gretje Dahl
    Hannah A. Swidergal

Adobe, Acrobat, and Reader are either registered trademarks or trademarks of Adobe Systems Incorporated in the United States and/or other countries. ASHRAE and BACnet are registered trademarks of the American Society of Heating, Refrigerating, and Air-Conditioning Engineers, Inc. The BTL Mark is a registered trademark of BACnet International. Echelon, i.LON, LNS, LON, LonMaker, LonMark, LonTalk, LonWorks, and Neuron are registered trademarks of Echelon Corporation in the US and other countries. EnOcean is a registered trademark of EnOcean GmbH. Green Building Rating System, Leadership in Energy and Environmental Design, and LEED are either registered trademarks or trademarks of the U.S. Green Building Council. Intel and Pentium are registered trademarks of Intel Corporation or its subsidiaries in the United States and other countries. JACE, Niagara Appliance, Niagara AX, Niagara Framework, Tridium, Vykon, and Vykon Security are registered trademarks of Tridium, Inc. MasterFormat is a registered trademark of the Construction Specifications Institute, Inc. MasterSpec is a registered trademark of the American Institute of Architects. Microsoft, Windows Vista, Windows XP, Windows 2000, Windows NT, Visio, and Internet Explorer are either registered trademarks or trademarks of Microsoft Corporation in the United States and/or other countries. Modbus is a registered trademark of Schneider Automation Inc. National Electrical Code and NEC are trademarks of the National Fire Protection Association, Inc., Quincy, MA 02169. Netscape is a registered trademark of Netscape Communications Corporation in the United States and other countries. Quick Quiz and Quick Quizzes are registered trademarks of American Technical Publishers, Inc. UL is a registered certification mark of Underwriters Laboratories Inc. WebCTRL is a registered trademark of Automated Logic Corporation. ZigBee is a registered trademark of ZigBee Alliance Corporation.

© 2009 by the National Joint Apprenticeship & Training Committee for the Electrical Industry and
American Technical Publishers, Inc.
All rights reserved

1 2 3 4 5 6 7 8 9 – 09 – 9 8 7 6 5 4 3 2

Printed in the United States of America

    ISBN 978-0-8269-2012-6

 This book is printed on 10% recycled paper.

Acknowledgments

Technical information and photographs were provided by the following companies, organizations, and individuals:

Alerton
Automated Logic Corporation
Cooper Bussmann
Datastream Systems, Inc.
Delta Controls
Distech Controls
Echelon Corporation
Ernst Eder
Gamewell-FCI
Honeywell International Inc.
Lutron Electronics, Inc.
Potter Electric Signal Co.
Tridium, Inc.
Bennie Tschirky
Viconics Inc.

Contents

Section 1. Building Automation

Chapter One — **Building Automation Interoperability** _____ 2
Building Automation Communication ▪ Automated Building Systems

Chapter Two — **Control Concepts** _____ 26
Control Strategies ▪ Control Logic ▪ Supervisory Control ▪ Building System Management

Chapter Three — **Data Communication** _____ 50
Data Communication ▪ Open System Interconnection (OSI) Model ▪ Network Architecture ▪ Media Types ▪ Common Building Automation MAC Layers

Section 2. LonWorks Systems

Chapter Four — **LonWorks System Overview** _____ 84
LonWorks Development ▪ LonWorks Technology

Chapter Five — **LonWorks Network Architecture and Infrastructure** _____ 100
LonWorks Network Architectures ▪ LonWorks Network Infrastructure ▪ Infrastructure Planning

Chapter Six — **LonWorks Nodes** _____ 118
LonMark Certification ▪ LonWorks Node Hardware Components ▪ LonWorks Node Software Components ▪ LonWorks Node Types

Chapter Seven — **LonWorks Network Programming** _____ 138
LonWorks Network Programming ▪ Network Variable Bindings ▪ Device Commissioning

Chapter Eight — **LonWorks Network Testing** _____ 162
Testing and Verifying the Network ▪ Optimizing Network Performance

Chapter Nine — **LonWorks Network Maintenance** _____ 178
Network Maintenance Tasks ▪ Network Documentation

Section 3. BACnet Systems

Chapter Ten — **BACnet System Overview** _____ 190
BACnet Systems ▪ Information Architecture ▪ System Architecture ▪ Testing and Certification

Building Automation: System Integration with Open Protocols

Chapter Eleven — **BACnet Transports and Internetworking** — **216**
Typical BACnet Physical Architecture • BACnet LAN Types • MS/TP Nodes and Token Passing • BACnet Network Layer • BACnet Over IP Infrastructures

Chapter Twelve — **BACnet Basic Objects and Core Services** — **246**
BACnet Objects • Basic Objects • Special Function Objects • Object Access Services • Remote Device Management Services

Chapter Thirteen — **BACnet Alarming, Scheduling, and Trending** — **270**
Change-of-Value Notification • Alarming • Scheduling • Trending

Chapter Fourteen — **BACnet Special Applications** — **290**
Special Application Objects and Services

Chapter Fifteen — **BACnet Installation, Configuration, and Troubleshooting** — **304**
Network Tools • Installation • Configuration • Troubleshooting

Section 4. Building System Integration

Chapter Sixteen — **System Integration** — **328**
Building Automation System Example • Control Scenario: Opening the Building on a Regularly Scheduled Workday • Control Scenario: Demand Limiting

Chapter Seventeen — **Cross-Protocol Integration** — **346**
Cross-Protocol Integration • Cross-Protocol Implementations

Chapter Eighteen — **Future Trends in Building Automation** — **362**
Industry Trends • Networking Trends • Open Protocol Trends • Control Strategy Trends • Automating Existing Buildings

Appendix — **379**

Glossary — **391**

Index — **399**

CD-ROM Contents

Using the CD-ROM • Quick Quizzes® • Illustrated Glossary • Flash Cards • Media Clips • ATPeResources.com

Features

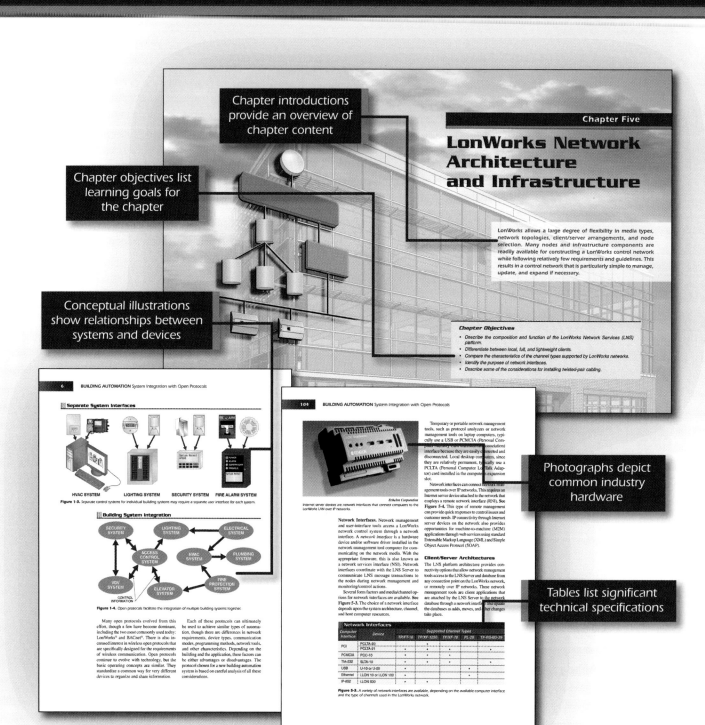

Building Automation: System Integration with Open Protocols

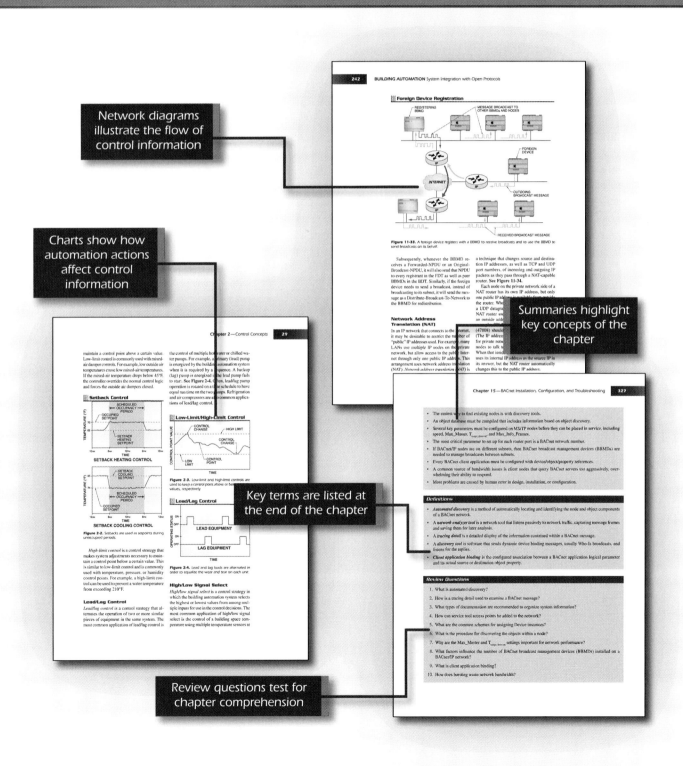

Introduction

Advanced building automation technologies include a decision-making ability within the individual control devices, which are linked by a common data communication network. These devices are known as smart or intelligent devices. All networks require a communication protocol that governs the electronic signals passed between devices to ensure that they are all speaking the same understandable language. If the structure of the protocol language is available to all manufacturers so that they can produce and market compatible control devices, then it is known as an open protocol.

Building Automation: System Integration with Open Protocols, the second book in a two-book series on building automation. The first book, *Building Automation: Control Devices and Applications,* addresses the basic functions of building systems and how devices are used to monitor and control these systems. This second book introduces the concepts of intelligent devices, automated control, and network communication utilizing open protocols. The two primary protocols for wired networks, LonWorks and BACnet, are described in detail, including information about their communication methods, information architecture, configuration, operation, and troubleshooting.

Building Automation: System Integration with Open Protocols provides a foundation of control concepts and network data communication in the first three chapters. After the LonWorks and BACnet sections, the final three chapters offer capstone coverage of previous chapter concepts and their relationships. The System Integration chapter includes a series of applications that illustrate the design, installation, and configuration of each protocol in various scenarios. Applications highlight the implementation differences between the protocols in different situations. The Cross-Protocol Integration chapter discusses strategies for incorporating multiple protocols together into a building automation system. The final chapter discusses the future of building automation, such as greater capabilities in system control and new technologies in network communication and protocol languages.

The Publisher

Contributing Writers

Building Automation: System Integration with Open Protocols

Mr. David Fisher is President of PolarSoft Inc., a Pittsburgh-based software company that specializes in BACnet software development and consulting. He was a charter voting member of ASHRAE's SPC 135P and has been very active in the development and authoring of the BACnet® standard since its inception. Mr. Fisher has over 35 years of experience in real-time software, human-interface design, and distributed direct digital control systems, and holds several patents for laboratory control systems and fiber-optic communications. Mr. Fisher attended Carnegie Mellon University, where he studied computer science and artificial intelligence.

Mr. Greg Powell is the Chief Technology Officer and CEO of Enerlon, a building services contractor in Los Angeles. Enerlon provides mechanical/electrical and network integration services for commercial, institutional, and industrial clients. Mr. Powell also delivers LonWorks training for building automation technicians and end users. Prior to Enerlon, Mr. Powell worked as senior network integration trainer for Echelon Corporation. He has trained over a thousand network integrators in the use of Echelon network tools and LonWorks control technology. Mr. Powell has a California teaching credential, is a certified LonMark professional, and holds contractor license classifications in HVAC, refrigeration, electrical, and general categories.

Mr. Jeremy J. Roberts is the Technical Director of LonMark International, a not-for-profit trade organization devoted to supporting the LonWorks networking platform. He has been working with LonWorks technology since 1993 and has headed the LonMark technical staff since 1998. He also provides LonWorks education through seminars, training, speaking engagements, and industry publications. Quarterly he authors the technical column in the LonMark magazine. Mr. Roberts holds an MBA degree in global management from the University of Phoenix and a BS degree in computer technology from Central Michigan University.

Mr. Chuck Sloup is a licensed mechanical engineer. During his career, he has worked as an application engineer for a controls contractor and a design engineer for a large engineering consulting firm specializing in hospitals, data centers, wet labs, cleanrooms, and pharmaceutical manufacturing facilities. He also owns a startup company focusing on advanced applications in controls optimization. He was president of the Nebraska Chapter of ASHRAE and was a participant in a committee that edited ASHRAE 90.1, a standard regarding energy use in buildings. Mr. Sloup holds a BS degree from the University of Nebraska.

NJATC staff contributors:
    Jim Simpson
    Assistant Director of Curriculum Development
    Technical Editor

    Marty Riesberg
    Director of Curriculum Development
    Technical Editor

Chapter One

Building Automation Interoperability

With ever-greater emphasis on energy efficiency and occupant comfort, building automation is becoming a significant factor in both new and existing commercial building operations. In particular, networked building automation is popular because it provides precise and reliable communication between control devices that can be associated with completely different building systems. It is even possible for control devices from different manufacturers to share information on the same network. Open protocols have contributed greatly to this trend, as they offer variety and flexibility that allow building owners to optimize every aspect of the building's control systems.

Chapter Objectives

- Compare the characteristics of proprietary and open protocols.
- Identify some of the leading open protocols used for building automation.
- Describe the features and benefits of control system interoperability.
- Differentiate between the network architectures used to integrate control devices together into a common building automation network.
- Evaluate some of the common strategies for integrating different building systems together within a control system.

BUILDING AUTOMATION COMMUNICATION

Advanced building automation technologies include decision-making ability within the individual control devices, which are linked by a common data communication network. **See Figure 1-1.** These devices are known as "smart" or "intelligent" control devices. These intelligent control systems are radically changing the building automation industry, allowing the integration of multiple building systems that results in lower installation costs, lower energy costs, and higher levels of comfort, safety, and security.

Building automation systems must structure and share information between control devices in a consistent and reliable way. A communication protocol governs the format, timing, and signals passed between devices to ensure that they are all speaking the same understandable language. A *protocol* is a set of rules and procedures for the exchange of information between two connected devices. The protocol is implemented in both the electronics and software of the control devices to assemble and interpret structured network messages that contain the detailed control information.

Each structured network message includes the identification of the shared variable and its value, plus other information. **See Figure 1-2.** Much of the content of a message is overhead, such as device-addressing information, message service type, error-checking information, and other parameters. The information is encoded and transmitted via a series of digital signals. Almost any type of information can be shared in this way. The standardization of protocols allows a variety of different control devices from different manufacturers to operate together, as long as they are all compatible with the same protocol.

> Automated controls are often used with other measures to maximize the energy efficiency of building systems. Changes in HVAC and lighting systems typically have the biggest impact on energy efficiency. For example, improving the insulation and sealing of a building envelope helps contain conditioned air, while light fixtures can be adjusted for optimal placement and use lamps with higher light output per watt ratings.

Building Automation

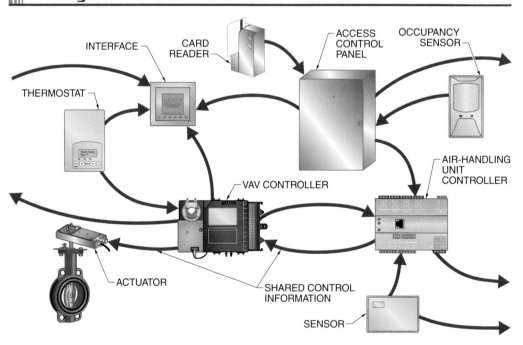

Figure 1-1. Networked building automation systems allow a variety of control devices from many different building systems to communicate and share control information.

Building automation systems using this type of communication are connected together in a way similar to a computer network. In this configuration, any device can communicate with any other device on the network. In fact, some systems can communicate over the same wiring and routing infrastructure as the building's local area network (LAN).

Proprietary Protocols

Protocols have been used by manufacturers for many years as part of proprietary building automation systems. A *proprietary protocol* is a communications and network protocol that is developed and used by only one device manufacturer. For example, early proprietary HVAC control systems were developed by many manufacturers. However, only the specific manufacturer's equipment and related proprietary software can be used with the system. The details of proprietary protocols are protected so that other manufacturers cannot market compatible devices.

Proprietary protocols also tend to be specific to certain building systems. Buildings utilizing proprietary protocols may include separate control systems for the HVAC, lighting, security, and life safety systems. It is not unusual for a building engineer to have separate workstations or interfaces for each of these systems, each requiring support from different vendors. **See Figure 1-3.**

End users typically are locked in to these vendors and have relatively few choices when it comes to adds, moves, or changes. Plus, these separate systems cannot easily share information without installing expensive and complex gateways or translation devices. Proprietary protocols lack the vendor competition (resulting in lower costs and greater variety) and potentially greater integration opportunities of open protocols.

Open Protocols

In the 1980s, a movement in the building automation industry started to create a system that allowed open access to information by devices using universal communication schemes. This introduces the idea of an open protocol. An *open protocol* is a standardized communications and network protocol that is published for use by any device manufacturer. Systems using open protocols must still use protocol-specific devices, but since the protocol standard is publicly available, the device can be made by any manufacturer. Without being tied to a specific manufacturer, open protocols have also fostered the expansion of automation to include nearly any building system. **See Figure 1-4.**

Network Messages

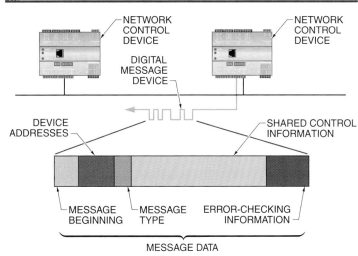

Figure 1-2. Network messages contain addressing, message type, error-checking, and other information, in addition to the control information to be shared.

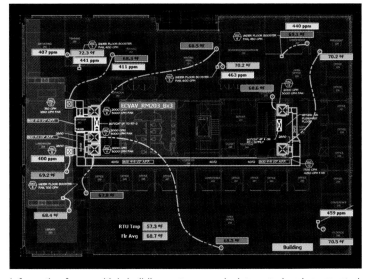

Information from multiple building systems may be integrated and represented in a single graphical interface.

Separate System Interfaces

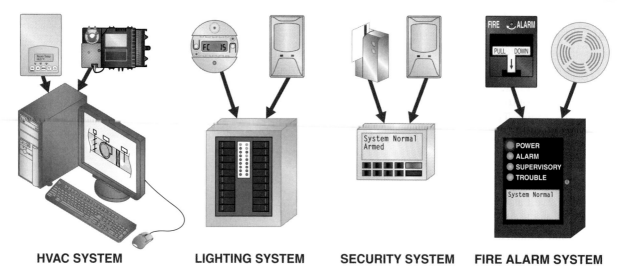

Figure 1-3. Separate control systems for individual building systems may require a separate user interface for each system.

Building System Integration

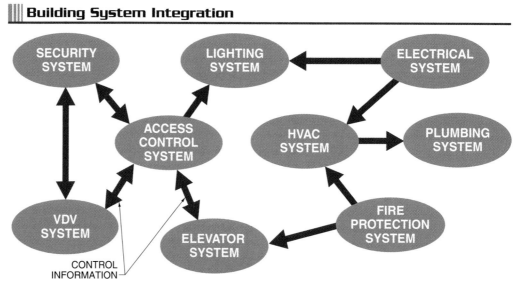

Figure 1-4. Open protocols facilitate the integration of multiple building systems together.

Many open protocols evolved from this effort, though a few have become dominant, including the two most commonly used today: LonWorks® and BACnet®. There is also increased interest in wireless open protocols that are specifically designed for the requirements of wireless communication. Open protocols continue to evolve with technology, but the basic operating concepts are similar. They standardize a common way for very different devices to organize and share information.

Each of these protocols can ultimately be used to achieve similar types of automation, though there are differences in network requirements, device types, communication modes, programming methods, network tools, and other characteristics. Depending on the building and the application, these factors can be either advantages or disadvantages. The protocol chosen for a new building automation system is based on careful analysis of all these considerations.

LonWorks Systems. LonWorks systems are based largely on a specially designed microcontroller chip, called the Neuron® chip, that is part of a LonWorks control device. The Neuron chip, and the associated LonTalk® communication protocol, were developed by Echelon Corporation as a way to embed intelligence into individual control devices as small, inexpensive electronics. The LonTalk protocol can be implemented on other microcontrollers, but the Neuron chip is the most common solution used by control device manufacturers. The ongoing support and development of the LonTalk protocol, as well as device conformance testing, is overseen by the independent LonMark International organization.

The Neuron chip manages the details of the LonTalk communication protocol, including the message transport, addressing, media access, and signaling tasks. **See Figure 1-5.** Depending on the control device's design, it may also manage the data organization and control logic. All control data is organized into standard variable types, with structures and formats defined in the protocol's standard. Each fundamental decision-making function of a control device's application program is represented as a block, which accepts input variables, performs some process or calculation, and provides output variables. Relatively simple devices contain few blocks, while complex controllers may contain many. The operation of each function block can be adjusted by changing its properties.

The sharing of control data is represented as bindings (connections) between input and output variables. These variables may be associated with the same function block, between two function blocks within the same control device, or between two function blocks in different control devices. When the connection is one of the former, the processing is internal to a control device. However, if the connection involves two separate control devices, the control information must be formatted and packaged into a network message and transmitted to the second device. **See Figure 1-6.** The LonWorks standard specifies the media, addressing, and other network requirements for efficient and reliable communication between the control devices.

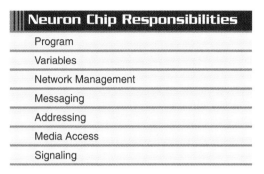

Figure 1-5. The Neuron chip manages many of the communication and application tasks in a LonWorks system.

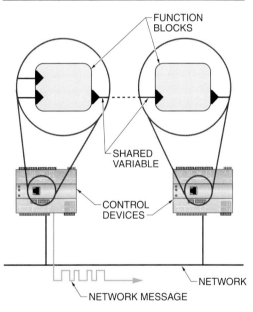

Figure 1-6. LonWorks control devices communicate over the network to share variables linking two of their function blocks.

LonWorks systems provide a fully developed platform for the design and implementation of a control system. This includes how the network is designed and how the control devices are commissioned and programmed. Software is available from Echelon for performing all of these tasks, though compatible third-party solutions are also available.

After two decades, the LonWorks platform has proven to be a major player in the building automation industry. Thousands of different products from hundreds of different manufacturers are available for nearly every control application found in commercial buildings. Millions of individual LonWorks-compatible control devices have been deployed worldwide.

BACnet Systems. The BACnet protocol was developed in 1987 by the American Society of Heating Refrigerating and Air-Conditioning Engineers (ASHRAE) as a data communication protocol for building automation and control networks. Even though the organization's focus is on HVAC applications, it made a conscious effort to develop a protocol that could potentially be used with any building system. It also maintains a process for continually collecting, evaluating, and implementing proposed changes to the protocol to improve its interoperations and expand its capabilities with new features.

The challenges facing BACnet-based automation systems are similar to LonWorks systems and to any other networked automation system. Control information must be organized into standardized units, which can then be shared over a network using standardized procedures. BACnet uses these two objectives to divide the goal of interoperability into more easily manageable tasks. **See Figure 1-7.**

Like LonWorks systems, BACnet maintains a highly structured information architecture that organizes information as software objects, each of which has a number of configurable properties. These objects represent not only control points, but also processes and control functions within the control device's application program. Information is identified in a hierarchical scheme by the identity of the device, the identity of the object within the device, the identity of the property of the object, and sometimes the identity of elements within the property. **See Figure 1-8.**

BACnet Information Hierarchy

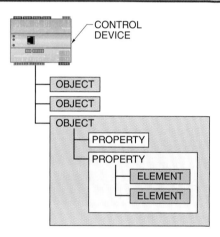

Figure 1-8. The BACnet information hierarchy divides control information into objects, each of which is composed of properties that may be composed of smaller elements.

This addressing is used to identify the sources and/or destinations of information to be shared between control devices. A variety of communication message types are defined by BACnet that standardize the desired actions requested by one control device of another device. In the language of the BACnet protocol, the objects are like nouns and these message types are like the grammar and structure of sentences.

The BACnet standard supports a variety of network types by which these messages are transmitted. These include both types that are unique to BACnet and types that are used by other applications of data networks. Each has a

BACnet Methodology

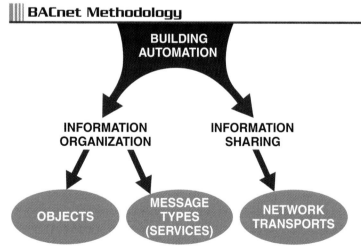

Figure 1-7. BACnet divides the challenges of building automation into two primary goals, which are addressed by BACnet-specific solutions.

different cost, performance, and other distinct characteristics. The selection allows integrators to choose the most appropriate network for each application.

BACnet is different from LonWorks in that the standard does not specify procedures or outcomes for installing, commissioning, or programming control devices. However, this has by no means prevented BACnet from being implemented in building automation systems worldwide. A number of solutions exist from third-party vendors to accomplish this integration.

Wireless Systems. The categorization of wireless building automation system protocols can be misleading. Many of the traditionally wired protocol systems, including LonWorks and BACnet systems, support wireless networks as one of their message transport options. They can even incorporate both wired and wireless segments on the same network. Therefore, they can technically be wireless systems, though they are not traditionally included in this category.

The protocols that are considered to be in the wireless protocol category, however, rely on the unique characteristics of wireless communication for their operation. Many use a networking technique that involves the relaying of messages between adjacent wireless control devices until the message reaches its ultimate destination. **See Figure 1-9.** The message content of wireless protocols is also reduced to accommodate the low-power and low-data-rate capabilities of wireless transmitters. Since "wireless" can apply to both communication and power wiring, wireless control devices must rely on self-contained power supplies, which limits the performance of the transmitters.

In situations that preclude network wiring, wireless systems can be an ideal solution. For example, wireless automation may be the best choice in an existing building if it is not feasible to run new network conductors throughout the building. Also, wireless systems eliminate improper wiring, connector, and termination issues as potential sources of communication problems.

Wireless Communication

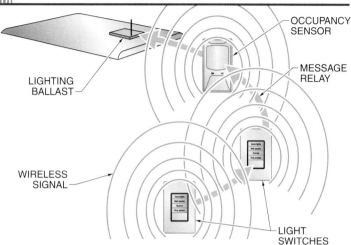

Figure 1-9. Wireless communication relies of the relaying of messages between adjacent control devices in order to reach the final destination.

However, wireless building automation systems are not as prevalent as wired systems. The reliability, security, and effectiveness of wireless systems is often questioned, though continual system advancements have addressed these concerns in many applications. Some of the more common wireless protocols, such as ZigBee®, are gradually making inroads into mainstream building automation applications.

Interoperability

Unlike proprietary systems, systems based on open protocols provide interoperability. *Interoperability* is the capability of network devices from different manufacturers and systems to interact using a common communication network and language framework. The concept of interoperability is common to many everyday systems. In control systems, interoperable devices from different manufacturers are able to work together in the same system. They speak and understand the same language, store control information in the same way, and use the same group of control actions. These are all defined in the protocol.

There are many different kinds of interoperations. For example, one device can announce the value of a control point to any other interested device, ask another device for information, or ask another device to perform an action. In order for these communications to be interoperable, each device must support the same network type and be fully compliant with the information architecture. There are two primary goals of interoperability: manufacturer and vendor integration, and building system integration.

Manufacturer and Vendor Integration. Since the protocol is open and available for any user, a manufacturer can develop a control device that can operate with any other manufacturer's device, as long as they both follow the specifications in the protocol's standard. Therefore, building automation systems using open protocols may consist of devices from a variety of manufacturers.

With multiple manufacturers offering compatible control devices, integrators can select the most appropriate device for each application. For example, a VAV controller from one manufacturer may be the best choice due to size and mounting reasons, while the thermostat with the desired temperature and humidity features are available from another manufacturer. Other factors in choosing a manufacturer's products include interoperability, cost, documentation, support, maintenance requirements, and availability.

The same approach applies to vendors offering integration, programming, maintenance, and troubleshooting support. Any vendor that has the appropriate training on an open protocol can work on the system, which provides additional choices for building owners.

The benefits of interoperability also allow an obsolete or faulty control device to be replaced by another from a different manufacturer. As long as the new device supports the same standard functions and features, the replacement is relatively a one-to-one procedure.

Building System Integration. Second, open protocols allow greater opportunities for integrating building systems together. For example, when a person enters a building, the access control system can share information about the entry and the person's identity with the lighting system, which turns on the person's office lights, and the HVAC system, which changes to an occupied mode for the person's work area. Open protocols can provide interoperability without being system-specific. For example, the standardized information infrastructure can be used to structure and share nearly any type of data, regardless of whether it is a temperature reading, damper position, or lighting level.

The combination of interoperability and the resulting variety in compatible vendors has fostered greater integration of building systems and more opportunities for automation. New controls manufacturers find it attractive to enter the control device market for a particular open protocol. Once invested in developing products for open protocols, some manufacturers expand outside of their traditional system markets. For example, a manufacturer of HVAC controls may leverage their experience in open protocol systems by adding lighting controls to their product line. The result is an increasing availability of control devices for all of the major building systems. Since they are all interoperable, these building systems can then be integrated together on a common control network.

Even if there is no need in a particular building for the systems to share information with each other, the integration of multiple building systems on a single automation system may still be desirable. **See Figure 1-10.** This allows control operations to be monitored from a unified operator interface, which is particularly desirable for a maintenance staff that is responsible for all of these very different building systems. The integrated system can also be used to log control data together.

Network Architectures

Integration and interoperability requires that all control devices reside on the same network. Messages may need to be able to travel between any two control devices, depending on the level of system integration. This can result in networks of different architectures and even the incorporation of different network types or protocols for some portions.

Integrated Interface

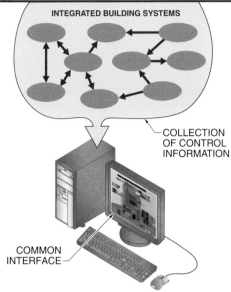

Figure 1-10. Interoperability allows a single user interface to access the information shared among integrated building systems.

Hierarchical Network Architectures

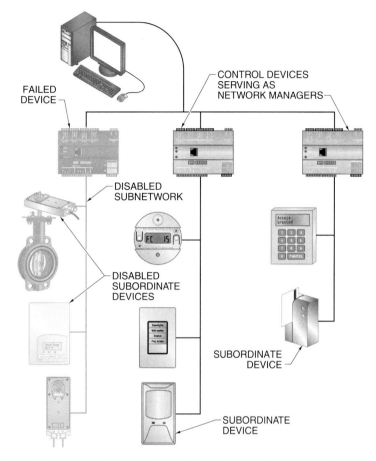

Figure 1-11. Hierarchical network architectures rely on network manager devices to organize the communication of subordinate devices below them.

Hierarchical Network Architectures. A *hierarchical network architecture* is a network configuration where control devices are arranged in a tiered network and have limited interaction with other control devices. One control device serves as a network manager for the devices attached to it in a lower tier. **See Figure 1-11.** These arrangements are also called supervised or master/slave network architectures, and network managers are also known as central or master controllers. Multiple network manager devices may be used to supervise many groups of subordinate devices.

The network manager typically includes applications for installing, configuring, and programming the network. Additional features often include scheduling, alarm, data logging, programmable control sequences, proprietary protocol drivers, and web-based user interface functions. The presence of the network manager may also be required for the subordinate devices to perform their control functions or communicate. Therefore, this architecture creates single points of failure that may render large portions of a control system, or even the entire system, inoperable if a network manager fails.

Several characteristics of hierarchical network architectures are generally considered disadvantages to this type of system. Implementing changes in a hierarchical system often requires proprietary software that can be complex and require specialized training. Network manager devices typically go off-line while program modifications are loaded, resulting in control system downtime. Network manager devices use polling to read sensor values, which is an inefficient use of the network and may miss important control events. Hierarchical architecture systems tend to be limited to specific building systems, making them challenging to integrate with other systems.

Some older control systems employed hierarchical architectures that required hardwired sensors and actuators to have home runs back to a master control panel. Some open-protocol-based systems support hierarchical architectures, but are typically implemented in flat network architectures when that arrangement is also supported.

Flat Network Architectures. A *flat network architecture* is a network configuration where control devices are arranged in a peer-to-peer way. This allows any device to communicate directly with any other device on the network without the need for network managers. **See Figure 1-12.** Control devices perform their local control functions independently while sharing data with other control devices on the network media.

Programming tools may be needed to initially commission and program the control network, but since the individual devices do not require any outside controllers to operate normally, the tool can then be disconnected from the network. It can also be reconnected as needed for moves, adds, or changes without affecting the operation of control devices unassociated with the changes.

A flat network architecture has no single point of failure. Individual control devices may still fail, but they typically do not disable any other devices in the process. A control device relying on information from a failed device is typically programmed to use default failsafe values until the problem is remedied.

Gateway Network Architectures. A *gateway network architecture* is a network configuration where a gateway is used to integrate separate control systems based on different protocols. **See Figure 1-13.** A *gateway* is a network device that translates transmitted information between different protocols. Gateway architectures are mergers of these separate control systems, which may be any combination of hierarchical and flat network architectures.

The systems may be any mix of open and proprietary protocols. System-level interoperability is achieved by translating the control information between the different control protocols. Corresponding control points must be mapped to each other during programming, typically requiring manufacturer-specific programming tools as well as knowledge of each protocol's format and structure.

AUTOMATED BUILDING SYSTEMS

Building automation may involve any or all of a building's systems. The level of sophistication of a building automation system is affected by the number of building systems integrated together. The most commonly automated building systems are the heating, ventilating, and air conditioning (HVAC) and lighting systems. However, any building system can potentially be automated and integrated, including electrical, plumbing, fire protection, security,

Flat Network Architectures

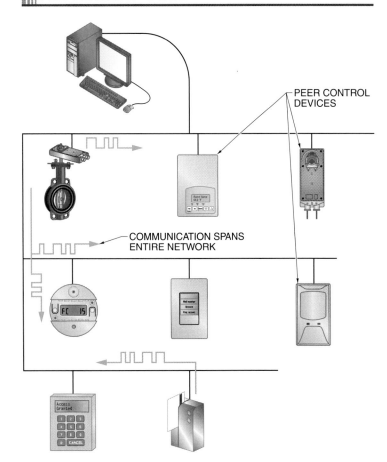

Figure 1-12. In flat network architectures, every control device can communicate directly with every other control device in a peer-to-peer way.

access control, voice-data-video (VDV), and elevator systems. It may also be possible to integrate specialized systems that are unique to certain buildings.

Electrical Systems

A building's electrical system distributes and controls the flow of electricity from the utility's service entrance to points of use throughout the building and surrounding property. In a typical commercial building, most electricity is used to operate the building's equipment, especially HVAC equipment and lighting fixtures, in order to maintain a comfortable indoor environment for the occupants. Other significant loads may include elevators, large computer networks, and material handling equipment.

Control of electrical systems involves minimizing the rate of electrical consumption while maintaining adequate operation of the equipment that relies on electrical power. It is becoming increasingly important to reduce energy consumption, both for financial and environmental reasons, but this should not be done at the expense of occupant comfort, security, productivity, and building operations. The application of intelligent controls helps balance these considerations.

> The system interoperability achieved in gateway network architectures is sometimes described as the "gluing together of islands of intelligence."

Gateway Network Architectures

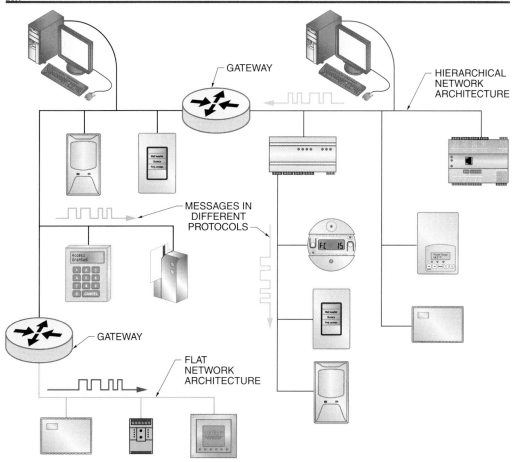

Figure 1-13. Gateway network architectures are used to connect networks using different communication protocols.

At many points within an electrical system, automation devices can be used to control the flow of electricity. The control of the electrical system involves ensuring a constant and reliable power supply to all building loads by managing building loads, uninterruptible power supplies (UPSs), and back-up power supplies. Sophisticated switching systems are used to connect and disconnect power supplies as needed to maintain building operation and occupant productivity.

One of the most important and common applications of building automation in electrical systems is demand limiting, also known as load shedding. This strategy monitors the system for high demand and/or high electrical rates (if rates change during the day) and shuts off or reduces loads on designated noncritical circuits in order to reduce the building's overall electrical consumption. **See Figure 1-14.** The most commonly affected loads are HVAC and lighting circuits.

Uninterruptible power supplies (UPSs) and back-up power supplies are employed to ensure reliable electrical power is always available, regardless of possible utility outages. These are especially important for enterprises relying on information technology and security systems. Building automation control devices are used to detect a power outage or power quality problem and initiate the transfer of some or all of the building's electrical circuits to an alternate power source, such as an on-site engine generator.

Similarly, alternate power sources can be managed by control devices in nonemergency situations for the benefit of the building and its owner. The electricity from power sources such as photovoltaic (PV) arrays or wind turbines can be monitored for output and power quality. For on-demand power sources, such as engine generators, an automation system can be programmed to operate them periodically in order to exercise or offset utility power during high-rate periods. In fact, utilities facing periods of especially high demand on their grid can be set up to communicate with and start up these distributed power sources, thus adding capacity to the grid. **See Figure 1-15.**

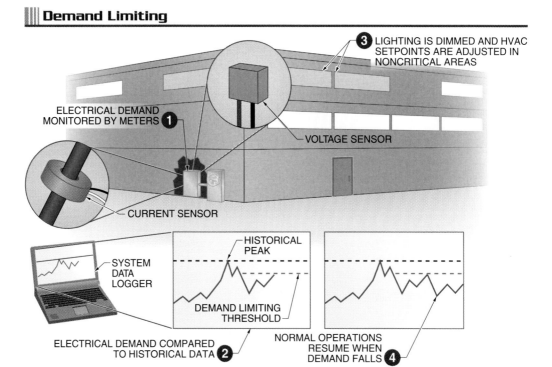

Figure 1-14. Demand limiting is a common application of electrical system automation and is used to reduce a building's rate of electrical consumption.

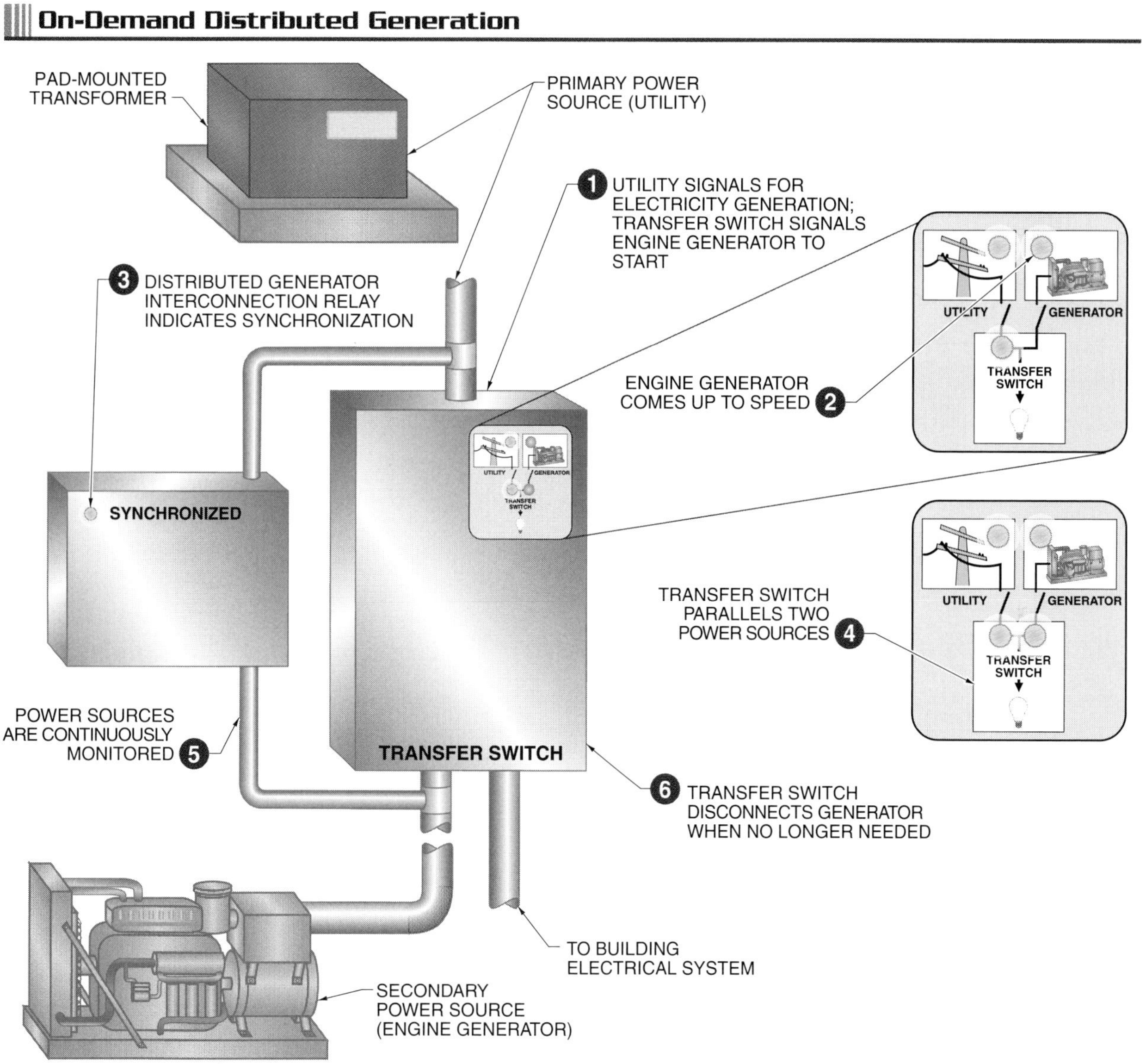

On-Demand Distributed Generation

Figure 1-15. Engine generators can be automated to respond to remote activation signals, such as those from a utility requesting extra electrical supply for the grid.

Electrical systems affect many other systems, such as lighting, security, and voice-data-video circuits since all of these use electrical power to operate. However, other than for purposes of load shedding, control of those systems is handled separately.

Lighting Systems

A lighting system provides artificial light for indoor and outdoor areas. Lighting is one of the single largest consumers of electricity in a commercial building. Therefore, lighting system controls are a major part of improving energy efficiency. Lighting system control involves switching OFF or dimming lighting circuits as much as possible without adversely impacting the productivity and safety of the building occupants or the security of the building.

As commercial buildings are typically occupied during daylight hours, a prime strategy in reducing indoor lighting use is to incorporate natural sunlight into the interior occupied areas. This must be done carefully to avoid adding to the HVAC system's cooling during the summer. Automated shading and skylight systems let in variable amounts of sunlight. These are controlled based on information from light level sensors in the occupied areas to maintain a minimum light level for the tasks in that area, but without excess heat from solar radiation. **See Figure 1-16.** When it is dark or cloudy outside, artificial lighting is used to supplement the available natural light, but it is dimmed as needed to keep electrical consumption low.

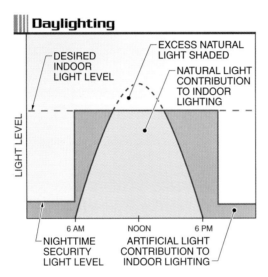

Figure 1-16. Daylighting applications control indoor lights and sometimes also shading to maintain optimal lighting levels with a maximum of natural light.

Lighting controls can automate the raising and lowering of motorized shades, or the opacity of window glass. Shading control is used to prevent too much natural light from entering the indoor space, which can cause glare on computer screens and raise the air temperature.

The other prime strategy in lighting control is based on occupancy, where the light fixture in a particular room or area is switched OFF when no one is in the space. Occupancy sensors deployed throughout the building can use one or more of a variety of technologies for detecting the presence or movement of a person in the area. The information from these occupancy sensors can also be shared with other building systems, such as HVAC and security systems, in an integrated automation system.

For outdoor lighting, lighting system controls focus on using schedules and light level sensors to switch lamps off during the day, when they are not needed. Light level sensors are particularly useful in ensuring that lighting is switched OFF promptly as soon as the sunrise provides adequate light, and switched back on at dusk. This can be more accurate than schedules alone, as the lengths of days change significantly during the year in northern regions.

Specialty lighting control systems can also produce custom lighting scenes for special applications. Lighting scenes are custom lighting designs that specify the light level of multiple lighting circuits within an area. For example, rooms typically used for instruction or presentations may include separate circuits of different types of lights, for general and spot lighting needs. Various combinations of lighting levels for each type can be saved as preset scenes, each of which can be selected from a single programmed button. **See Figure 1-17.**

HVAC Systems

The HVAC system is a building system that controls a building's indoor climate. With the lighting system, HVAC systems are typically the largest consumers of electricity in a commercial building. HVAC systems are controlled to operate at optimum energy efficiency while maintaining desired environmental conditions for the comfort and health of building occupants. The characteristics of indoor air that are monitored and/or controlled—temperature, humidity, circulation, filtration, pressurization, and ventilation—are largely interrelated. A change in one variable often affects at least one of the others. Therefore, HVAC systems can be the most complicated systems to control. They employ sophisticated control logic that must be carefully designed and tuned (calibrated). A well-automated HVAC system, however, can operate efficiently and with little manual input from occupants or maintenance personnel.

Lighting Scenes

"SET UP" SCENE

"MEETING" SCENE

"PRESENTATION" SCENE

"UNOCCUPIED" SCENE

© *2008 Lutron Electronics, Inc.*

Figure 1-17. Lighting scenes store the programmed levels of multiple lighting circuits so that certain combinations can be accessed with a single command.

Typical HVAC system operation involves controlling fans, dampers, and other control devices until the measured environment within a conditioned area matches the desired setpoints. **See Figure 1-18.** However, the setpoints can be changed dynamically according to a variety of control strategies. Most commonly, setpoint schedules are used to lower the heating or cooling loads when the entire building is unoccupied, such as at night and on weekends.

Schedules can also be used to control smaller areas, such as individual rooms. For example, if a conference room is typically used only at certain times on certain days, that schedule can be programmed into the HVAC controllers. One group of setpoints is used while the room is occupied (and to prepare the room for occupancy, since it takes some time to change room conditions), while another is used for conserving energy while the room is not occupied. If meeting times change, dynamic scheduling integrated with the occupants' software can be used to adjust the room's HVAC schedule. Setpoint changes can also be triggered by occupancy, access authorization, or any other parameter shared by other automated building systems.

HVAC control systems can be programmed for actions in specialized scenarios. For example, if smoke is detected in the building, the HVAC system is automatically shut down to avoid spreading the smoke hazard, or even used to help exhaust the smoke to outside the building. **See Figure 1-19.** Stairways can be pressurized to keep smoke out of this vital escape route.

HVAC System Control Devices

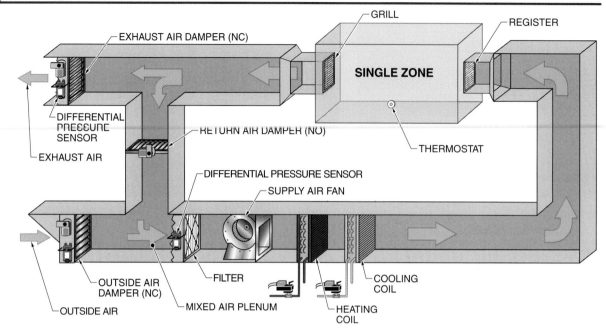

Figure 1-18. HVAC systems involve many control devices to condition and distribute air throughout a building.

Smoke Control Modes

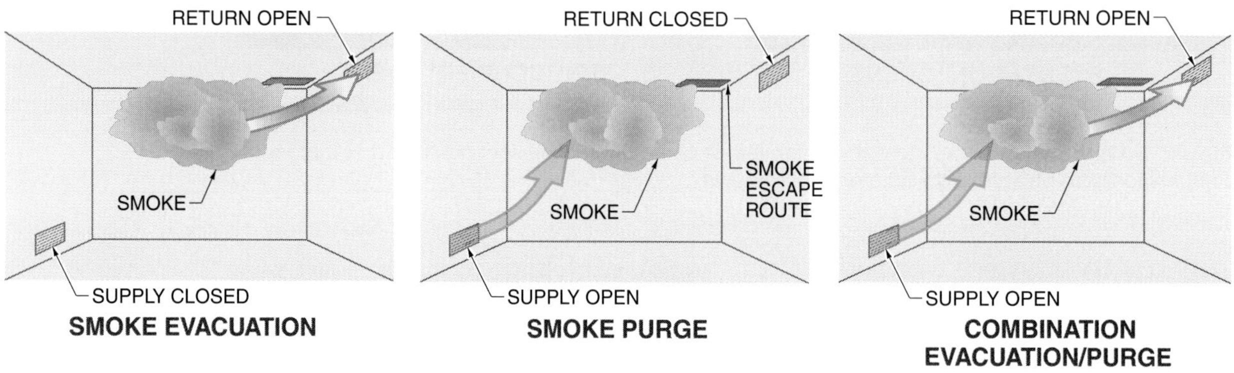

Figure 1-19. HVAC systems can initiate special sequences during fire alarms to help contain or disperse smoke.

Plumbing Systems

A plumbing system uses a system of pipes, fittings, and fixtures to convey a water supply and remove wastewater and waterborne waste. It is not common for automation systems to be integrated with the plumbing system other than for the specific uses of water in other building systems (such as boilers or fire suppression systems). However, there are a few applications in which certain parts of a commercial building's plumbing system are controlled. These applications typically involve maintaining adequate water-supply pressure and temperature.

Pressure management systems use controls to boost low-water-supply pressure. Pressure may be low due to the water source or the location of the building within a municipal water system. Sensors monitor water pressure and operate pumps as needed to maintain a consistent and adequate pressure. Pressure management may be needed at all times, so this automation application is not likely to be controlled by schedules.

Temperature management systems ensure that sufficient hot water is available quickly, at least at certain fixtures. Associated controls are used to operate hot water circulation loops to maintain hot water close to the fixtures. **See Figure 1-20.** However, since domestic hot water is probably not needed when the building is unoccupied, this system can be controlled by the same schedules used to adjust the HVAC, lighting, and other systems for nights, weekends, or holidays.

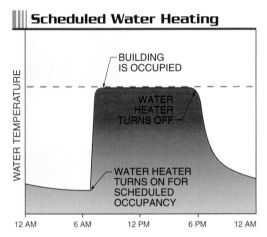

Figure 1-20. Water heating can be automated according to an anticipated occupancy schedule, reducing unnecessary energy use.

Fire Protection Systems

A fire protection system may include both fire alarm systems and fire suppression systems. Fire protection systems automatically sense fire hazards, such as smoke and heat, and alert building occupants to the dangers via strobes, horns, and other devices. They also monitor their own devices for any wiring or device problems that may impair the system's proper operation during an emergency. Fire suppression systems attempt to control the spread of a fire by applying a suppression agent in the area.

Fire protection systems are highly regulated due to their role as a life safety system. For this reason, fire alarm systems are only integrated with other building systems through the fire alarm control panel (FACP). **See Figure 1-21.** The fire-sensing devices and output devices are on their own network, which is connected only to the FACP. This means that fire protection systems are not typically integrated with building automation networks in the same way as some other building systems.

However, many FACPs include special features that enable integration with other automated systems. Some include special output connections that can be used to share fire alarm signals with other systems that have special functions during fire alarms. For example, HVAC and elevator controllers may each have a fire alarm input that is used to activate special operating modes. FACPs may also control fire doors, which are either unlocked (if normally locked) to ensure accessible escape routes, or closed (if normally held open) to keep fire and smoke from spreading quickly through hallways.

Gamewell-FCI

Fire alarm control panels (FACPs) are available for different size systems and with different types of interfaces.

Security Systems

A security system protects the building and/or property by detecting and deterring intrusion, theft, and vandalism. Security systems are similar to fire protection systems in that they are typically implemented as separately designed package systems and allow connection with other building systems only at the security control panel.

Fire Alarm Systems

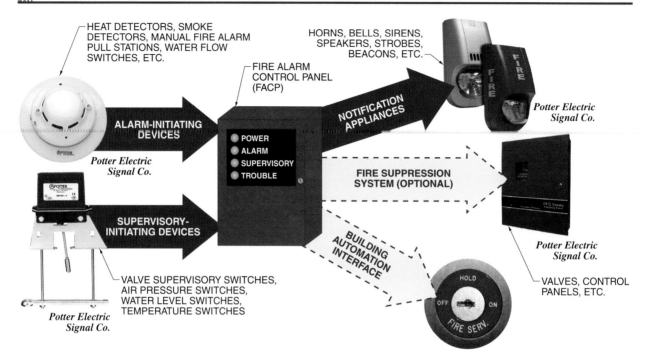

Figure 1-21. Fire alarm systems are typically connected to a building automation network only through the fire alarm control panel (FACP).

Honeywell International Inc.
Security systems are often designed to communicate with central offices, which are staffed 24 hr a day, in order to alert building owners and local authorities to any security incidents.

Security systems make heavy use of occupancy sensors, which can also be used to control other systems based on the presence of persons in the area. It is sensible to want a single group of occupancy sensors to share detection information with all of these systems, but this may or may not be possible. Being on a shared network could make a security-based control device an attractive target for a computer-based attack by an intruder. Sensors intended for systems with low security importance, such as lighting, may not have the encryption and authentication features appropriate for working with security systems. This may not always be the case, but this should be considered when integrating security systems. In either case, the control panel typically provides special communication and simple output connections that can be used to initiate special control sequences in other systems, such as unoccupied modes, access, and security monitoring with surveillance systems. **See Figure 1-22.**

Access Control Systems

An access control system is used to allow only those with proper credentials access to a specific building, area, or room. Authorized personnel use keycards, access codes, or other means to verify their right to enter the restricted area. **See Figure 1-23.** Access control systems are often confused with security systems, as both protect areas through the use of personnel identification. However, while security systems are designed to detect intruders inside the protected area, access control systems are focused on strictly controlling the passage of persons into (and sometimes out of) the protected area. Furthermore, access control systems have features for specifying different access rights for each person, detailing which areas they may access and at which times.

Access control systems, through their control panels, can be integrated with other systems in much the same way as fire protection systems and security systems. In fact, due to their similarities, access control and security systems are often highly integrated by the manufacturer and may share the same control panel, keypads, and configuration software.

When a person's credentials have been verified by the access control system, the control panel can initiate the subsequent response of other systems, such as disarming a security alarm and directing the surveillance system to monitor the person's entry. This may be through simple output connections at the control panel, or via protocol-based messages.

Voice-Data-Video (VDV) Systems

Voice-data-video (VDV) systems encompass a variety of individual systems, all related to the transmission of information. Voice systems include telephone and paging systems. Data systems include computer and control device networks, but can also carry voice or video information that has been encoded into compatible data streams. Video systems include closed-circuit television (CCTV) systems. Strictly speaking, the building automation network itself is a VDV system, though the roles of VDV systems in this context are typically as separate systems that can be controlled by the devices that communicate on the building automation network.

Security Control Panel Connections

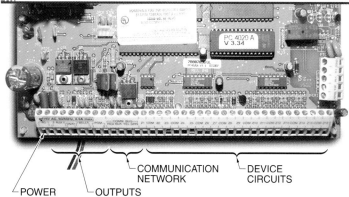

Figure 1-22. Like fire alarm systems, security systems are interfaced with a common building automation communication network at the control panel.

Access Control Authentication

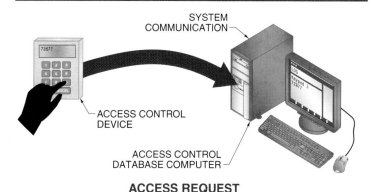

ACCESS REQUEST

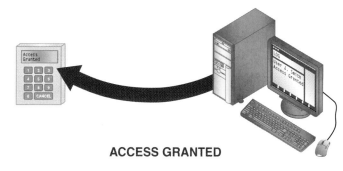

ACCESS GRANTED

Figure 1-23. Access control devices communicate with central database computers to authenticate users.

Closed-circuit television (CCTV) systems can be triggered by other building systems, such as security or access control systems, to record short clips of video.

Sound-Masking Technologies

A relatively new commercial building system technology is sound masking. Sound-masking systems work by gently raising the ambient background sound in an office environment with a comfortable, uniform sound that provides speech privacy for the occupants in the space. When paging is incorporated into the system, the message overrides the masking sound to deliver a clear, quality signal.

The sound-masking company Lencore has taken this technology to the next level by integrating sound-masking systems with networked building automation systems. This adds all of the advantages of open-protocol building automation networking to the system features, including flexible control strategies, easy expansion, and remote access. The system can be calibrated and adjusted through a networked open protocol interface.

Lencore's Sound Manager user interface centralizes the system management and can access the system either on-site or remotely. Through this web browser interface, users can initiate the system, change settings, control the system wirelessly, run diagnostics, and set timer functions. Settings for individuals, groups, or the entire system can be adjusted for volume, contour, and equalization of masking, paging, and music functions.

The system uses Echelon's i.LON® Internet server platform to allow remote connectivity into the system and integrate the system to existing LonWorks systems. This web server acts as the network conduit and relays information from the Sound Manager software to the system's operating platforms and speaker channels.

The most common applications of controlling a VDV system are those for fire/security or access control purposes. For example, a telephone line may be seized by the fire alarm or security systems in order to alert the local authorities to the alarm in the building. The fire alarm or security system may even be able to transmit information about the location and size of the problem. This helps responders to act more quickly to the alarm.

Also, a CCTV system that is used for surveillance may be controlled by the security or access control systems to monitor and record certain areas where people have been detected. If the surveillance cameras support pan-tilt-zoom (PTZ) control, information on the desired field of view can be transmitted by the security or access control devices to the camera controller or the camera itself. **See Figure 1-24.**

Elevator Systems

An elevator system is a conveying system for transporting people and/or materials vertically between floors in a building. Elevator systems operate efficiently, largely on their own, and with their own control devices, but can accept inputs from other systems to add call signals or modify their operating modes.

Elevators are usually called to a particular floor by a person pressing a call button in the elevator lobby. This initiates a call signal that is added to the elevator controller's queue, which determines the next floor in the elevator's itinerary. However, similar call signals can also be added to the queue via building automation systems. For example, an access control system admitting a person into the building may anticipate their need for an elevator by transmitting a message to the elevator controller to add a call for that floor. **See Figure 1-25.** The light in the call button is typically lit automatically by the controller, indicating that the call has already been received. By the time the person reaches the elevator lobby, the elevator car may already be there waiting.

Another example of elevator system integration is with fire alarm systems. Since elevators can be dangerous to use during a fire, the elevator cars must be removed from service if the fire alarm is activated. An output on the fire alarm control panel (FACP) is connected to the elevator controller. In the event of a fire alarm signal, the elevator controller switches over to a fire service mode and disables the elevator car.

Pan-Tilt-Zoom (PTZ) Camera Control

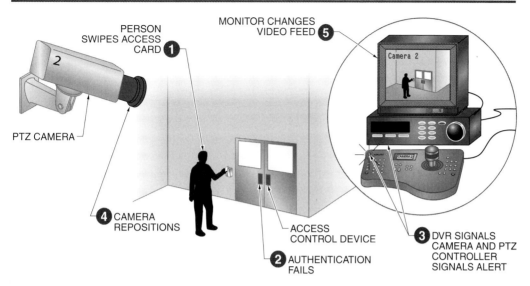

Figure 1-24. Video surveillance systems are often integrated with access control systems to document access attempts. Depending on the scenario, pan-tilt-zoom (PTZ) cameras can be controlled by information from the access control system to focus on a particular area.

Automatic Elevator Call

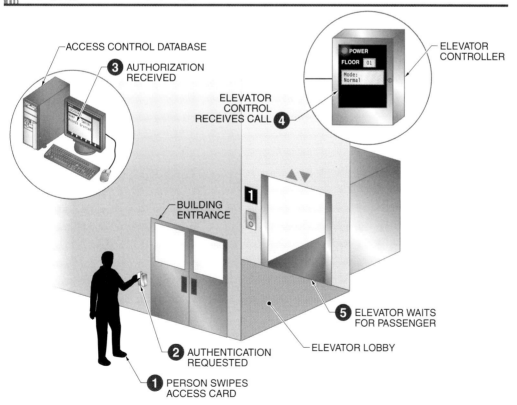

Figure 1-25. Integrated building automation systems can be used to add calls to the elevator system controller from other building systems, such as the access control system.

Summary

- Advanced building automation technologies include decision-making ability within the individual control devices, which are linked by a common data communication network.
- A communication protocol governs the format, timing, and signals passed between devices to ensure that they are all speaking the same understandable language.
- Systems using open protocols must still use protocol-specific devices, but since the protocol standard is publicly available, the device can be made by any manufacturer.
- LonWorks systems are based largely on a specially designed microcontroller chip, called the Neuron® chip, that is part of a LonWorks control device.
- BACnet maintains a highly structured information architecture that organizes information as software objects, which each have a number of configurable properties.
- The protocols that are considered to be in the wireless protocol category rely on the unique characteristics of wireless communication for their operation.
- In control systems, interoperable devices from different manufacturers are able to work together in the same system. They speak and understand the same language, store control information in the same way, and use the same group of control actions.
- Since an open protocol is available for any user, a manufacturer can develop a control device that can operate with any other manufacturer's device, as long as they both follow the specifications in the protocol's standard.
- Open protocols allow greater opportunities for integrating building systems together.
- In a hierarchical network architecture, a control device serves as a network manager for the devices attached to it in a lower tier, which have limited interaction with other control devices.
- A flat network architecture allows any device to communicate directly with any other device on the network without the need for network managers.
- Building automation may involve any or all of a building's systems. The level of sophistication of a building automation system is affected by the number of building systems integrated together.
- One of the most important and common applications of building automation in electrical systems is demand limiting, also known as load shedding.
- The prime strategies in reducing indoor lighting use are to incorporate natural sunlight into the interior occupied areas and to use occupancy to switch the light fixtures in a particular room or area OFF when no one is in the space.
- HVAC system setpoints can be changed dynamically according to a variety of control strategies. Most commonly, setpoint schedules are used to lower the heating or cooling loads when the entire building is unoccupied, such as at night and on weekends.
- Fire alarm, security, and access control systems are only integrated with other building systems through their main control panels.
- The security system control panel typically provides special communication and simple output connections that can be used to initiate special control sequences in other systems, such as unoccupied modes, access, and security monitoring with surveillance systems.
- Access control systems have features for specifying different access rights for each person, detailing which areas they may access and at which times.

- The most common applications of controlling a VDV system are those for fire/security or access control purposes.
- Elevator systems operate efficiently, largely on their own, and with their own control devices, but can accept inputs from other systems to add call signals or modify their operating modes.

Definitions

- A *protocol* is a set of rules and procedures for the exchange of information between two connected devices.
- A *proprietary protocol* is a communications and network protocol that is developed and used by only one device manufacturer.
- An *open protocol* is a standardized communications and network protocol that is published for use by any device manufacturer.
- *Interoperability* is the capability of network devices from different manufacturers and systems to interact using a common communication network and language framework.
- A *hierarchical network architecture* is a network configuration where control devices are arranged in a tiered network and have limited interaction with other control devices.
- A *flat network architecture* is a network configuration where control devices are arranged in a peer-to-peer way.
- A *gateway network architecture* is a network configuration where a gateway is used to integrate separate control systems based on different protocols.
- A *gateway* is a network device that translates transmitted information between different protocols.

Review Questions

1. What is the role of a protocol in building automation?
2. What are the primary differences between proprietary and open protocols?
3. Compare the basic approaches for data structuring and decision-making organization of LonWorks and BACnet systems.
4. Why is the categorization of "wireless protocols" potentially unclear?
5. How does interoperability affect the integration of different manufacturers and vendors?
6. How does interoperability affect the integration of building systems?
7. What are the advantages of flat network architectures over hierarchical network architectures?
8. How can building automation be used to reduce electricity consumption?
9. Describe the two primary strategies for reducing electricity consumption through lighting control.
10. How are separate fire alarm, security, access control, and elevator systems typically integrated with a building automation system?

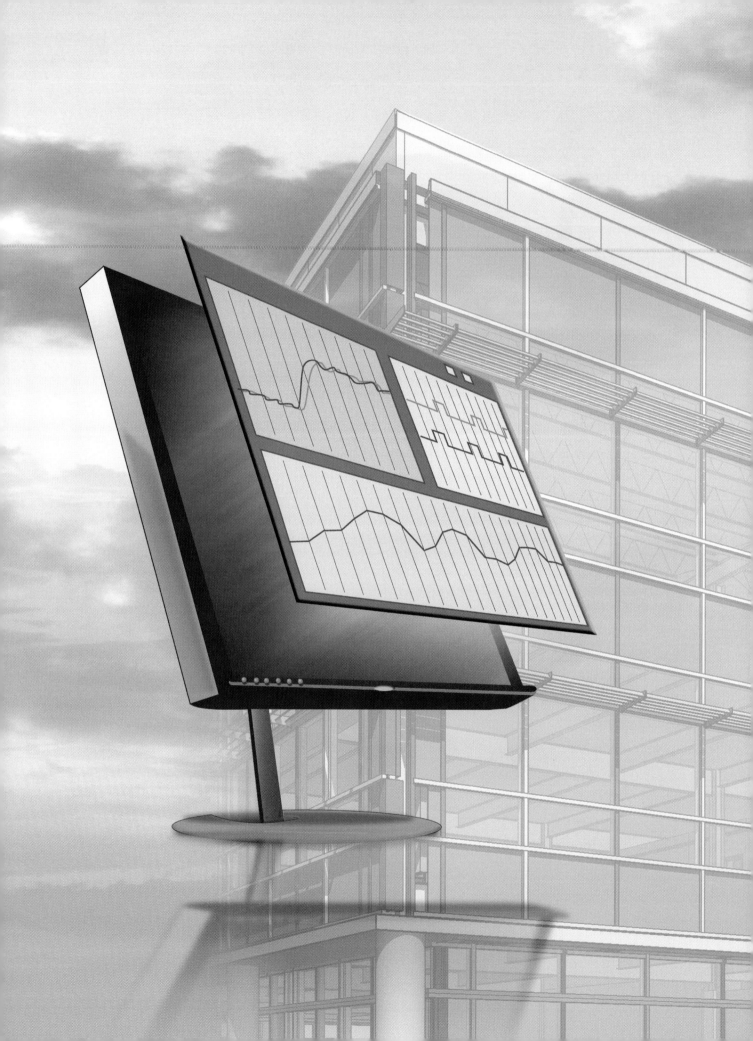

Chapter Two

Control Concepts

Control strategies and logic are not new to networked building automation systems, but have evolved from earlier types of building control systems. However, while the control concepts are similar, the electronic and distributed intelligence features of modern building automation systems allow for much greater precision and reliability than earlier systems. It is also generally easier to program and manage parameters for multiple control functions and use the data gathered by the building automation system for other uses.

Chapter Objectives

- Compare the control strategies that can be used to provide the optimum building system operation.
- Compare the algorithms used to make output decisions based on inputs.
- Identify the parameters and guidelines involved in calibrating control logic algorithms.
- Evaluate the use of supervisory control strategies for the overall operation of the building automation system.
- Describe how building automation systems can be used to help accomplish building management goals.

CONTROL STRATEGIES

A *control strategy* is a method for optimizing the control of building system equipment. The optimum outcome is one that fulfills the requirements of the sequence of operations, such as maintaining a comfortable indoor environment or providing adequate lighting, while minimizing energy use, manual interaction, and equipment wear and tear. There are often multiple strategies that can be used to achieve this, each with a slightly different method for controlling the energy-using equipment in a building. Strategies are chosen based on the particular layout of the building; the type, number, and location of the control devices; how the building will be used; and the building owner's priorities.

Setpoint Control

Setpoint control is a control strategy that maintains a setpoint in the system. **See Figure 2-1.** A *setpoint* is the desired value to be maintained by a system. Setpoint control is the most common building automation control strategy. The setpoint can be one of many controlled variables, such as temperature, humidity, pressure, light level, dewpoint, and enthalpy. The setpoint and the desired stability are programmed into a building automation controller. For example, if a building automation system is required to maintain a temperature of 72°F in a building, the 72°F temperature is the setpoint of the building automation system.

A *control point* is a variable in a control system. For example, a temperature sensor in a building space measures a temperature of 74°F. The control point is the indoor air temperature, which currently equals 74°F. At any given time, a control point may differ from the setpoint. *Offset* is the difference between the value of a control point and its corresponding setpoint.

Setback Control

Most setpoints are meant for when the building space is occupied. However, it is sometimes necessary to maintain certain conditions when the building is unoccupied. Setback control uses setpoint values that are active during the unoccupied mode of a building automation system. **See Figure 2-2.** A *setback* is the unoccupied heating or cooling setpoint. For example, if a heating setpoint is lowered from 70°F during the day to 65°F at night, then the setback heating setpoint is 65°F. If a cooling setpoint is raised from 74°F during the day to 85°F at night, then the setback cooling setpoint is 85°F.

Setback control is commonly used with building space temperature setpoints. This strategy saves energy by reducing the heating and/or cooling load when a building is unoccupied, but still prevents excessively hot or cold temperatures. It also reduces the time needed to reach the occupied setpoint.

Reset Control

Reset control is a control strategy in which a primary setpoint is adjusted automatically as another value (the reset variable) changes. For example, when the outside air temperature falls to a certain point, the water heating temperature setpoint is automatically reset to a higher setpoint. This ensures that the boiler is able to provide the hot water needed to effectively heat the space. A *setpoint schedule* is a description of the amount a reset variable resets the primary setpoint.

Low-Limit/High-Limit Control

Low-limit and high-limit controls ensure that a control point remains within a certain range. **See Figure 2-3.** *Low-limit control* is a control strategy that makes system adjustments necessary to

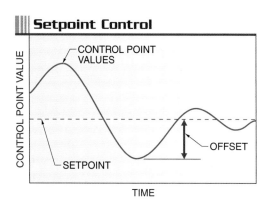

Figure 2-1. Control systems make adjustments in order to minimize the offset between a control point and its setpoint.

maintain a control point above a certain value. Low-limit control is commonly used with mixed-air damper controls. For example, low outside air temperatures cause low mixed-air temperatures. If the mixed-air temperature drops below 45°F, the controller overrides the normal control logic and forces the outside air dampers closed.

Setback Control

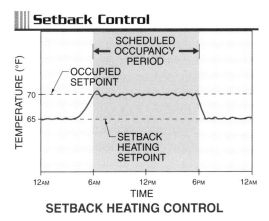

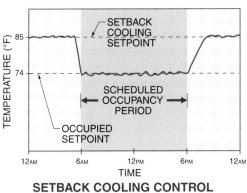

Figure 2-2. Setbacks are used as setpoints during unoccupied periods.

High-limit control is a control strategy that makes system adjustments necessary to maintain a control point below a certain value. This is similar to low-limit control and is commonly used with temperature, pressure, or humidity control points. For example, a high-limit control can be used to prevent a water temperature from exceeding 210°F.

Lead/Lag Control

Lead/lag control is a control strategy that alternates the operation of two or more similar pieces of equipment in the same system. The most common application of lead/lag control is the control of multiple hot water or chilled water pumps. For example, a primary (lead) pump is energized by the building automation system when it is required by a sequence. A backup (lag) pump is energized if the lead pump fails to start. **See Figure 2-4.** Often, lead/lag pump operation is rotated on a time schedule to have equal run time on the two pumps. Refrigeration and air compressors are also common applications of lead/lag control.

Low-Limit/High-Limit Control

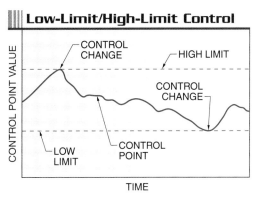

Figure 2-3. Low-limit and high-limit controls are used to keep a control point above or below certain values, respectively.

Lead/Lag Control

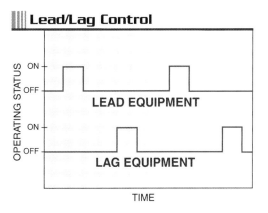

Figure 2-4. Lead and lag loads are alternated in order to equalize the wear and tear on each unit.

High/Low Signal Select

High/low signal select is a control strategy in which the building automation system selects the highest or lowest values from among multiple inputs for use in the control decisions. The most common application of high/low signal select is the control of a building space temperature using multiple temperature sensors at different locations within the zone. The highest signal represents the warmest area and the lowest signal represents the coolest area. The building automation system uses these signals and reset control to determine the setpoint. The highest signal may be used to reset a cooling function to satisfy the warmest space. The lowest signal may be used to reset a heating function to satisfy the coolest space.

Averaging Control

Averaging control is a control strategy that calculates an average value from multiple inputs, which is then used in control decisions. Averaging control is used with a group of sensors that may include high and low values that do not accurately represent the overall conditions in a building. For example, a temperature sensor located in a foyer senses outside air temperatures whenever a door is open and returns the coldest or warmest space temperatures from among a group of sensors in the area. **See Figure 2-5.** It would not be desirable for this sensor to control the heating or cooling reset setpoint most of the time. Instead, an average value from several temperature sensors is more representative of zone conditions. Correct placement of the averaging control sensors is required for the best results.

CONTROL LOGIC

The decisions that controllers make to change the operation of a building system involve control logic. *Control logic* is the portion of controller software that produces the calculated outputs based on the inputs. There are many different ways in which these decisions can be made, depending on the inputs used to make the decisions and the algorithms used to produce the results. An *algorithm* is a sequence of instructions for producing the optimal result to a problem. The decision-making process is described with a control loop. A *control loop* is the continuous repetition of the control logic decisions. Control systems are categorized as either open-loop control or closed-loop control.

Open-Loop Control

An *open-loop control system* is a control system in which decisions are made based only on the current state of the system and a model of how it should work. An example of an open-loop control system is a controller that turns a chilled water pump ON when the outside air temperature is above 65°F. The controller has no feedback to verify that the pump is actually ON. **See Figure 2-6.** *Feedback* is the measurement of the results of a control action by a sensor.

Averaging Control

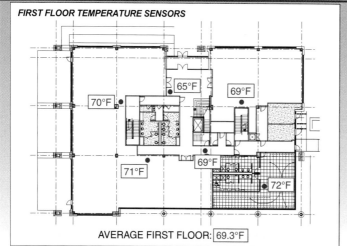

Figure 2-5. Averaging is an effective way to manage areas with multiple sensors because it moderates the effect of very high or low readings from one sensor.

Open-Loop Control

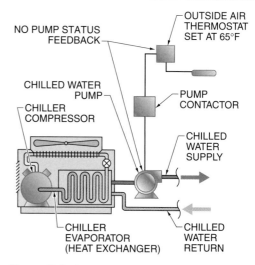

Figure 2-6. Open-loop control makes changes to a system without receiving feedback on the system's actual state.

Open-loop control requires perfect knowledge of the system and assumes there are no disturbances to the system that would otherwise change the outcome. There is no connection between the controller's output and its input.

The most common example of open-loop control in a building automation system is based on time schedules. Time-based control is a control strategy in which the time of day is used to determine the desired operation of a load. Time-based control turns a load ON or OFF at a specific time, without knowledge of any other factors that may affect the need for that load to operate. For example, open-loop time-based landscape irrigation control may activate the sprinklers based on a schedule. However, if it had recently rained, the system is overwatering the landscape and wasting water. Without a moisture sensor to provide any input based on the system output, the system has only open-loop control.

Closed-Loop Control

To address the limitation of open-loop control, most essential control loops include feedback. This makes them closed-loop. A *closed-loop control system* is a control system in which the result of an output is fed back into a controller as an input. For example, a thermostat controls the position of a valve in a hot water terminal device to maintain an air temperature setpoint. The thermostat in the building space provides the feedback of the air temperature that is used to continually adjust the hot water valve. **See Figure 2-7.**

However, a malfunction of one component in a closed-loop control system results in other components within the system having an incorrect value or position. This can cause problems such as uncomfortable indoor environments, wasted energy, or even equipment damage.

Control Algorithms

Algorithms are used to determine the necessary output value based on the inputs, enabling a building automation system to achieve a high level of accuracy. Control algorithms are selected when a control device is initially installed and configured. To achieve proper control, the correct algorithm must be selected and accurate setpoints and other parameters must be input into the device. Common algorithms used in building automation systems include two-position, proportional, integral, derivative, and adaptive control algorithms. Each algorithm has different characteristics of accuracy, stability, and response time.

Closed-Loop Control

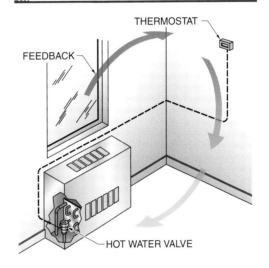

Figure 2-7. Closed-loop control makes decisions based on information fed back into the controller from the system.

Two-Position Control. Two-position control is the simplest control algorithm. A *two-position control algorithm* is a control algorithm in which the output assumes only one of two positions, which is switched when the input reaches certain setpoints. The two positions can be ON/OFF, open/closed, high/low, or any other combination with only two possible fixed outputs. The result of two-position control is that the controlled variable responds quickly, but continuously fluctuates slightly above and below the setpoint. **See Figure 2-8.**

For example, a two-position valve on a steam radiator opens or closes based on the temperature in the building space it serves. The setpoint is 72.0°F. If the temperature drops below the setpoint, the valve is opened, which raises the temperature. As it rises above the setpoint, the valve is closed, which allows the space to cool. The temperature in the space stays close to the setpoint, but can be above or below the setpoint at any time.

Two-Position Control

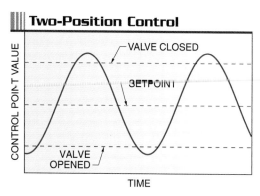

Figure 2-8. Two-position control maintains a control point value near a setpoint by alternating an actuator between two extremes, such as fully open and fully closed.

A deadband is used to prevent short cycling. A *deadband* is a range of values in which no control actions are made. *Short cycling* is the rapid alternation of a two-position control output in order to keep a controlled variable within a narrow range. For example, without a deadband, the valve would open when the temperature falls to 71.9°F and close when it rises to 72.1°F. **See Figure 2-9.** This very small range would cause the valve to activate constantly, which can damage the mechanical components. Plus, a range this small is not necessary to maintain occupant comfort. Instead, a deadband of 70°F to 74°F keeps the temperature within an acceptable range without excessive valve action. Therefore, the valve opens when the temperature reaches 70°F and closes at 74°F. A deadband may be specified as either a range (70°F to 74°F) or a differential from the setpoint (72°F ±2°F).

Due to system response times and thermal lag, the actual control point value, such as space temperature, can extend slightly outside the deadband. For example, when the valve is closed at the upper end of the deadband, the residual heat in the system continues to raise the space temperature slightly, and then the temperature begins to fall. Likewise, the lag in the application of heat when the valve is opened allows the space temperature to continue to fall slightly before it rises again.

Deadband

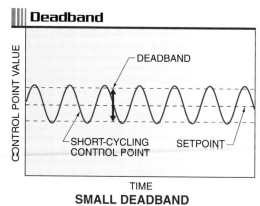

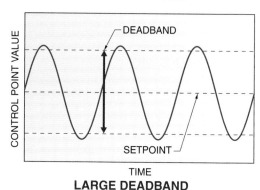

Figure 2-9. The size of the deadband programmed into a controller affects how frequently the actuator cycles between the two positions.

Proportional Control Algorithms. A *proportional control algorithm* is a control algorithm in which the output is in direct response to the amount of offset in the system. **See Figure 2-10.** For example, a 10% increase in room temperature results in a cooling control valve opening by 10%. Proportional controllers output an analog signal, which requires compatible actuators.

The algorithm is based on the setpoint offset and a desired proportion (throttling) parameter. Proportional control systems have a lower tendency to undershoot or overshoot than other algorithms, but may not offer precise control. They are used successfully in most applications, but may be inaccurate if not set up properly. When the system reaches the setpoint, the controller outputs a default actuator position, typically a 50% setting. However, a load may require a different position when at the setpoint, resulting in increased offset and energy use.

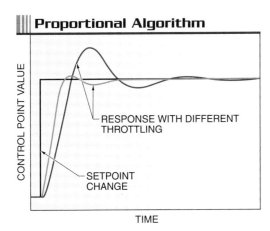

Figure 2-10. In order to bring a control point to a new setpoint, a proportional algorithm adjusts the output in direct response to the current offset.

Integral Control Algorithms. An *integral control algorithm* is a control algorithm in which the output is determined by the sum of the offset over time. Integration is a function that calculates the amount of offset over time as the area underneath a time-variable curve. **See Figure 2-11.** This offset area is then used to determine the output needed to eliminate the offset. The time period used for the calculation changes the results. Integral control algorithms tend to move the system toward the setpoint faster than proportional algorithms.

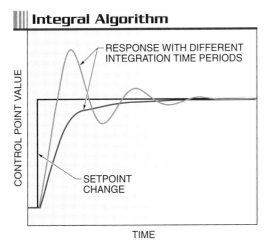

Figure 2-11. In order to bring a control point to a new setpoint, an integral algorithm adjusts the output according to the sum of the offsets in a preceding period.

However, since the integral is responding to accumulated errors from the past, it can cause the present value to overshoot the setpoint, crossing over the setpoint and creating an offset in the other direction. Proportional/integral (PI) control is the combination of proportional and integral control algorithms. This combination is generally more stable and accurate than the integral-only algorithm.

Derivative Control Algorithms. A *derivative control algorithm* is a control algorithm in which the output is determined by the instantaneous rate of change of a variable. **See Figure 2-12.** The rate of change is then used to determine the output needed to eliminate the offset. As the input approaches the setpoint, then the output change is reduced early to allow the input to coast to the setpoint. If the input moves rapidly away from the setpoint, extra change is applied to the output to maintain the setpoint. The amount of derivative control in an algorithm affects the overall response. However, this algorithm amplifies noise in the signal, which can cause the system to become unstable.

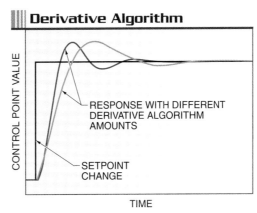

Figure 2-12. In order to bring a control point to a new setpoint, a derivative control algorithm adjusts the output according to the rate of change of the control point.

Sophisticated control technologies have increased expectations for accuracy and precision. In the past, a tolerance of ±2°F was acceptable. With electronic systems, a much greater precision (such as ±0.5°F) is possible. Control system accuracy is determined by the capabilities of the controllers, sensors, and actuators, and the quality of the system design and tuning.

Proportional/integral/derivative (PID) control is the combination of proportional, integral, and derivative algorithms. **See Figure 2-13.** The offset is calculated from feedback from the building system that is used as an input. The offset is then used in separate proportional, integral, and derivative calculations. The results of the three separate calculations are added together to determine the output value. The relative proportions of each algorithm are controlled by gain multipliers.

PID Control

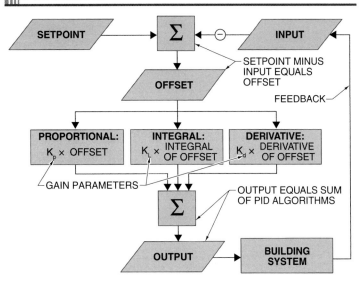

Figure 2-13. PID control systems include contributions from the proportional, integral, and derivative algorithms.

The PID combination improves stability and precise control. The derivative control algorithm moderates the effects of integral control algorithm, which is most noticeable close to the controller setpoint. It reduces the magnitude of the overshoot produced by the integral component. Only extremely sensitive control applications require PID control. Proportional/integral (PI) control is normally sufficient to achieve a setpoint.

Calibration (Tuning)

Control logic, especially PI and PID control algorithms, must be carefully calibrated by a building automation technician to ensure accuracy and stability. *Calibration* (tuning) is the adjustment of control algorithm parameters to the optimal values for the desired control response. The parameters are the relative contributions (gains) of each algorithm to the final output decision, plus any parameters used within each algorithm calculation. **See Figure 2-14.** The control logic must be calibrated during commissioning of the system, and routinely checked and adjusted if the response is incorrect. Each system responds differently and must be calibrated individually. The following are several guidelines by which experienced control professionals can determine the best parameter values:

1. Calibrate the control loop during a heavy load demand, which satisfies the setpoint under the most adverse conditions.
2. Calibrate the control loop using proportional control only, disabling the integral and derivative algorithms. The proportional control algorithm sets the control system in the desired range. Therefore, a control loop cannot be calibrated at all if it cannot be calibrated approximately using only proportional control. Continue the calibration process only after achieving good control with proportional-only algorithm.
3. Once the proportional control algorithm is stable, double the throttling parameter.
4. Adjust the gain of the integral control algorithm. Start with long integration times, which provide stable conditions. Short integration times result in quick responses but increase the chance of cycling.
5. Shorten the integration time slowly, checking the system response. As the times are shortened, the system becomes unstable. Lengthen the time until the system retains stability.
6. When the integral algorithm has been calibrated in a stable manner, the derivative control algorithm can be adjusted to ensure a quick response in the event of a rapid load change. A small gain of the derivative algorithm makes the system stable but react slowly. A larger gain makes the system unstable but react quickly. When increasing the derivative algorithm gain, also increase the integration time.

Effects of Increasing PID Gain Parameters				
Algorithm	Rise Time	Overshoot	Settling Time	Steady-State Error
Proportional	Decreases	Increases	Small Change	Decreases
Integral	Decreases	Increases	Increase	Eliminates
Derivative	Small Decreases	Decreases	Decreases	None

Figure 2-14. Calibration involves adjusting the relative contributions of the proportional, integral, and derivative algorithms to quickly stabilize a control point at a setpoint.

Control loops can be calibrated manually by introducing disturbances into the system and adjusting variables accordingly. Software is also available for automating the process.

When properly calibrated, control loops provide optimal control of building systems. The optimal control behavior varies depending on the application. Some logic must not allow an overshoot of the output beyond the setpoint if, for example, it would create an unsafe situation. Other processes must minimize the energy expended in reaching a new setpoint. Generally, long-term stability is required and the response must not oscillate for any combination of conditions and setpoints.

Hunting

If the control-loop parameters are not well calibrated, the control logic can be unstable. The output may diverge from the setpoint or hunt excessively. *Hunting* is an oscillation of output resulting from feedback that changes from positive to negative. **See Figure 2-15.** Positive feedback increases the offset, while negative feedback decreases the offset. The alternation between the two causes the output to oscillate above and below the setpoint. In some cases, the oscillation can worsen over time. Then, the only limits to the extremes in oscillation are saturation or mechanical limits.

Adaptive Control Algorithms

An *adaptive control algorithm* is a control algorithm that automatically adjusts its response time based on environmental conditions. This results in increased accuracy and stability, and it is simpler to implement. Adaptive control algorithms are the most sophisticated control algorithms because they are a self-calibrating form of PID control.

See Figure 2-16. Adaptive control algorithms require less calibration because they can adjust their parameters to load changes or incorrect programming. Not all control devices provide adaptive control algorithms.

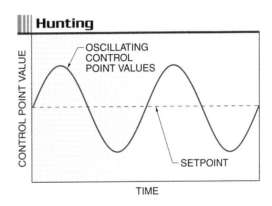

Figure 2-15. Hunting is an oscillating response that does not quickly settle at the setpoint.

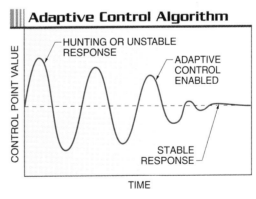

Figure 2-16. An adaptive control algorithm can automatically determine the best calibration parameters to bring a hunting or unstable response back to the setpoint.

A deadband is the range between two setpoints in which no control action takes place. Deadband reduces energy use and frequent cycling of the system.

Adaptive control is often used with air-handling unit dampers. If the outside air temperature is very close to the return air temperature, the outside air damper must move a lot to change the mixed-air temperature. However, during winter, small adjustments of the outside air damper can impact the mixed-air temperature significantly. Adaptive control algorithms adjust the tuning parameters as needed to account for these types of seasonal operational changes.

SUPERVISORY CONTROL

Modern electronic control devices include the control logic and programming to efficiently operate individual building equipment. These networked control systems do not rely on centralized controllers for normal building system operation. However, some control functions affect the overall operation of the entire building automation system and may use centralized (supervisory) controllers or interface software for their configuration.

A *supervisory control strategy* is a method for controlling certain overall functions of a building automation system. Supervisory control strategies typically override the control logic decisions of individual control devices. For example, a lighting controller's inputs may indicate the need to increase the lighting level in a building area. However, if a supervisory control strategy indicates that the lighting in the area should be OFF, the local control is overridden and the lighting is turned OFF.

Multiple supervisory control strategies can be integrated together into an overall approach for efficiently controlling the same loads during the same period. In fact, this is very common. Each control strategy has a priority relative to the others, so that the highest priority strategy overrides all others.

Life Safety Control

Life safety control is a supervisory control strategy for life safety issues such as fire detection and suppression. Life safety control strategies have the highest priority of all control strategies. A building automation technician must be familiar with life safety system wiring, software, and codes.

Scheduled Control

Scheduled control is a supervisory control strategy in which the date and time are used to determine the desired operation of a load or system. Based on programmed schedules, this strategy turns loads ON or OFF, adjusts setpoints, or changes the occupancy states of building zones, which are then used by the control system to control building systems. Scheduled control strategies are among the most basic and common supervisory control strategies for building automation systems.

Many different types of scheduled control strategies were created to manage a variety of operating situations. Building automation technicians set up these schedules during system commissioning, though most systems provide a user-friendly way to make future schedule changes.

Seven-Day Scheduling. *Seven-day scheduling* is the programming of time-based control functions that are unique for each day of the week. **See Figure 2-17.** Seven-day programming is common, but some building automation systems use a 5+2 schedule. A 5+2 schedule system recognizes Monday through Friday (5 days) as normal workdays with the same daily schedule, with Saturday and Sunday (2 days) treated separately. For maximum efficiency and flexibility, building automation systems should provide the capability to program each day independently, which can accommodate the specific needs of any building.

Seven-Day Scheduling	
Day	Normal Occupancy Schedule
Sunday	none
Monday	07:00 – 18:00; 20:00 – 22:00
Tuesday	07:00 – 20:00
Wednesday	07:00 – 18:00
Thursday	07:00 – 20:00
Friday	07:00 – 16:00
Saturday	10:00 – 18:00

Figure 2-17. Seven-day scheduling allows occupancy periods to be programmed individually for each day of the week.

Time Compensation

Special attention is required to address certain aspects of local time when developing scheduled control strategies. Changes to local time due to daylight saving and time zones may require additional programming.

Building automation systems provide software mechanisms for automatic daylight saving time changeover, if necessary. The system is programmed with the changeover dates, and the local time throughout the system is changed automatically at the correct moment.

Building automation systems can also operate within a group of buildings, even if they are located in different time zones. However, this introduces the possibility of complicating the scheduled control strategies for devices with different local times. Building automation systems typically solve this problem by assigning an offset time value, in either positive or negative hours, to each device. The devices are all synchronized with a standard time, such as Coordinated Universal Time (UTC), and each applies its time offset to determine local time. Schedules may then be shared between devices according to the standard time without causing problems.

Daily Multiple Time Period Scheduling. *Daily multiple time period scheduling* is the programming of time-based control functions for atypical periods of building occupancy. With this function, building systems or individual loads can be scheduled to operate during multiple independent time periods. For example, the normal scheduled operating hours of a rooftop unit in a commercial building are 8 AM to 5 PM. However, the building is also used for a continuing education class from 8 PM to 10 PM three times a week, so this time period is added to the schedule. Daily multiple time periods can be programmed for certain hours and for certain days. Building automation system software typically provides for several separately programmable time periods per day.

Holiday and Vacation Scheduling. *Holiday and vacation scheduling* is the programming of time-based control functions during holidays and vacations. This is typically used to reduce loads or turn them OFF completely since the building is expected to be unoccupied. This overrides the normal occupancy schedules that would have operated the equipment.

A comprehensive yearly operation calendar is required when programming holiday and vacation scheduling. A *permanent holiday* is a holiday that remains on the same date each year. A *transient holiday* is a holiday that changes its date each year. For example, New Year's Day, which is on January 1 each year, is a permanent holiday. Memorial Day, which falls on the last Monday in May of each year, is a transient holiday.

Timed Overrides. A *timed override* is a control function in which occupants temporarily change a zone from an UNOCCUPIED to OCCUPIED state. During this period, the controller uses the setpoints for the OCCUPIED state. The state reverts back to UNOCCUPIED after a programmed time period elapses. Time overrides provide a quick response to unanticipated changes in building occupancy. A timed override can be activated by a pushbutton or other type of user input. **See Figure 2-18.** Activating the input a second time within the override period may cancel the timed override. Some building automation systems record the amount of time spent in the override mode each month. This information can be used to investigate ways to improve the normal scheduled control functions so that overrides are needed less frequently.

Timed Override

Figure 2-18. Timed override inputs can include pushbuttons for occupants to temporarily add a new occupancy period to the control system.

Optimum start/stop control determines the best time to operate HVAC system equipment in order to exactly fulfill setpoints during occupancy periods.

Temporary Scheduling. *Temporary scheduling* is the programming of time-based control functions for a one-time temporary schedule. Temporary schedules are commonly associated with a specific calendar date with unique occupancy needs, accommodating for a specific event in a building without using a timed override. Temporary schedules take priority over normal time schedules. At the end of a temporary schedule, it is erased and the normal time schedules resume. Temporary scheduling is commonly used for regularly scheduled weekly or monthly events.

Alternate Scheduling. *Alternate scheduling* is the programming of more than one unique time schedule per year. Alternate scheduling is commonly used during seasonal changes in building operations. For example, a retail business may use alternate scheduling during the holiday shopping season in December. One time schedule may extend from January through November. Beginning on the day after Thanksgiving, the alternate time schedule would take effect, overriding the yearly schedule until the end of the holiday shopping season.

Optimum start/stop control relies on scheduled periods of expected occupancy and cannot account for unexpected occupancy, such as arriving early or visiting the building on the weekends. Access events can be used to trigger unscheduled operation at occupancy setpoints.

Schedule Linking. *Schedule linking* is the association of loads within the building automation system that are always used during the same time. For example, when a rooftop unit is energized for a particular zone, the lighting load for that zone is also energized. Schedule linking enables both loads to be energized simultaneously from the same schedule.

Optimum Start/Stop Control

Schedules do not always represent the actual operation of a load, but instead the period of occupancy. For HVAC systems in particular, there is a lag between the start of load operation and the reaching of a setpoint. Therefore, it is necessary to determine when the loads need to operate in order for the setpoint conditions to be fully achieved during the occupied period. **See Figure 2-19.** Optimizing these start and stop times fulfills this requirement without operating the loads any more than necessary. This maximizes energy savings while maintaining comfort levels.

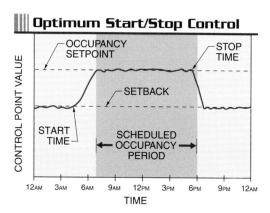

Figure 2-19. Optimum start/stop control determines the best actual start and stop times for HVAC equipment to meet the occupancy requirements.

Optimum Start Control. *Optimum start control* is a supervisory control strategy in which the HVAC load is turned ON as late as possible to achieve the indoor environment setpoints by the beginning of building occupancy. The actual start times of the HVAC equipment are calculated based on building and other conditions, and may change daily. The current outside air temperature influences the heating or cooling load, and the current indoor air temperature and setpoint determine the temperature change required within a building space. Two methods used in building automation systems to determine the actual start time are adaptive control and estimation control.

Adaptive start time control is a process that adjusts the actual start time for HVAC equipment based on the building temperature responses from previous days. This method tries to achieve the optimum start time for each day. A 7-day, 10-day, or 14-day history is commonly used to determine the success rate of previous start times and as a guide for adjusting new calculations. **See Figure 2-20.** Adaptive start control is the most common optimum start method used in building automation systems.

Estimation start time control is a process that calculates the actual start time for HVAC equipment based on building temperature data and a thermal recovery coefficient. A disadvantage of this control method is that the estimated coefficient can be calculated or input incorrectly, causing the HVAC system to start late so that the building space temperature is not at the setpoint when occupancy begins.

Thermal Recovery Coefficients. A *thermal recovery coefficient* is the ratio of a temperature change to the length of time it takes to obtain that change. A thermal recovery coefficient is expressed in temperature degrees per unit of time, such as degrees Fahrenheit per minute (°F/min). Thermal recovery coefficients are used to calculate the actual start time of HVAC systems in commercial buildings. For example, a rooftop unit may start operation at 7 AM for building occupancy at 9 AM. During this time, the indoor temperature increases from 60°F to 72°F. Therefore, it takes 120 min to increase the temperature 12°F. The thermal recovery coefficient is 0.1°F/min (12°F/120 min).

Thermal recovery coefficient values can also be used as indicators of HVAC equipment efficiency and/or mechanical problems. For example, an HVAC system having a thermal recovery coefficient of 0.2°F/min in one month and 0.1°F/min the next month may require a filter replacement or preventive maintenance.

| Adaptive Start Time Control ||||
Previous Attempts	Start Time	Time When Setpoint Reached	Start of Occupancy
1	05:18	07:01	07:00
2	05:17	07:02	07:00
3	05:16	06:59	07:00
4	05:17	07:03	07:00
5	05:15	06:57	07:00
6	05:17	07:05	07:00
7	05:15	06:53	07:00
8	05:20	07:03	07:00
9	05:30	07:10	07:00
10	05:10	06:50	07:00

Figure 2-20. A record of the start-up responses from several previous days is used to help calculate the best start and stop times.

Optimum Stop Control. *Optimum stop control* is a supervisory control strategy in which the HVAC load is turned OFF as early as possible to maintain the proper building space temperature until the end of building occupancy. For example, in winter, optimum stop supervisory control allows the building temperature to gradually decline until the end of occupancy. Optimum stop supervisory control is commonly limited to a specific length of time, such as 15 min or 30 min, and may be implemented independently of optimum start supervisory control.

Duty Cycling Control

Duty cycling control is a supervisory control strategy that reduces electrical demand by turning OFF certain HVAC loads temporarily. When HVAC units are oversized, they can be cycled in this way without adversely affecting building space temperature. The ON cycles of the loads are staggered so that only one load is operating at a time. For example, if two HVAC units that operate simultaneously have an electrical demand of 15 kW each, the electrical demand is 30 kW. By implementing duty cycling control, the operation of each unit is alternated so that the electrical demand is 15 kW. **See Figure 2-21.**

Duty cycling control is effective for reducing electrical demand and the associated expenses. However, duty cycling increases motor wear due to the frequent motor starts. Duty cycling also results in the loss of temperature control and loss of ventilation during the duty cycle's OFF time. In addition, the cycling of HVAC equipment may cause excessive noise.

when loads are shed. The building automation system must have a method of measuring the electrical power at the building service.

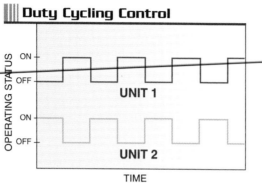

Figure 2-21. Duty cycling control reduces peak electrical demand by alternating the operation of multiple units.

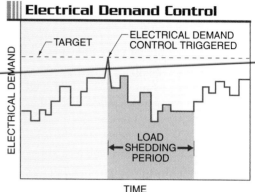

Figure 2-22. Electrical demand control sheds low-priority loads if the building's demand reaches a certain target.

Duty cycling is especially used in commercial buildings that have a large number of electric baseboard heaters or small exhaust fans. The duty cycling sequence for the baseboard heaters can be programmed to alternate around the building. For example, a heater in one office is duty cycled for a short period, followed by another on the opposite side of the building.

Electrical Demand Control

Commercial building consumers are typically charged for the highest period of electrical power demand for the month. Therefore, controlling the operation of electrical loads to lower the electrical demand can significantly lower utility bills. *Electrical demand control* is a supervisory control strategy designed to reduce a building's overall electrical demand. This strategy turns OFF certain electric loads in a way that reduces building electrical demand without adversely affecting occupant comfort or productivity. **See Figure 2-22.**

A *shed load* is an electric load that has been turned OFF for electrical demand control. A *restored load* is a shed load that has been turned ON after electrical demand control. Loads are shed (turned OFF) as the building electrical demand increases to a specific limit (target). The electrical demand decreases below the target

Shed Tables. A *shed table* is a table that prioritizes the order in which electrical loads are turned OFF. Building automation systems commonly provide low-priority and high-priority shed tables. A *low-priority load* is a load that is shed first for electrical demand control. A *high-priority load* is a load that is important to the operation of a building and is shed last when demand goes up. High-priority loads are restored (re-energized) first when the electrical demand drops. A thorough analysis of the electrical loads in a building must be performed to document the priority of each load. Not all building loads have to be included in shed tables. Some loads are essential to the efficient operation of a commercial building and cannot be shed.

When building electrical demand is above the target, the loads in the low-priority shed table are shed in order. If the building electrical demand stabilizes or drops, the loads are restored in reverse order. However, if the electric demand continues to increase, all loads in the low-priority shed table are shed and the building automation system begins to shed loads in the high-priority shed table. The low-priority shed table is often referred to as first OFF/last ON. The high-priority shed table is often referred to as last OFF/first ON.

Demand Control Considerations

The type of mechanical system significantly affects the demand control strategy, due to the interactions between the various pieces of equipment. For example, the chiller in a variable-air-volume (VAV) system with a chilled water plant would appear to be a large load that is easily shed temporarily. However, it is possible that shedding this load could have the opposite effect. If the chiller is taken out of service, the chilled water temperature in the supply to the air-handling unit (AHU) rises. In order to provide adequate cooling, the VAV box and AHU fans must move far more air and run at a much higher speed. Because the electrical demand of fan motors increases significantly at higher speeds, the resulting overall building electrical demand may actually rise. The whole building, its pattern of usage, and the type and potential interaction of the HVAC system equipment must be considered when prioritizing shed loads.

Rotating Priority Load Shedding. Load shedding reduces electrical demand in a commercial building, but can also create problems. For example, if the first load in a low-priority shed table is an electric water heater, it is always the first load to be shed. If the building has frequent high electrical demand periods, this load is often OFF and not available for use. The solution is to rotate the loads within the shed table. *Rotating priority load shedding* is an electrical demand control strategy in which the order of loads to be shed is changed with each high electrical demand condition.

In rotating priority load shedding, if load 1 is the first load shed for one high electrical demand period, load 2 is the first load shed for the next high electrical demand condition. **See Figure 2-23.** Some systems rotate both low- and high-priority shed tables while other systems rotate one shed table and leave the other shed table fixed.

Shedding Strategies. Electrical demand control features often include additional parameters that can be programmed for efficient control of shed loads.

The building automation software typically includes timers for the maximum and minimum shed times. The maximum shed time timer causes a load to be restored after it has been shed for a certain length of time. This load is restored regardless of the electrical demand in a commercial building at the time. The maximum shed time timer is commonly used with loads that are essential to the building operation. These loads would not be involved in electrical demand control without the maximum shed time timer.

The minimum shed time timer ensures that a shed load cannot be restored until a specific time period has elapsed. The minimum shed time timer reduces the possibility that a load is cycled ON and OFF repeatedly by the building automation system.

Electrical demand control can also be programmed to shed certain loads only while other loads are operating. For example, the cooling compressor of a package unit can be turned OFF during supply fan operation. This allows air to be circulated, providing some relief while the compressor is OFF. Demand is still reduced because the compressor uses more electrical power than the supply fan.

Rotating Priority Load Shedding

Shed Order	Shed Loads				
	Cycle 1	Cycle 2	Cycle 3	Cycle 4	Cycle 5
1	Lighting Circuit 5	Lighting Circuit 7	VAV Unit 12	Fan 2	Pump 7
2	Lighting Circuit 7	VAV Unit 12	Fan 2	Pump 7	Lighting Circuit 5
3	VAV Unit 12	Fan 2	Pump 7	Lighting Circuit 5	Lighting Circuit 7
4	Fan 2	Pump 7	Lighting Circuit 5	Lighting Circuit 7	VAV Unit 12
5	Pump 7	Lighting Circuit 5	Lighting Circuit 7	VAV Unit 12	Fan 2

Figure 2-23. The order in which equal-priority loads are shed can be rotated so that no one load is always the first to be turned OFF.

Electrical demand control can be programmed to temporarily change the setpoint when electrical demand is high. Instead of shedding a load, the normal setpoints are changed by a certain number of degrees. For example, a rooftop unit is programmed for a normal cooling setpoint of 72°F. However, 4°F is added to the cooling setpoint when the building is in a high electrical demand period. The new, temporary setpoint is 76°F. This setpoint change causes the unit compressor to shut OFF. If the temperature in the building space changes outside of the new setpoint, the compressor turns ON regardless of the building's electrical demand. This feature can also be programmed for heating applications.

Electrical Demand Targets. An effective electrical demand control strategy requires accurate monthly electrical demand targets. Load shedding is reduced if electrical demand targets are high and increased if electrical demand targets are low. **See Figure 2-24.** The development of accurate electrical demand targets requires experience in evaluating prior electric bills. After the targets are developed and the actual building electrical demand is evaluated, the targets may require adjustment. The electrical demand targets can be lowered incrementally over a period of time until the optimum level is reached.

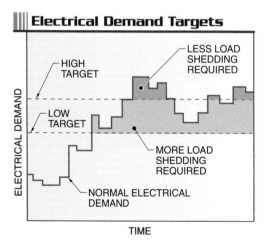

Figure 2-24. The choice of electrical demand target affects how much load shedding is required.

BUILDING SYSTEM MANAGEMENT

The operational data gathered by a building automation system can also be used by maintenance technicians to help manage building system equipment. The collection of information by a single system helps operators and technicians efficiently monitor abnormal conditions, document system performance, and perform preventive maintenance.

Alarming

Alarming is the detection and notification of abnormal building conditions. The most common alarms are associated with temperature sensors. Alarms can also indicate abnormal levels of humidity and pressure, or failure of a fan or pump. Alarms quickly alert maintenance personnel to equipment failures and/or problems.

Alarm Classification. Most modern building automation systems allow the operator to classify alarms into different categories. The most common alarm-classification categories are critical alarms and noncritical alarms. Critical alarms concern devices vital to proper operation of the building. Critical alarms are reported immediately to multiple maintenance personnel for a quick response. Noncritical alarms concern elements of building operation that are not vital. Noncritical alarms may or may not be reported immediately because a quick response is not necessary to the proper operation of the equipment or building.

Alarm Setpoints. Alarms can be set up to monitor most inputs and outputs of a building automation system. However, an operator monitoring many inputs and outputs may be overloaded with information. For example, a dangerous condition may result if a critical alarm is overlooked among a large number of noncritical alarms, even though alarms are classified to help ensure that operators address critical alarms promptly.

The building automation system is programmed with alarm setpoints. An *alarm setpoint* is the control point value that should trigger an alarm. Alarm differentials are used to prevent an alarm from quickly changing state between alarm and no-alarm status. An *alarm differential* is the amount of change required in a variable for an alarm to return to normal after it has been in alarm status. This is similar to deadband. **See Figure 2-25.** For example, an alarm differential of 2°F to 4°F is normally used for temperature sensors. If a temperature sensor is set to alarm at 80°F, a 2°F alarm differential does not allow the alarm to change to a no-alarm (normal) status until the temperature value drops to 78°F or below. Most building automation systems also have an alarm time-delay feature. The time delay gives equipment a short time to operate before a change in alarm status.

system performance, calibrate a control loop during commissioning, certify or publicize energy-efficient buildings, troubleshoot equipment problems, and other uses. Data trending functions can record temperature, humidity, power, and any other equipment or building condition variable. **See Figure 2-26.** Any control point used by the building automation system can be recorded. Multiple inputs and outputs can be recorded simultaneously, which can show relationships between control points, such as the outside air temperature and heating/cooling equipment operation.

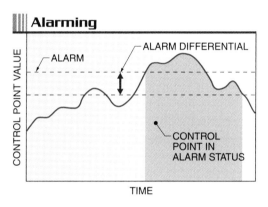

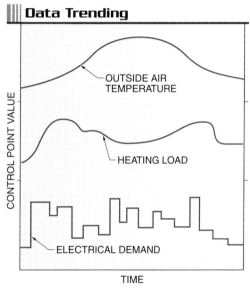

Figure 2-25. A control point is put into alarm status if it reaches a certain value, and does not return to normal status until it falls below (or rises above if it is a low limit alarm) the alarm differential.

Figure 2-26. Data trending records values of many control points over a long period. This information is saved for later analysis or troubleshooting.

Alarm Notification. Alarm conditions can also be configured with a type of notification. Building automation systems typically provide a number of ways to notify personnel when an alarm is triggered, such as status lights, buzzers, pop-up windows, prerecorded telephone messages, e-mail, and text messages. Alarm information is also often output to printers, which provide a permanent record of the event. Most building automation systems allow different categories of alarms to be reported through different methods.

Data Trending

Data trending is the recording of past building equipment operating information. This information can then be used to predict future

The control point values are recorded at a specific time interval or minimum change. The time interval used to record these values is programmable. A time interval of 20 min is commonly used for long-term data trending. A minimum change setting indicates how much the value must change before a new entry is made into the data log. Long-term data trending is used by a technician to view temperatures and other values that occur overnight or during weekends. For troubleshooting equipment problems, the time interval is often shortened to 1 min or 2 min in order to diagnose abnormal equipment operation.

Electronic meters at the electrical service or other points within the electrical system can provide control point information recorded for data trending purposes.

Data trending can be started and stopped at specific times and dates. For example, there may be complaints that a building space is excessively cold when occupants arrive in the morning. A data trend is created to record the building space temperature, outside air temperature, and equipment ON/OFF status beginning before occupancy and ending at midmorning. The data trend interval time is 5 min. The results of this data trend are used to change the actual start time of a unit or correct mechanical equipment problems. Data trends can also be used to make a decision regarding the purchase or replacement of equipment. Data trend records can be imported into spreadsheet software to create graphs and charts.

Preventive Maintenance

Preventive maintenance is scheduled inspection and work that is required to maintain equipment in peak operating condition. Corrective (breakdown) maintenance is performed after equipment has failed. A preventive maintenance program is typically less expensive and causes less downtime than corrective maintenance.

> Graphical interfaces are often software based, but can also be used in hardware-based human-machine interfaces (HMIs). These interfaces are small computers that are specifically designed to gather, process, and display system data. They include communication ports for collecting control point data from the automation network and a display screen for showing these values.

Many building automation systems can be integrated with computerized maintenance management systems (CMMS). When integrated, the separate CMMS software uses operational data from the building automation system to help manage preventive maintenance tasks, such as generating work orders based on certain control point information. **See Figure 2-27.** For example, a work order may be automatically generated for lubricating a piece of equipment based on the number of hours it has operated as logged by the building automation system. Other values that are commonly used to trigger preventive maintenance work orders are the number of motor starts or the pressure drop across filters.

Predictive maintenance is the monitoring of wear conditions and equipment characteristics in comparison to a predetermined tolerance to predict possible malfunctions or failures. Predictive maintenance attempts to detect equipment problems with vibration analysis and other methods. If the building automation system includes the necessary sensors, it may be possible to also integrate that data with predictive maintenance software.

Graphical Interfaces

Many building automation systems use graphical interfaces to visually communicate building and equipment conditions. Building systems and major equipment are represented in illustrations and icons, which are often easier to understand. **See Figure 2-28.** Illustrations can be created by the building automation technician or other vendor. Some systems use photographs of equipment. The actual temperature, humidity, pressure, or status values are superimposed onto the graphics.

Graphical interface software falls into two categories, thin client or thick client. Thin client interfaces do not require the use of any software other than a standard web browser. These interfaces are available on nearly any computer. Thick client interfaces are software programs specifically designed to display information about a building's systems. Thick client software is required on a computer before using the interface.

Chapter 2—Control Concepts   45

Preventive Maintenance Software

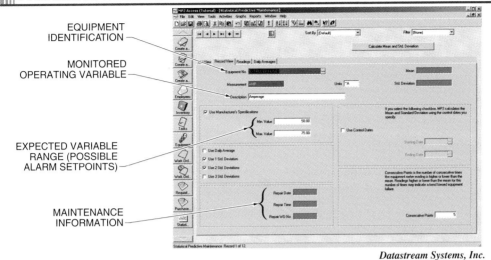

Figure 2-27. Some preventive maintenance software can be integrated with building automation systems to use control points for equipment information such as status, run time, and current draw.

Graphical Interfaces

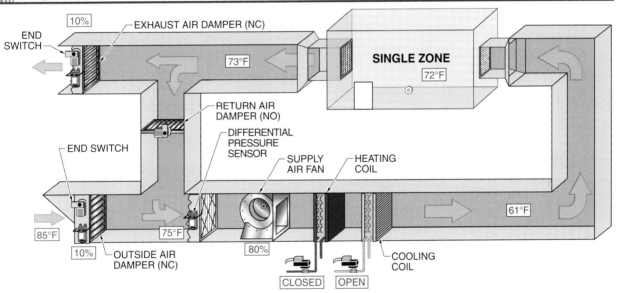

Figure 2-28. Graphical interfaces help convey important system information at a glance.

Summary

- There are often multiple strategies that can be used to fulfill the requirements of the system while minimizing energy use, manual interaction, and equipment wear and tear.
- Setpoint control is the most common building automation control strategy.
- Setup and setback are setpoint values that are active during the unoccupied mode of a building automation system.

- Low-limit and high-limit controls ensure that a control point remains within a certain range.

- Averaging control is used with a group of sensors that may include high and low values that do not accurately represent the overall conditions in a building.

- The decisions that controllers make to change the operation of a building system involve control logic.

- Open-loop control requires perfect knowledge of the system, and assumes there are no disturbances to the system that would otherwise change the outcome. There is no connection between the controller's output and its input.

- Closed-loop control includes feedback.

- Common algorithms used in building automation systems include proportional, integral, derivative, and adaptive control algorithms. Each algorithm has different characteristics of accuracy, stability, and response time.

- Calibration involves adjusting the relative contributions of each algorithm to the final output decision, plus any parameters used in each algorithm calculation.

- If the control loop parameters are not well calibrated, the control logic can be unstable.

- Adaptive control algorithms are the most sophisticated control algorithms because they are a self-calibrating form of PID control.

- Supervisory control functions affect the overall operation of the entire building automation system and may use centralized controllers or interface software for their configuration.

- Based on programmed dates and times, scheduled control turns loads ON or OFF, adjusts setpoints, or changes the occupancy states of building zones, which are then used by the control system to control building systems.

- Schedules do not always represent the actual operation of a load, but instead the period of occupancy.

- Optimum start and stop times fulfill setpoint requirements during the occupancy period without operating the loads any more than necessary.

- Electrical demand control turns OFF certain electric loads in a way that reduces building electrical demand without adversely affecting occupant comfort or productivity.

- Alarms are classified to help ensure that operators address critical alarms promptly.

- Data trending information can be used to predict future system performance, calibrate a control loop during commissioning, certify or publicize energy-efficient buildings, troubleshoot equipment problems, and other uses.

- Many building automation systems can be integrated with computerized maintenance management systems (CMMS), which use operational data from the building automation system to help manage preventive maintenance tasks, such as generating work orders based on certain control point information.

Definitions

- A *control strategy* is a method for optimizing the control of building system equipment.
- *Setpoint control* is a control strategy that maintains a setpoint in the system.
- A *setpoint* is the desired value to be maintained by a system.
- A *control point* is a variable in a control system.
- *Offset* is the difference between the value of a control point and its corresponding setpoint.

- A *setback* is the unoccupied heating or cooling setpoint.
- *Reset control* is a control strategy in which a primary setpoint is adjusted automatically as another value (the reset variable) changes.
- A *setpoint schedule* is a description of the amount a reset variable resets the primary setpoint.
- *Low-limit control* is a control strategy that makes system adjustments necessary to maintain a control point above a certain value.
- *High-limit control* is a control strategy that makes system adjustments necessary to maintain a control point below a certain value.
- *Lead/lag control* is a control strategy that alternates the operation of two or more similar pieces of equipment in the same system.
- *High/low signal select* is a control strategy in which the building automation system selects the highest or lowest values from among multiple inputs for use in the control decisions.
- *Averaging control* is a control strategy that calculates an average value from multiple inputs, which is then used in control decisions.
- *Control logic* is the portion of controller software that produces the necessary outputs based on the inputs.
- An *algorithm* is a sequence of instructions for producing the optimal result to a problem.
- A *control loop* is the continuous repetition of the control logic decisions.
- An *open-loop control system* is a control system in which decisions are made based only on the current state of the system and a model of how it should work.
- *Feedback* is the measurement of the results of a control action by a sensor.
- A *closed-loop control system* is a control system in which the result of an output is fed back into a controller as an input.
- A *proportional control algorithm* is a control algorithm in which the output is in direct response to the amount of offset in the system.
- An *integral control algorithm* is a control algorithm in which the output is determined by the sum of the offset over time.
- A *derivative control algorithm* is a control algorithm in which the output is determined by the instantaneous rate of change of a variable.
- *Calibration* is the adjustment of control algorithm parameters to the optimal values for the desired control response.
- *Hunting* is an oscillation of output resulting from feedback that changes from positive to negative.
- An *adaptive control algorithm* is a control algorithm that automatically adjusts its response time based on environmental conditions.
- A *supervisory control strategy* is a method for controlling certain overall functions of a building automation system.
- *Life safety control* is a supervisory control strategy for life safety issues such as fire detection and suppression.
- *Scheduled control* is a supervisory control strategy in which the date and time are used to determine the desired operation of a load or system.

- *Seven-day scheduling* is the programming of time-based control functions that are unique for each day of the week.
- *Daily multiple time period scheduling* is the programming of time-based control functions for atypical periods of building occupancy.
- *Holiday and vacation scheduling* is the programming of time-based control functions during holidays and vacations.
- A *permanent holiday* is a holiday that remains on the same date each year.
- A *transient holiday* is a holiday that changes its date each year.
- A *timed override* is a control function in which occupants temporarily change a zone from an UNOCCUPIED to OCCUPIED state.
- *Temporary scheduling* is the programming of time-based control functions for a one-time temporary schedule.
- *Alternate scheduling* is the programming of more than one unique time schedule per year.
- *Schedule linking* is the association of loads within the building automation system that are always used during the same time.
- *Optimum start control* is a supervisory control strategy in which the HVAC load is turned ON as late as possible to achieve the indoor environment setpoints by the beginning of building occupancy.
- *Adaptive start time control* is a process that adjusts the actual start time for HVAC equipment based on the building temperature responses from previous days.
- *Estimation start time control* is a process that calculates the actual start time for HVAC equipment based on building temperature data and a thermal recovery coefficient.
- A *thermal recovery coefficient* is the ratio of a temperature change to the length of time it takes to obtain that change.
- *Optimum stop control* is a supervisory control strategy in which the HVAC load is turned OFF as early as possible to maintain the proper building space temperature until the end of building occupancy.
- *Duty cycling control* is a supervisory control strategy that reduces electrical demand by turning OFF certain HVAC loads temporarily.
- *Electrical demand control* is a supervisory control strategy designed to reduce a building's overall electrical demand.
- A *shed load* is an electric load that has been turned OFF for electrical demand control.
- A *restored load* is a shed load that has been turned ON after electrical demand control.
- A *shed table* is a table that prioritizes the order in which electrical loads are turned OFF.
- A *low-priority load* is a load that is shed first for electrical demand control.
- A *high-priority load* is a load that is important to the operation of a building and is shed last when demand goes up.
- *Rotating priority load shedding* is an electrical demand control strategy in which the order of loads to be shed is changed with each high electrical demand condition.
- *Alarming* is the detection and notification of abnormal building conditions.
- An *alarm setpoint* is the control point value that should trigger an alarm.

- An *alarm differential* is the amount of change required in a variable for an alarm to return to normal after it has been in alarm status.
- *Data trending* is the recording of past building equipment operating information.
- *Preventive maintenance* is scheduled inspection and work that is required to maintain equipment in peak operating condition.
- *Predictive maintenance* is the monitoring of wear conditions and equipment characteristics in comparison to a predetermined tolerance to predict possible malfunctions or failures.

Review Questions

1. How does setup/setback control help optimize building operations?
2. When may averaging control be needed?
3. What is the difference between open-loop and closed-loop control?
4. How are control algorithms used to produce output values?
5. What is modified when calibrating a control algorithm?
6. What is the advantage of seven-day scheduling?
7. Why might the optimum start/stop times be different from the occupancy schedule?
8. How is electrical demand control used to reduce utility costs?
9. How can data trending be used to record and analyze control point information?
10. How might a building automation system improve the efficiency of preventive maintenance programs?

Chapter Three

Data Communication

Data communication includes a large number of technologies and concepts for communicating digital data between two or more entities on a network. This information is displayed to users and/or used to control the action of some machine. A familiar example is the computer networks used to access the Internet, though data communication is necessary in a variety of applications, including the type of machine-to-machine communication used in building automation applications.

Chapter Objectives

- Describe the role of protocols in ensuring accurate and reliable network communication.
- Identify common fields and their arrangement in different message frames.
- Differentiate between the roles of the seven layers of the OSI Model.
- Compare the functions of different network devices.
- Compare the characteristics of different network topologies.
- Identify the media types used in building automation networks.
- Describe the characteristics of the common building automation MAC layers.

DATA COMMUNICATION

The role of any data communications technology is to facilitate the exchange of information in predictable and reliable ways. Like the many forms of communication, many methods have been developed for allowing machines to communicate to other machines. Machines that are involved in network communication are known as nodes. **See Figure 3-1.** A *node* is a computer-based device that communicates with other similar devices on a shared network.

A *signal* is the conveyance of information. Symbols are used to group signals together, often as binary bits representing the concept of zero and one. Such groups of zeroes and ones can also be thought of as a number. **See Appendix.** A code equates symbols or numbers with real ideas that often depend on context. The protocol's rules must define what kinds of signals and symbols are used to convey information, and the structure, encoding, and content of messages that may be exchanged.

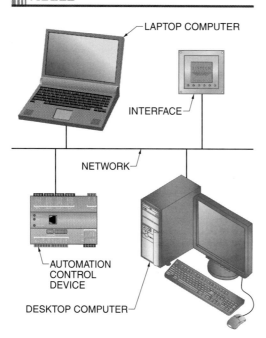

Figure 3-1. Nodes are computer-based devices that are connected together on a data communications network.

Everyday Protocols

Though they may seem limited to computer networks, the concept of standardized protocols is common, even within a person's everyday life. People use protocols all the time. For example, to control traffic flow and avoid accidents, drivers follow the rules of traffic signs and signals. Traffic lights use colored lights for signaling instructions to drivers. Which light is illuminated indicates a simple code: red means "stop" and green means "go." The standardized signaling rules also require that only one solid light is on at a time and that the red light is always on top. These rules form a commonly understood protocol for managing vehicle traffic.

Communication Protocols

Before any two nodes can effectively communicate with each other, they must agree on a protocol to govern the manner in which they exchange information. A *protocol* is a set of codes, message structures, signals, and procedures implemented in hardware and software that permits the exchange of information between nodes. In other words, a protocol is a collection of rules that enable the nodes to exchange information in reliable and repeatable ways.

Signaling

Signaling is the use of electrical, optical, and radio frequency changes in order to convey data between two or more nodes. Since computer-based devices deal most effectively with binary information (zeros and ones), most signaling methods involve the sending and receiving of streams of binary bits. Typically, a digital signal uses the presence and absence of voltage or light, or radio frequency manipulation, to represent 1 and 0. **See Figure 3-2.** For example, a common signaling scheme uses a 5 VDC electrical level to represent 1 and a 0 V electrical level to represent 0.

Signaling between nodes is accomplished with transceivers. A *transceiver* is a hardware component that provides the means for nodes to send and receive messages over a network. The word "transceiver" is a combination of "transmitter" and "receiver" since it handles both sending and receiving messages. Each node on the network must have a transceiver for communication.

Signaling

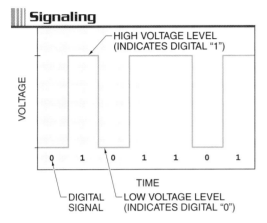

Figure 3-2. Electrical signaling commonly uses changes between two voltage levels to indicate the ones and zeroes of a digital signal.

Each signaling method has characteristics that make it more appropriate for certain situations. Some methods operate faster but require equipment that is more expensive and harder to install or that has distance limitations. For these reasons, there is no "one size fits all" solution, and many methods have evolved to suit different circumstances.

Bandwidth. *Bandwidth* is the maximum rate at which bits can be conveyed by a signaling method over a certain media type. A *media type* is the specification of the characteristics and/or arrangement of the physical conductors or electromagnetic frequencies used for digital communication. Although there are many measures that can be used to express bandwidth, for data communications networks, it is common to describe bandwidth in terms of some quantity of bits per second (bit/s). Depending on the signaling technology, the bandwidth may be thousands of bits per second (kbit/s), millions of bits per second (Mbit/s), or even billions of bits per second (Gbit/s).

As a rule, the speed and performance (bandwidth) of a given signaling method is proportional to the cost. **See Figure 3-3.** The higher-speed, higher-performance technologies typically cost more. Also, the maximum distance of total network length tends to be inversely proportional to speed, and consequently cost. Lower-performance technologies generally allow greater distances for wiring without degradation of the information. Higher performance technologies generally have more severe distance limitations. It is sometimes possible to extend the maximum distance with signal repeaters, but these also increase cost.

Effects of Bandwidth

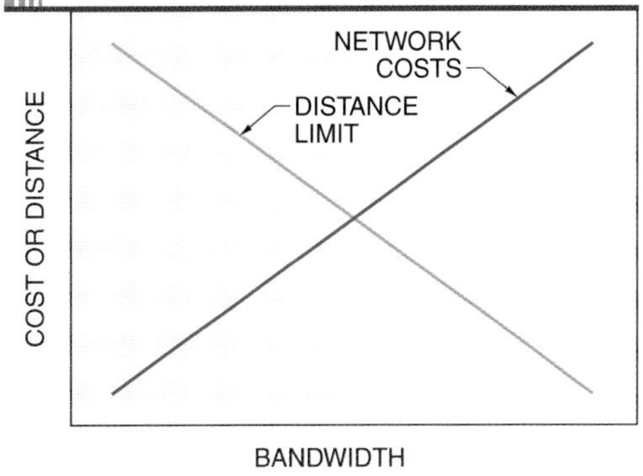

Figure 3-3. When comparing different data communication technologies, the speed of the network tends to be proportional to its cost and inversely proportional to the maximum network distance.

Throughput. *Throughput* is the actual rate at which bits are transmitted over a certain media at a specific time. While bandwidth is the maximum amount of data that can be transmitted over a media type, throughput is the actual measure of data that is transmitted over a specific network route, on a certain media, and at a specific time. Factors that affect throughput include the type of data being transmitted, the network topology, the number of network users, and network congestion. Throughput can never exceed the bandwidth of a media type; it is always less than or equal to bandwidth.

> The concepts and rules of data communication are common to all computer-based networks, including office LANs of personal computers and building automation networks of intelligent controllers. Networks can be as small as two connected devices and as large as the global Internet.

Bandwidth

Bandwidth, throughput, and latency terminology can be confusing and easily used incorrectly. The distinctions between the meanings of these terms can be made clearer with an analogy. For example, a highway can be thought of as the communications media, such as conductors, through which bits (vehicles) travel from one point to another at a fixed speed, such as 55 mph.

The bandwidth of the highway is represented by the number of lanes times the vehicle speed. With more lanes, the highway has a greater capacity for vehicle traffic. However, the actual number of vehicles using the highway (throughput) is less than or equal to its traffic capacity (bandwidth). During rush hour, the throughput is very high, approaching 100% bandwidth utilization. At other times, the throughput can be significantly less.

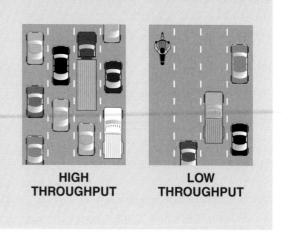

HIGH THROUGHPUT **LOW THROUGHPUT**

Latency. *Latency* is the time delay involved in the transmission of data on a network. In one-way communication, latency is the difference between when a message is transmitted and when it is received. In two-way communication, latency also includes the delay required for the receiving node to respond to the message, such as sending a reply to the message. Increased latency decreases throughput as a fraction of the total bandwidth capacity of the signaling method.

Signaling Directions. Data communication systems are defined partly by the signaling directions enabled by the node and media types. **See Figure 3-4.** *Simplex communication* is a system where data signals can flow in only one direction. These systems are often employed in broadcast networks, where the receivers do not send any data back to the transmitter.

Most computer and node networks utilize duplex communication, which includes two variations. *Half-duplex communication* is a system where data signals can flow in both directions, but only one direction at a time. Once a node begins receiving a signal, it must wait for the transmitter to stop transmitting before it can reply. An example of a half-duplex system is a set of "walkie-talkie"-style two-way radios, where one person must indicate the end of transmission before the other can reply.

Signaling Directions

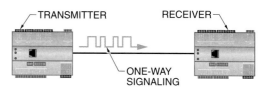

SIMPLEX COMMUNICATION

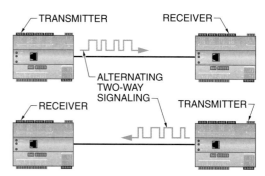

HALF-DUPLEX COMMUNICATION

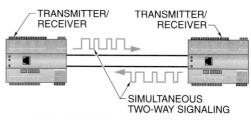

FULL-DUPLEX COMMUNICATION

Figure 3-4. Different types of communication arrangements allow signals to be transmitted in one or both directions.

Full-duplex communication is a system where data signals can flow in both directions simultaneously. For example, landline telephone networks are full-duplex systems because callers can speak and be heard at the same time. Full-duplex systems often use separate sets of conductors, one for each direction, to accomplish this. Full-duplex communication improves throughput, since there are no collisions that require retransmission. Full bandwidth is available in both directions and nodes do not require media access methods because there is only one transmitter for each twisted pair.

Message Frames

All networks send individual messages in discrete units that may be called frames or packets, though these terms have slightly different meanings. A *packet* is a collection of data message information to be conveyed. A *frame* is a packet surrounded by additional data to facilitate its successful transmission and reception by delineating the start and end (or length) of the packet. Frames can have somewhat different structures depending on the message protocol, but most have similar parts. **See Figure 3-5.**

Each frame has a beginning and an end that mark the frame using a special sequence of bits. The names of the beginning and end markers are not standardized, but there are several common terms. The beginning marker is often referred to as the header, start-of-frame, or preamble, and the end marker is often referred to as the trailer, end-of-frame, or postamble. Between the frame markers, the data is structured to have meaning.

Almost all sections of a frame are transmitted as a sequence of octets. An *octet* is a sequence of 8 bits. These are sometimes also referred to as bytes. In this context, the term byte is usually accurate but is a less precise term because it has been used for groupings of other than 8 bits in older computer systems. Octet, by definition, is always a group of 8 bits, so it is always an accurate term for this context.

> It is important to pay close attention to the numbering systems used in data communication contexts. The binary nature of data sharing lends itself to non-decimal (base 10) numbering systems, such as hexadecimal (base 16). Numbers in these systems may be easily confused, so hexadecimal numbers are typically distinguished with the prefix 0x. For example, the hexadecimal number 0x38 is equal to the decimal number 56. Other, less common hexadecimal designations include h, #, and hex.

Message Frames

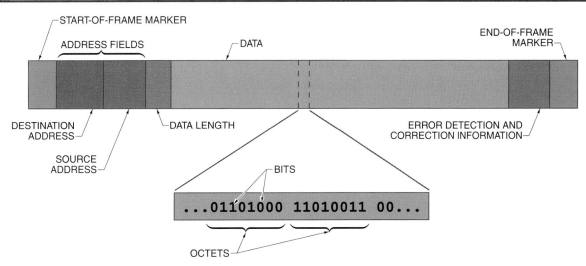

Figure 3-5. Message frames are like envelopes for sharing data between nodes. Frames include defined fields for addresses, data type and length, and error-checking information, in addition to the actual data payload.

Frame Fields. Information is conveyed within the boundaries of a frame using logical groupings called fields. Most protocols use fields that are made up of one or more sequential octets. Frames typically include several fields arranged in a certain order.

The two primary fields or groups of fields are address and data information. The address octets encode where a frame is supposed to be going and where it came from. Messages may vary in length, so a portion of this information usually also indicates the length of the data area. In some protocols, though, the length must be implied from the receipt of the end-of-frame marker.

In many cases, frames also contain error-checking information that can be used to detect if the data has been altered or corrupted during transmission. This field is typically called a frame-check sequence (FCS). There are many different schemes for detecting errors using mathematical algorithms, such as a cyclical redundancy check (CRC). Some sophisticated methods can not only detect, but also correct, some errors.

Addresses. The address portion of a frame identifies the sender and the recipient of the message. To prevent confusion, each node on a network segment must have some unique address. Different protocols and frame types use different lengths of the address field. The length of the node address defines the number of unique network addresses, which determines the maximum number of nodes that may communicate on a network. **See Figure 3-6.** For a single-octet address, the maximum number of nodes is only 256 (2^8), which is the number of unique numbers in one octet. For Ethernet, which uses a 6-octet address, the maximum number of nodes is much larger: 281,474,976,710,656 (2^{48}).

Some address schemes reserve a portion of the possible addresses for special uses like broadcasting and multicasting. A *broadcast* is the transmission of a message intended for all nodes on the network. Some networks allow nodes to be members of one or more logical groups, which have multicast addresses. A *multicast* is the transmission of a message intended for multiple nodes, which are all assigned to the same multicast group. A multicast is only received by nodes that are configured to be members of that particular multicast group. A given node may receive messages addressed specifically to its unique node address, a broadcast address, or to a multicast group that it is a member of.

Address Length	
Number of Octets	Number of Unique Network Addresses
1	256
2	65,536
3	16,777,216
4	4,294,967,296
5	1,099,511,627,776
6	281,474,976,710,656

Figure 3-6. The number of octets reserved for node addresses in the frame determines how many unique addresses (nodes) the network can support.

Sometimes a network message must be directed to a certain application program on a node, so a node-only address is not sufficient. In this case, a port number is added to identify the application destination. A *port* is a virtual data connection used by nodes to exchange data directly with certain application programs on other nodes. The most common ports are TCP and UDP ports, which are used to exchange data between computers on the Internet.

Ports are identified by a 16-bit number, which is often specified following the node address. For example, TCP port 80 is used to share hypertext transfer protocol (HTTP) information with web browsers. Some port numbers are officially assigned to certain applications, while others are available for any use. By using port numbers, messages for different applications on the same node can be received and used efficiently within the node.

Segmentation. There is a fixed maximum size for the payload in the data portion of a frame, though this varies between different technologies. Typically, the limit is in the 512 octet to 1500 octet range. However, some data communication applications, including building automation, require relatively large data payloads. Therefore, nodes that need to send more data than the limit allows must break the data stream into segments. At the destination, the segments are reassembled back into one large data unit.

Endianness

The octets in a frame are typically represented graphically as beginning at the left and ending at the right, but this may not reflect the order in which the octets and bits within a field are transmitted. This order in which data is transmitted is an example of one of the rules that make up a protocol. Endianness is the protocol rule that determines from which end a series of bits or octets are transmitted. A big endian protocol transmits the most significant octet (the "big end") first. A little endian protocol transmits the least significant octet (the "little end") first.

For example, the decimal number 367 is 0x016F in hexadecimal, which is composed of the octets 0x01 and 0x6F. The most significant octet is 0x01, so it is transmitted first in a big endian scheme, which follows the left-to-right order of the graphic representation of the frame. In the little endian scheme, 0x6F is the least significant octet, so it is transmitted first. The effect of this is that bits within frame fields are transmitted from right-to-left, even though the fields are graphically arranged left-to-right.

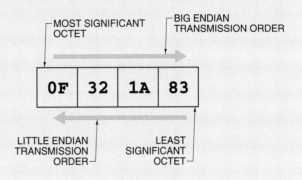

Segmentation is a protocol mechanism that controls the orderly transmission of large data in small pieces. **See Figure 3-7.** Each segment must be marked with a sequence number so that the receiving end knows where each piece goes in the fully assembled data unit. This is also important because pieces may arrive out of order.

In most cases, the sender wants assurance that all of the pieces have made it to the destination. Each piece that is successfully received can be individually acknowledged by the receiver. However, this approach is costly in terms of time and network bandwidth. Instead, most protocols allow groups of segments to be acknowledged all at once. This reduces the number of acknowledgments required, but retains the efficiency of resending only failed segments.

Media Access

Every protocol must define a method for managing when nodes can transmit to each other. This determines how they access the medium of the network. If more than one node transmits at the same time, the messages can collide. A *collision* is the interaction of two messages on the same network media, which can cause data corruption and errors. **See Figure 3-8.** As the traffic of messages on the network increases, the problem of collisions worsens. Several methods have been developed to resolve this problem.

Master/Slave. In master/slave networks, the master node controls the message traffic. Slave nodes transmit only when granted permission by the master. The failure of the master makes all communication impossible. Historically, this has been the most commonly used method of media access control in building automation systems.

However, implementing this method can be risky because failure of a single node, the master, blocks communication between the remaining functional nodes. In response to this issue, media access methods have been developed for peer-to-peer networks, where each node has equal rights and responsibilities.

Contention. Contention is a media access control method that is used with peer-to-peer networks. The contention method allows a node access to the medium at any time, but the node must choose the best time to begin transmitting. The node listens for activity on the medium and waits for silence before starting to transmit, trying to avoid collisions.

However, two nodes may detect silence and then begin transmitting at the same time, causing collisions. There are various contention-based media access control schemes that try to address this possibility. Some can detect the collision and immediately stop transmitting. The node then waits a short time and attempts to transmit again.

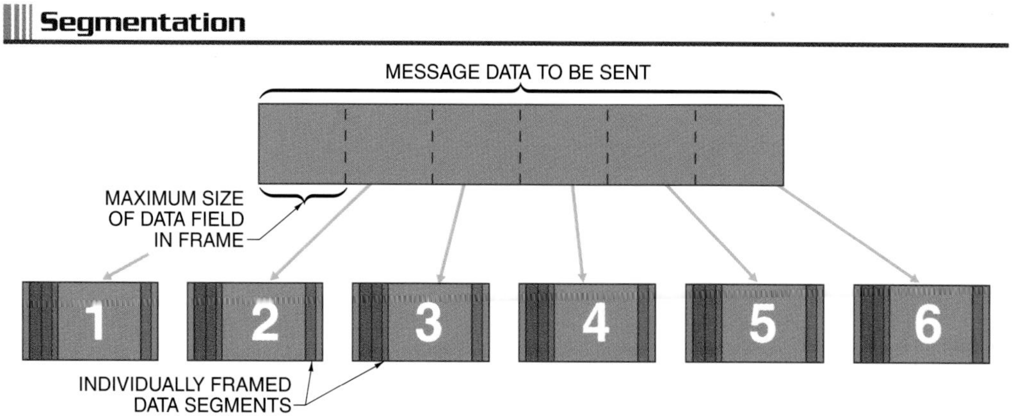

Segmentation

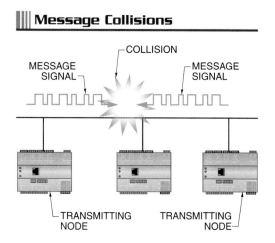

Figure 3-7. Segmentation breaks large messages into smaller pieces that are sent in individual message frames and reassembled at the receiving side.

Message Collisions

Figure 3-8. Message signals can collide on the network medium if two nodes transmit at the same time.

Contention schemes degrade in performance as traffic increases. More traffic causes more collisions, retries, and transmission failures. Also, due to the random-length delays, one cannot predict exactly how long it will take a message to reach its destination. Because of this, the contention method is called nondeterministic.

Token Passing. The token-passing method requires nodes to receive a token message before being allowed to access the medium. **See Figure 3-9.** Each node voluntarily passes ownership of the token to the next node so that the medium is shared in a predictable manner. Nodes must wait until they receive the token before they can transmit. Once a node has the token, it may transmit before passing the token again.

The advantage of this orderly technique is that the performance of the network is guaranteed. Since the worst-case transit time for the token can be calculated, this kind of network is called deterministic. However, the disadvantage is that a portion of the network's available bandwidth is wasted on passing the token. Also, a node must wait for the token before transmitting, even when traffic is light. The token-passing method must also be capable of dealing with nodes entering and leaving the network, as well as instances when the token is lost.

OPEN SYSTEMS INTERCONNECTION (OSI) MODEL

The International Organization for Standardization (ISO) has led an effort to develop a wide assortment of data communications standards that foster the design and implementation of open computer systems that can be interconnected for a variety of applications. The first open-system standard approved by the ISO was the Basic Reference Model (ISO 7498), also known as the Open Systems Interconnection (OSI) Model. The *Open Systems Interconnection (OSI) Model* is a standard description of the various layers of data communication commonly used in computer-based networks. The purpose of this standard was to create a framework that could be used as the basis for defining standard communication protocols.

Token Passing

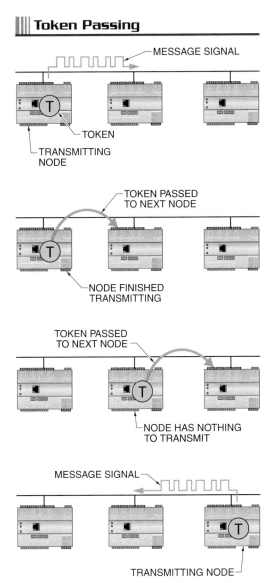

Figure 3-9. Token passing shares the right to transmit between cooperating nodes. The nodes pass the token throughout the network until it is received by a node that needs to transmit.

The concept behind the OSI Model is to divide the very complex problem of computer-to-computer communication into several smaller pieces. Each piece has a carefully defined function to perform, and a well-defined interface through which it interacts with the other pieces. This arrangement sometimes even allows replacement of one piece by another, providing the same functions without changing the rest of the communication hardware and software. This is very similar to the way computer programs are divided into subroutines and functions.

The OSI Model arranges these functional pieces in a hierarchy of layers. **See Figure 3-10.** Each layer provides services to higher layers and relies on the services provided by lower layers. A *protocol stack* is a combination of OSI layers and the specific protocols that perform the functions in each layer. This is also known as a protocol suite. Protocol stacks may include some or all of the OSI layers.

Each computer has its own protocol stack. Parallel layers in each node are called peer layers. Peer layers communicate with each other through their own matching protocol. In the sending computer, messages flow down one protocol stack, and each layer may add data to the message according to its specific protocol. **See Figure 3-11.** At each layer, the data portion of the message is made up of the original message plus the cumulative overhead from all of the higher layers. Although it is possible for the data itself to be transformed, for example through encryption or translation, most often a layer simply adds a new header and/or trailer before passing the message down to the next layer. The final message is then transmitted across the physical medium to the receiving computer, where it then passes up its protocol stack. Each layer strips out the portion of the message data that is meant for it and passes the remainder on.

Protocol Terminology

When discussing the organization of functions according to the OSI Model, the term "protocol" may become confusing. In general terms, a protocol is the set of rules and procedures that permit the reliable exchange of information between computers. However, this definition can apply to both the low-level protocols managing just one aspect of the communication and the group of these protocols that are used together to form a complete communication solution. The latter are more accurately called protocol stacks or protocol suites, though use of the word "protocol" alone for these technologies is not uncommon.

Open Systems Interconnection (OSI) Model

Layer	Type	Function
Application	End-to-end communication	Interfaces with user's application program
Presentation	End-to-end communication	Converts codes, encrypts/decrypts, and reorganizes data
Session	End-to-end communication	Manages dialog and synchronizes data transfers
Transport		Provides error detection and correction, segmentation, and reliable end-to-end transmission
Network		Manages logical addressing and determines routing between nodes
Data Link	Point-to-point communication	Manages physical addressing and orderly access to physical transmission medium
Physical		Transmits and receives individual bits on the physical medium

Figure 3-10. The OSI Model breaks the complex procedure of data communications between network nodes into seven different layers, each with specific responsibilities.

Protocol Stack

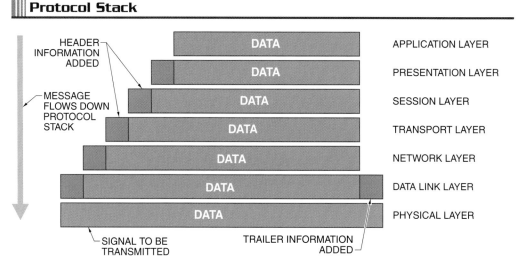

Figure 3-11. Each OSI layer in a protocol stack adds information, usually in the form of a header, to a message before passing it down to the next layer.

Physical Layer

The physical layer is the lowest layer in the OSI Model. The *physical layer* is the OSI Model layer that provides for signaling (the transmission of a stream of bits) over a communication channel. No frame headers or trailers are added to the data by the physical layer protocol. The protocols for the physical layer define the signaling and wiring rules for communications, such as how a binary 0 or 1 is to be represented. For example, a voltage above a certain level represents a 1 and a voltage below another level represents a 0. Physical layer protocols also define the types of physical media to be used to transmit the data (such as twisted-pair conductors, coaxial cable, optical fiber, or infrared transmission), the connectors to be used, and the allowable network arrangements.

Data Link Layer

The *data link layer* is the OSI Model layer that provides the rules for accessing the communication medium, uniquely identifying (addressing) each node, and detecting errors produced by electrical noise or other problems. Some data link protocols also provide a way to correct transmission errors.

MAC Layer. In the OSI Model, the physical and data link layer protocols form a special combination known as the MAC layer. The *MAC layer* is a sublayer of the OSI Model that combines functions of the physical and data link layers to provide a complete interface to the communications medium. This interface presents each message as a collection of bits that is destined for a particular node, which is identified by its MAC address. **See Figure 3-12.** A *MAC address* is a node's address that is based on the addressing scheme of the associated data link layer protocol. Incoming messages received by a MAC layer are similarly presented to other layers as a collection of bits that came from a particular MAC address.

The MAC layer's job is to structure the data into the appropriate form suitable for the particular underlying framing and signaling technology, gain access to the signaling medium, and then transmit the message (or receive incoming ones). There are many kinds of MAC layer technologies, each using very different schemes for organizing data into frames, detecting and correcting errors, arbitrating media access, and signaling.

Local Area Networks (LANs). A *local area network (LAN)* is the infrastructure for data communication within a limited geographic region, such as a building or a portion of a building. **See Figure 3-13.** A LAN utilizes a particular MAC layer type. A LAN does not specify anything about the content of the messages or the way in which the messages are encoded. Higher-layer protocols provide these specifications. A single LAN can carry messages from several incompatible higher-layer protocols simultaneously. For example, TCP/IP, NetBEUI, and IPX/SPX messages can coexist on the same LAN. In effect, a LAN is a collection of computers using the same MAC layer on a common network.

Network Layer

The *network layer* is the OSI Model layer that provides for the interconnection of multiple LAN types (MAC layers) into a single internetwork. The LANs may be distinct because of differences in the LAN protocols or because there is a need to segregate address spaces. An *address space* is the logical collection of all possible LAN addresses for a given MAC layer type. For example, an 8-bit MAC address allows for a maximum of 256 possible addresses, and a 48-bit address allows for more than 280 trillion addresses in the address space.

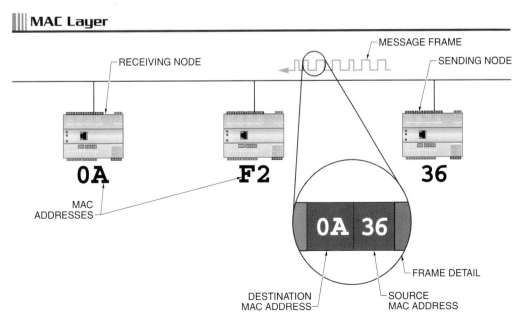

Figure 3-12. The MAC layer is an interface to the communication medium and handles addressing.

Local Area Networks (LANs)

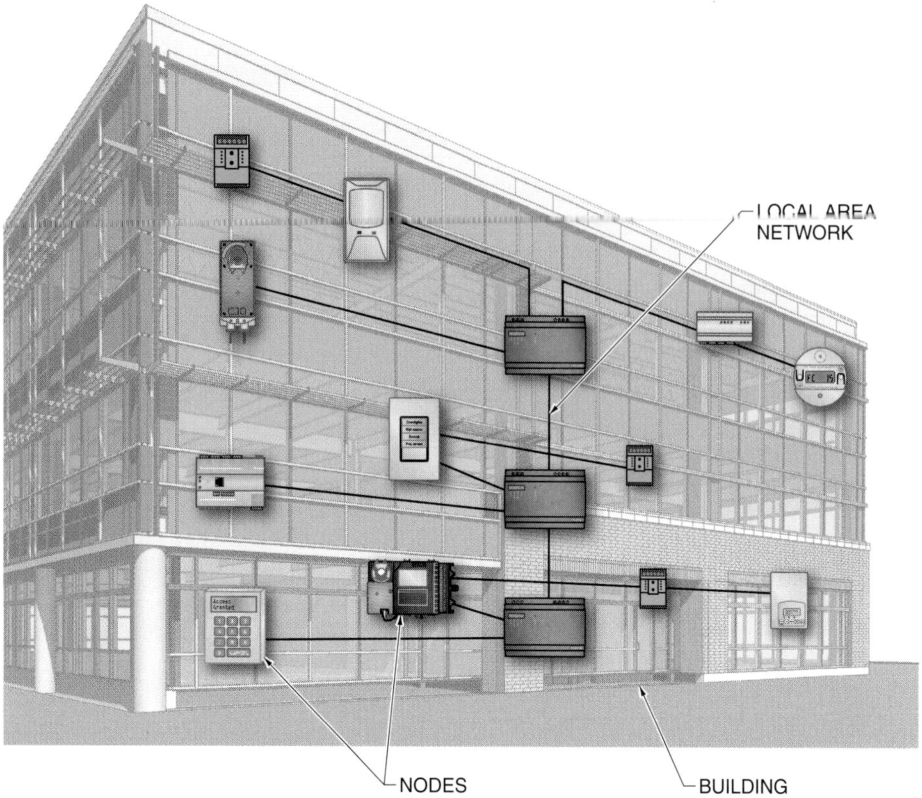

Figure 3-13. Local area networks (LANs) use the same MAC layer technologies and are limited to a small geographic area, like a building.

Some internetworks allow for multiple paths between computers on different LANs. The network layer protocol is primarily responsible for managing the delivery of messages through different possible routes. *Routing* is the process of determining the path between LANs that is required to deliver a message. **See Figure 3-14.** If there are multiple route choices, the network layer protocol may decide on a route based on certain criteria such as the cost, reliability, and speed of the routes. Sometimes network layer protocols provide message segmentation, in which case they are also responsible for message reassembly. Because nodes that perform routing functions have limited resources, some routes may become congested if there is a large amount of message traffic. Under these conditions, the network layer protocol may also be responsible for flow control and coordination with other routing nodes to manage and alleviate congestion.

Transport Layer

The transport layer is the first of the upper layers of the protocol stack. The *transport layer* is the OSI Model layer that manages the end-to-end delivery of messages across multiple LAN types. The lower layers (the physical, data link, and network layers) address only point-to-point protocol issues. This distinction is only relevant in internetworks, which transmit messages across multiple LANs.

The transport layer provides end-to-end error detection and correction, which may rely on network layer services. The transport layer protocol may also provide segmentation and reassembly of long messages, and a variety of levels of quality-of-service (QOS). QOS may be characterized in terms of throughput, transmit delay, residual error rate, and failure probabilities. The user of the transport service is guaranteed a particular QOS that is independent of any changes in the underlying network service or its quality.

Routing

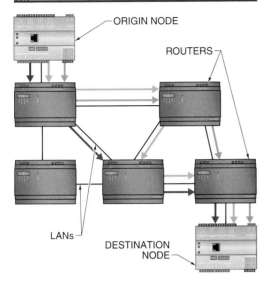

Figure 3-14. If network connections allow more than one route between the sending and receiving nodes, the network layer handles the routing choices.

Session Layer

The *session layer* is the OSI Model layer that provides mechanisms to manage a long series of messages that constitute a dialog. **See Figure 3-15.** The transfer of large files is an example of a dialog. The session layer organizes the exchange of data into a series of dialog units. Checkpoints between dialog units provide the ability to resynchronize the communication at intermediate points after a communication failure rather than at the start of the transaction.

Dialog Sessions

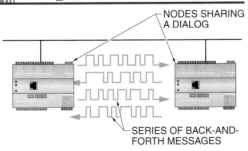

Figure 3-15. The session layer protocol is responsible for back-and-forth communications between two nodes, known as dialogs.

> The application layer provides services to application processes. Examples of application layer protocols include HTTP, FTP, TELNET, SOAP, IRC, SMTP, IMAP, and POP. The application process must still then provide the user interface, which is not part of the OSI Model.

Presentation Layer

The *presentation layer* is the OSI Model layer that provides transformation of the syntax of the data exchanged between application layer entities. The presentation layer protocol permits each local application to represent information using its own syntax but still share information with other applications. The protocol converts this local syntax into one of many possible transfer syntaxes. The appropriate transfer syntax to use is negotiated by the peer-presentation entities. This ensures that the data exchanged can be interpreted appropriately by the two application layer entities. Also, data compression and encryption is typically done at the presentation layer.

Application Layer

The application layer is the endpoint of the OSI Model. The *application layer* is the OSI Model layer that provides communication services between application programs. A specific type of application program interfaces with the application layer in order to participate in communications between itself and peer application programs in other nodes. There are many possible applications for computer-to-computer communication, so the details of a particular application are not part of the OSI Model. One can think of the application layer as a choice between different possible applications at a given endpoint, not unlike a telephone extension at a company telephone number.

There are many well-known applications as well as unique or proprietary applications that fit the OSI Model. An application protocol is a specific kind of communication, with its own rules, that communicates between peer entities at the application layer. For example, an Internet browser communicates to a website using the application protocol hypertext transport protocol (HTTP).

OSI Model Analogy

The OSI Model divides the process of each network transmission into a list of successive tasks, assigning a layer of responsibility for each one. This simplifies the implementation of a particular function and makes it somewhat independent of the other layers. However, the distinction between the responsibilities of the individual layers can be confusing. To understand each layer's function, it is often useful to apply the concepts in the OSI Model to a type of communication that is easier to relate to. For example, consider the following situation in which important messages are exchanged between company offices.

Sending Side	Layer	Receiving Side
The head of a company decides to send an important message to a colleague in another city. The message is dictated to an assistant as a letter.	**APPLICATION**	The colleague receives the message.
The assistant transcribes the dictation into written text.	**PRESENTATION**	The assistant reads the letter and decides how to get the message to the addressee, such as transcribing it to e-mail or reading it aloud over the phone.
The assistant puts the letter in an envelope marked "Urgent" and gives it to the mailroom to send.	**SESSION**	The assistant takes the letter out of the envelope.
The mailroom decides how to send the letter. Since it is urgent, the envelope is given to a courier company.	**TRANSPORT**	The mailroom removes the inner envelope from the courier envelope and delivers it to the correct office.
The courier company determines the best route to the destination city. It places the envelope in a larger courier envelope and hands it off to the workers who load courier packages on airplanes.	**NETWORK**	The routing office sees that the letter has reached its destination city and delivers the envelope to the destination company.
The courier workers tag the courier envelope with the code for the destination city, pack it into a box, and load it onto the appropriate airplane.	**DATA LINK**	The box is unloaded and the courier envelope is unpacked and given to the courier company's local routing office.
The airplane flies to the destination city.	**PHYSICAL**	The airplane arrives in the destination city.

Collapsed Architectures

The strengths of the OSI Model are its generality and adaptability. It can be applied to almost any data communication application between computers located anywhere in the world. However, it provides for a rich set of functionality that may not be necessary for every application. Since each function and each layer protocol adds size to the message, which reduces the efficiency of the system, it makes sense to consider adapting the model to a particular application by eliminating unnecessary functionality. A *collapsed architecture* is a protocol stack that does not include layers that are not needed for the application. In most collapsed architectures, some of the functionality of the missing layers may be provided by protocols in the other layers. This is common for most applications.

A good example of a collapsed architecture is the BACnet protocol for building automation and control networks. **See Figure 3-16.** The few aspects of the presentation, session, and transport layers that are needed by BACnet nodes are instead implemented as part of the application layer.

Collapsed Architectures

[Figure showing Full 7-Layer Architecture vs. Collapsed Architecture]

Figure 3-16. If a protocol does not need to implement each layer of the OSI Model, its collapsed architecture eliminates the unneeded layers.

Protocol Data Units

Each successive OSI Model layer adds fields of information to the beginning and end of a frame that are needed to fulfill its part of the transmission. These can be thought of as envelopes within envelopes. It is sometimes necessary to refer to one of these "envelopes" and its contents, which is known as a protocol data unit. A *protocol data unit* is the portion of the frame containing fields belonging to a certain OSI Model layer and the layers above it. **See Figure 3-17.**

An application protocol data unit (APDU) is the information to be shared between node applications. Fields are added to the beginning of this block of data to ensure that the information is sent to the correct node without error. Each successive protocol data unit is named for the lowest layer that contributed frame fields to the unit. For example, the network protocol data unit (NPDU) includes the APDU, network layer headers, and any presentation, session, and transport layer headers (if used). The NPDU is the envelope that directs an application message payload to its intended recipient.

Likewise, the NPDU is the payload of the link protocol data unit (LPDU), and the LPDU is the payload of the MAC protocol data unit (MPDU). The MPDU consists of the entire frame except the start-of-frame and end-of-frame markers, and error-checking information. Regardless of the MAC layer technology, all MPDUs share four conceptual characteristics: destination and source MAC addresses, a special broadcast address, a payload data length, and a payload that may contain any amount of data, up to some maximum size.

> The payload of one protocol data unit, sometimes called a service data unit (SDU), is the entirety of the enclosed protocol data unit. For example, the LPDU is the payload of the MPDU, which is then the payload of the physical layer protocol data unit. The physical layer protocol data unit includes the entire frame.

Protocol Data Units

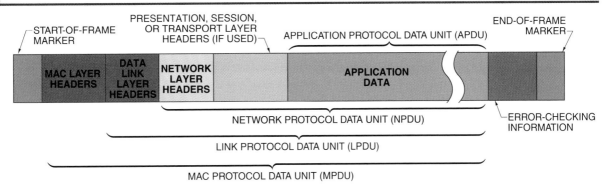

Figure 3-17. Protocol data units are named for the highest OSI Model layer information included in the unit.

NETWORK ARCHITECTURE

In the simplest network, two or more nodes share a segment. A *segment* is a portion of a network where all of the nodes share common wiring. **See Figure 3-18.** Segments are usually limited to certain lengths, depending on the characteristics of the particular physical media type. *Network architecture* is the physical design of a communication network, including the network devices and how they connect segments together to form more complex networks.

Segment

Figure 3-18. A segment is a section of network wiring that is continuous at the physical layer.

Network Devices

Network devices can be used to extend segments, connect segments together, and transition between different types of physical media, for example, from twisted-pair to optical fiber. A *logical segment* is a combination of multiple segments that are joined together with network devices that do not change the fundamental behavior of the LAN.

Repeaters. Physical layer protocols often place restrictions on the length of an individual segment. A *repeater* is a network device that amplifies and repeats the electrical signals, providing a simple way to extend the length of a segment. **See Figure 3-19.** It does not understand addresses and cannot filter or route traffic. A repeater forwards all traffic, regardless of validity. This means that corrupted packets or noise may also be repeated. There is a slight delay, or latency, from the time the signal arrives to when it is repeated. Since these delays are cumulative, there may be a limit to the number of repeaters that can be used to extend the segment. Repeaters operate at the physical layer of the OSI Model.

Repeaters

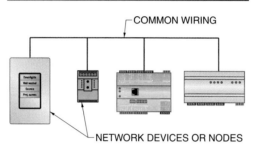

Figure 3-19. A repeater amplifies and repeats message signals at the physical layer.

Network Device Symbols

Network devices are represented in network diagrams by symbols that indicate their respective functions. The telecommunications industry has not standardized these symbols, but many are used so commonly that they have become de facto standards. Variations are commonly used to denote devices with special features or a combination of network functions.

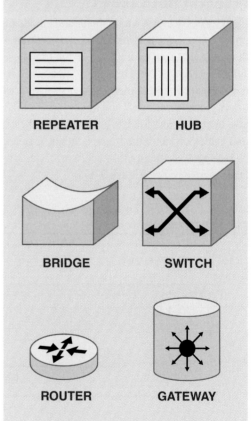

Hubs. A hub is a multiport repeater, so it also operates at the physical layer of the OSI Model. A *hub* is a network device that repeats messages from one port onto all of its other ports. When a hub detects a signal on one of its ports, it retransmits the signal through all the other ports of the hub. **See Figure 3-20.** Like repeaters, there may also be practical limits on the number of hubs that can be used along any given path because of the cumulative latency. Physical constraints on the electrical characteristics of the signaling technology may also limit the maximum distance of wiring from a node to a hub, or from hub to hub, thus further limiting the total length of a network.

Hubs

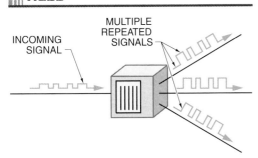

Figure 3-20. Hubs are multiport repeaters that amplify and repeat all signals that they receive.

Bridges. A *bridge* is a network device that joins two LANs at the data link and physical layers. **See Figure 3-21.** When two LANs are bridged, the address of each node must remain unique in the extended network. Bridges provide traffic filtering based on destination address. The locations of nodes on the network are learned as the bridge receives messages on its ports.

A bridge can be used to extend the length of a LAN as a traffic management device, or to connect two different topologies or physical media. For example, a bridge can connect a LAN of standard (thick) Ethernet (10BASE5) with a LAN of fiber optic Ethernet (10BASE-FL).

The placement of a bridge within a network is very similar to that of a repeater. The key difference is that a bridge understands data link layer addressing and can use that knowledge to provide filtering. A repeater simply amplifies and retransmits the electrical signals.

Bridges

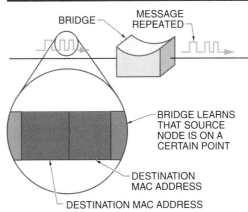

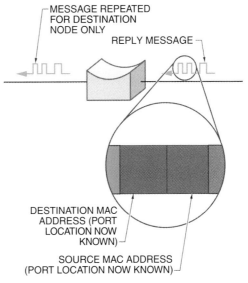

Figure 3-21. Bridges operate at the data link layer. They operate as hubs until they learn the locations of nodes by reading the source addresses; then they forward messages only to the port associated with the known address.

Switches. A switch is essentially a multiport bridge. A *switch* is a network device that can forward messages selectively to one of its other ports based on the destination address. A switch is designed for a particular kind of message frame. It can read addresses from the frames and learn which addresses belong to which ports. Once learned, the switch can immediately repeat the message to the correct port. **See Figure 3-22.** If the destination node's port is unknown, or the destination is a broadcast address, it can repeat the message on all ports.

Switches

Internetwork

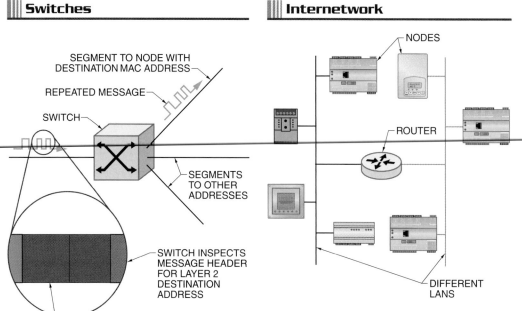

Figure 3-22. Switches forward messages selectively based on addresses in the data link layer.

Figure 3-23. Routers connect LANs together into internetworks and provide routing, segmentation, reassembly, and flow control.

With contention-based MAC layers, switches allow simultaneous communication between multiple port pairs without collisions. Also, since the switch has a dedicated point-to-point connection to each node, it can simultaneously transmit and receive to that node. Because signal-timing constraints are generally imposed on signals rather than frames, this can help to mitigate constraints on timing and distance and significantly reduce collisions even in high-traffic situations.

Routers. Two or more LANs are connected together into a single unified network with a router. A *router* is a network device that joins two or more LANs together at the network layer and manages the transmission of messages between them. The LANs may have fundamentally different physical, electrical, signaling, or behavior characteristics, or the LANs may be similar and the router only fills a logical requirement to separate the network segments. LANs joined with routers form an internetwork. **See Figure 3-23.** An *internetwork* is a network that involves the interaction between LANs through routers at the network layer. The most famous example of an internetwork is the Internet.

A router continuously listens for messages on all of its ports and isolates network traffic by selectively forwarding network messages based on the destination address. When the router receives a message on one LAN segment, it inspects the message for the destination address and completeness. If the message is destined for a node on another LAN segment, it repeats the message on the port for the proper LAN segment. **See Figure 3-24.** If the message is destined for a node within one LAN segment only, the router does not repeat it. Routers also validate messages and do not forward them if they are corrupted. All of these functions help reduce excess communication traffic that can negatively impact network performance.

The router does not need to understand, or even look at, the content of a message. It only needs to know the source and destination of each message. This is analogous to the post office delivering a letter while knowing only the destination and return addresses printed on the envelope. The logical segments that represent each LAN each have a separate address, much like the town or city in a mailing address.

Messages between nodes on an internetwork may have many possible paths. The network protocol determines the best route for the message to take. This may involve multiple hops across routers from one LAN to the next, but the network layer usually only needs to know how to direct messages to the nearest intermediate destination. A distinct address for each network provides for the possibility that two nodes have the same MAC address on their respective networks.

Configured routers are the most common type of router. These forward only valid packets to segments defined in the internal routing tables. Routing tables are updated by the network management tool when segments are added to the network design. Alternatively, learning routers dynamically update their internal routing tables based upon the source addresses of incoming message packets. When learning routers are reset, such as during a power cycle, they must relearn the subnet locations, which may cause an initial flood of network traffic.

A *firewall* is a router-type device that allows or blocks the passage of packets depending on a set of rules for restricting access. Modern firewalls include stateful handling, which recognizes the progression of states in a typical TCP conversation and uses the state as part of the rules for determining legitimate conversations between nodes inside and outside of the firewall.

Gateways. The primary goal of peer-to-peer networks is for each node to be able to communicate effectively with its peers, which typically involves nodes that use the same application protocol. However, there are many situations where more than one protocol is implemented within the same network. This is very common in building automation systems.

When two or more application protocols are being used to perform similar functions, a gateway can be used to integrate them together into one communication system. A *gateway* is a network device that translates transmitted information between different protocols. **See Figure 3-25.**

A gateway is similar to a router in that its job is to connect to two or more LANs and manage the necessary communication between them. However, a gateway has an additional and much more complex task. It cannot simply repeat the message on the other LAN because the application language or message content must be translated conceptually from one protocol into the other. The gateway has two independent protocol stacks, and possibly even physically separate and different types of physical media. The gateway makes a logical connection between peer entities at the application layer. **See Figure 3-26.**

Routers

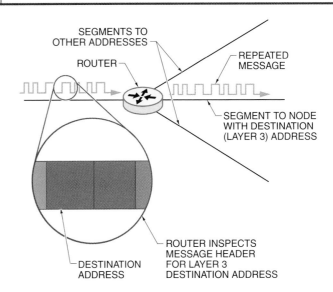

Figure 3-24. Routers forward messages selectively based on addresses in the network layer, providing transitions between LANs and managing traffic between groups of nodes.

Gateways

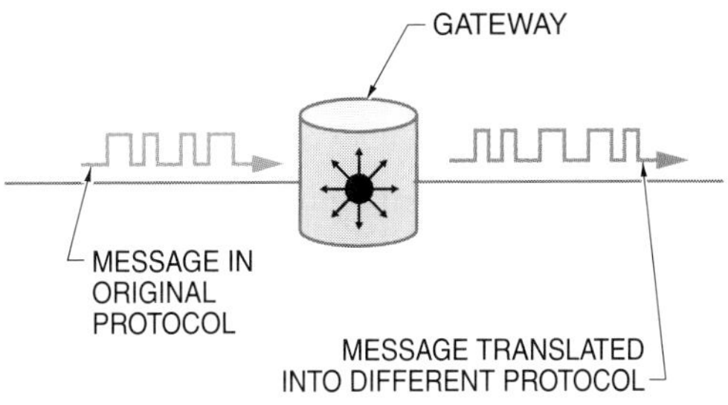

Figure 3-25. Gateways translate between application protocols that have similar functionalities.

Protocol Translation

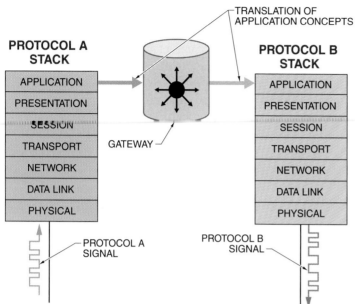

Figure 3-26. Translating between information concepts in two different protocols requires a complete protocol stack for each protocol.

The two application protocols must have common features between them so that it is possible for the gateway to integrate them together. This usually involves the logical association between some concept in one protocol and a similar or equivalent concept in another protocol. *Mapping* is the process of making an association between comparable concepts in a gateway. Some protocols are rich in functionality while others are simpler. When creating gateways between protocols, the protocol designer may sometimes be forced to simulate functionality or to emulate the functions of a more sophisticated device in order to have successful integration between protocols.

Because there are so many possibilities, the OSI Model falls short where gateways are concerned and does not define or model how they might work. As a result, most gateways are very specific to particular application protocols that share common concepts.

Network Topology

Nodes can be connected together in many different ways. In some cases, the signaling method affects the methods, locations, and shape of the connections. *Topology* is the shape of the wiring structure of a communications network. There are four primary physical topologies used in network technologies: bus, star, ring, and mesh. In some instances, technologies can use a mixture of these topologies by using network devices to connect them together.

Bus Topology. A *bus topology* is a linear arrangement of networked nodes with two specific endpoints. **See Figure 3-27.** This topology is often depicted as a line connected to nodes with short wiring stubs, or drops. Since most signaling methods use at least two conductors, this topology can also be drawn as a ladder shape.

Bus Topology

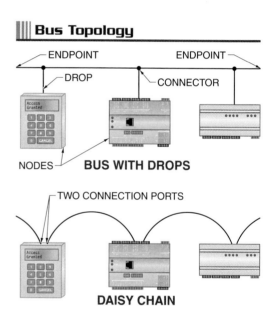

Figure 3-27. Bus topologies can be built by stubbing short connections to nodes off a continuous bus, or by daisy-chaining the nodes together.

However, the extra conductor lengths and connectors required for this arrangement can cause unwanted changes in electrical impedance, so this wiring method is rarely practical. A more common configuration of bus topology is a daisy chain. A *daisy chain* is a wiring implementation of bus topology that connects each node to its neighbor on either side. Nodes that can be used in bus topology typically include two connection ports or sets of terminals to facilitate daisy chaining.

Installing a bus topology network is relatively simple and it is easy to troubleshoot because segments can be easily isolated in order to locate failures. However, a disadvantage to bus topology is that it is sensitive to node failures and wiring problems. A single short or open connection along the bus can disable all or some of the other nodes. Also, the expansion of a bus-topology segment can be challenging because the daisy-chain configuration must be opened to add new nodes.

Star Topology. A *star topology* is a radial arrangement of networked nodes. At the center of a star topology is a hub, which includes multiple ports with which to connect to each node. **See Figure 3-28.** Each node has a dedicated connection to a hub port, which overcomes a principal limitation of bus topologies by providing isolation of individual nodes from others if a node or wiring failure occurs. The hubs can also be thought of as an inverted tree structure.

Ring Topology. A *ring topology* is a closed-loop arrangement of networked nodes. The network forms a loop, with each node having access to both ends of a segment. **See Figure 3-29.** Some ring networks may also use hubs along the ring, similar to star topology. These hubs facilitate connections to other rings. The principle advantage to the ring topology is that the failure of any one node does not compromise the rest of the nodes on the ring because there are always two paths to each node.

Ring Topology

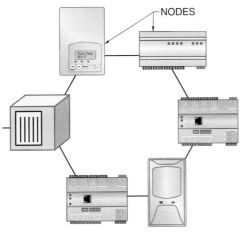

Figure 3-29. Ring topologies are versions of buses where the ends are both connected to a hub.

Star Topology

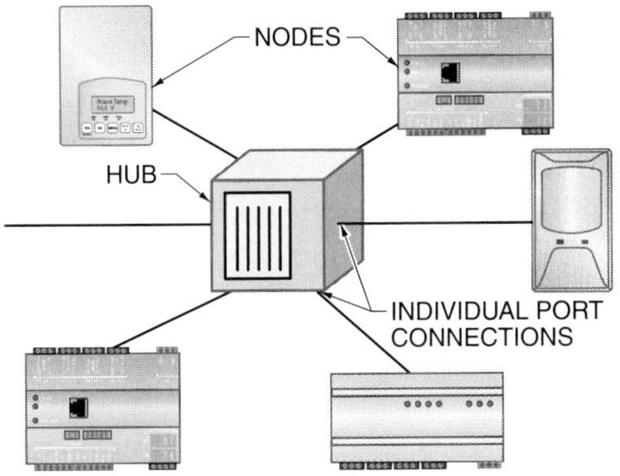

Figure 3-28. Star topologies use hubs to isolate segments and provide each node with a dedicated port.

Hubs can be connected to other hubs to form a collection of stars. The hubs may be daisy chained in a bus structure, though this arrangement is vulnerable to the same failure issues as other bus networks. Alternatively, it is quite common to connect hubs in hierarchies. Not only can less expensive twisted-pair wiring be used, but an individual failure is isolated to its segment (and any segments below it).

Rings are like bus topologies that have both ends connected together or to a hub. Therefore, like buses, the individual nodes can be connected to the ring through either drops or being daisy chained.

Mesh Topology. Mesh topologies are primarily utilized in wireless networks. Each wireless transmitter has a transmission range defined by a circular limit around the node. The nodes in a wireless network are all part of the same radio frequency grouping and can potentially all receive each other's transmitted messages. In practice, though, this would require relatively high-power transmissions for the farthest nodes to communicate with each other. Instead, the mesh topology takes advantage of the fact that many of the nodes are physically close to each other and the overlapping circles of adjacent nodes' transmission ranges form a mesh. A *mesh topology* is an interconnected arrangement of networked nodes. **See Figure 3-30.** Any node's radio transmissions must only have the range to reach the nearest other nodes. Each node is connected to multiple other nodes, and messages can travel in a variety of paths to reach destinations.

Mesh Topology

Figure 3-30. The overlapping transmission ranges of wireless nodes allow the nodes to form a mesh topology, which allows multiple connections between most nodes and their neighbors.

The strength of a mesh network is that each node can operate as a simple repeater. If a node receives a message that is not for itself, it passes the message on to its neighbor. Through multiple repeats, the message reaches its destination. Because multiple nodes participate in the meshing, the topology has a built-in redundancy that provides greater reliability and self-healing properties to the network in the event of node failures. Wireless nodes can also easily establish communication links with nearby nodes. In this way, the range of the network is easily extended and the power requirements for each node are reduced.

Free Topology. A *free topology* is an arrangement of nodes that does not require a specific structure and may include any combination of buses, stars, rings, and meshes. **See Figure 3-31.** A common implementation of free topology connects a variety of separate topologies together into a bus configuration. The bus portion is typically designed as a high-speed backbone network that can quickly route messages to individual network segments.

Free topology offers flexibility during installation, as there is no defined structure to the segment wiring. In addition, future expansion of the network is as simple as adding new nodes anywhere along any segment. However, troubleshooting physical wiring issues with a free topology network can be difficult, especially if there is no wiring documentation. Also, since free topology networks have no defined structure, installers can easily exceed a wire-length limitation for the signaling method.

Termination

Electrical signals can reflect from the ends of conductors and travel back along the length of a network segment, which can cause packet collisions and corrupted data. Segment termination is used to avoid this problem. A *terminator* is a resistor-capacitor circuit connected at one or more points on a communication network to absorb signals, avoiding signal reflections. The required values of the resistors and capacitors depend upon segment type and topology.

Network Tools

The programming and management of a communication network requires software network tools. A *network tool* is a software application that runs on a computer connected to a network and is used to make changes to the operation of the nodes on a network. Network tools are also commonly known as network management tools.

This software is used to define information-sharing relationships, change node communication settings, analyze problems, and perform other management functions. The software saves a representation of the network and its operation on the local computer and loads the necessary settings onto the nodes over the network when changes are made. Once the network is operating, the network tool is not required until changes are needed.

MEDIA TYPES

There are many common types of media used to convey signals for computer-based networks. Media types represent the physical layer characteristics of a particular method of signaling and those physical media (cabling) that may be used with them. These fall primarily into three categories: copper conductors, optical fiber, and radio frequency. Generally, a particular medium is used because it has one or more desirable characteristics such as noise immunity, low power, distance, cost, reliability, or speed.

Copper Conductors

By far, the most common physical media type for computer-based networks, including building automation networks, are copper conductors. The most common types of copper cabling are twisted-pair cable and coaxial cable.

Twisted-pair cable is a multiple-conductor cable in which pairs of individually insulated conductors are twisted together. **See Figure 3-32.** The twisting keeps the conductors a uniform distance apart, which enhances the impedance characteristics and improves noise immunity. Twisted pairs may be shielded with a foil or braided metal wrap around each pair or around the whole group. Twisted-pair cables with different numbers of conductor pairs or types of shielding are readily available. The number of conductors and shielding requirements are determined by the type of signaling to be conveyed with the cable. Category 5 (CAT5) cable, named for its performance specification, is a common twisted-pair cable type that provides four twisted pairs within a common cable jacket.

Coaxial cable is a two-conductor cable in which one conductor runs along the central axis of the cable and the second conductor is formed by a braided wrap. The two conductors are separated by a plastic or foam insulating material. The two conductors share the same central axis, making them "co-axial." This type of cable provides uniform impedance and good capacitance characteristics.

Free Topology

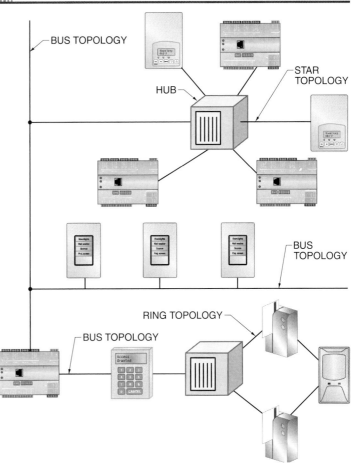

Figure 3-31. Free topology is any mixture of bus, star, ring, and mesh topologies.

Twisted–Pair Cable

Figure 3-32. Twisted-pair cables are the most common type of copper conductors for building automation networking.

TIA-232. The TIA-232 (also known as RS-232) signaling standard defines the electrical characteristics of a type of signaling circuit to be used for computer-to-computer communication based on serially transmitted sequences of data bits. The standard does not define frame structure, only physical characteristics like electrical voltage levels, timing, the slew-rate of signals, and capacitance.

The signals themselves are +15 VDC and −15 VDC, relative to a common ground, although the standard requires receivers to distinguish signals as low as 3 VDC. A negative voltage of −3 VDC to −15 VDC is a logical 1 and is defined as marking or mark level. A positive voltage of 3 VDC to 15 VDC is a logical 0 and is defined as spacing or space level.

The standard defines 20 different signals, but only 9 of them are commonly used. Each uses a different pin in the connector. **See Figure 3-33.** The signals fall into two groups: data and status. The Transmitted Data and Received Data signals use serially transmitted sequences of logical 1/0 bits of data in various framing schemes. The status signals each have logical true/false levels.

TIA-232 Signals

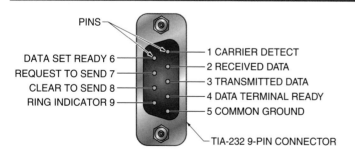

Figure 3-33. TIA-232 signals typically use a 9-pin connector, which carries the standardized data and modem control signals.

In practice, the modem control signals are often not required or used. A valid TIA-232 connection may use only the Transmitted Data, Received Data, and Common Ground connections. Because of the DC voltage level, the distance is also limited by the aggregate voltage drop caused by wire impedance. TIA-232 signaling is typically only used for direct point-to-point connections between two computers.

TIA-485. Like TIA-232, TIA-485 signaling is used to serially transmit sequences of data bits. The standard defines the physical characteristics of the signaling, such as electrical voltage levels, timing, slew-rate of signals, and capacitance. However, TIA-485 signaling is based on a differential voltage across two conductors relative to a common ground connection, which transmits the binary data. One polarity represents a logic 1 level and the opposite polarity represents logic 0. **See Figure 3-34.** Because the signal is differential, it is much more immune to noise than simple voltage-based signaling like TIA-232.

TIA-485 Signals

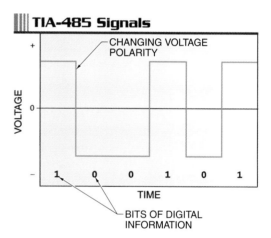

Figure 3-34. TIA-485 signals use two wires (and no ground) to transmit data based on the voltage polarity between the two wires.

The voltage difference must be at least 200 mVDC, but any voltages between +12 VDC and −7 VDC will accurately transmit bits. Although TIA-485 can be used at very high transmission speeds, it is most often applied in automation systems at speeds of 156 kbit/s or lower. Below 100 kbit/s, the signals can be transmitted for a distance of up to 1200 m (about 4000′).

Only one node may transmit at a time. Each node's transceiver contains both a differential driver and a receiver connected to the same two wires. When a node needs to transmit, it enables the appropriate pin, causing the driver to begin transmitting.

In automation systems, TIA-485 signaling is often used as a two-wire, half-duplex, multipoint serial connection. TIA-485 circuits use twisted-pair conductors, with or without a shield conductor. TIA-485 circuits are always arranged in a daisy-chain bus topology. No connector type is specified by the standard.

TP/FT. Twisted-pair/free topology (TP/FT) is a signaling technology that is only used with LonTalk devices. This type of signaling is a differential Manchester encoded signal for serially transmitted data. **See Figure 3-35.** This media allows free topology, though better performance is available for bus topologies. The data rate is 78 kbit/s for up to a maximum distance of 500 m (about 1640′) under a free topology, or up to 2700 m (about 8858′) if a bus topology with double terminations is strictly enforced.

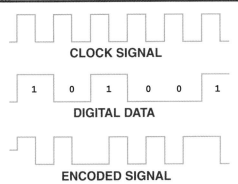

Figure 3-35. Differential Manchester encoding combines a synchronizing clock signal and a data signal into a self-clocking bit stream.

TP/XF. The TP/XF is a twisted-pair technology that is used only with LonTalk devices. The signal is a transformer-isolated differential Manchester encoded signal for serially transmitted data. There are two forms: TP/XF-78 and TP/XF-1250. Both use twisted-pair wiring. The TP/XF-78 can transmit at 78 kbit/s up to a maximum distance of 1400 m (about 4593′). The TP/XF-1250 can transmit at 1250 kbit/s up to a maximum distance of 130 m (about 426′).

Fiber Optics

Fiber optics is a form of signaling based on light pulses to convey signals. A glass or plastic fiber acts like a pipe to convey light over great distances. **See Figure 3-36.** The ends of each fiber are polished and fitted with a standard connector.

Figure 3-36. Glass or plastic fibers can carry optical signals over large distances and are immune to many electrical problems.

Fiber optic cables are used in pairs: one for transmitting and one for receiving. A fiber optic transmitter converts electrical signals into light pulses using a light emitting diode (LED) or laser diode. The receiver converts the light signals back into electrical signals with a photodiode. Fiber optics are always used point-to-point. Multipoint networks with fiber optics require bridging hubs that convert between the optical signals and electrical signals.

The characteristics of optical fiber signaling allow for very high speeds, about 14 Tbit/s over 160 km of fiber. In automation systems, though, much slower speeds and shorter distances are typically used. Because the signaling is optical, fiber-optic transmissions are immune to electrical noise, transients, grounding, and lightning. This makes fiber-optic signaling a superior choice for signaling between buildings, especially over long distances.

Radio Frequency

Wireless signaling uses radio waves as the medium for conveying signals. *Radio frequency signaling* is a communications technology that encodes data onto carrier waves in the radio frequency range. Signals are conveyed by modulating (changing) the characteristics of a radio wave carrier. A radio wave of a known frequency is used as a baseline, and then digital data is mixed with it to produce a new wave with the data encoded into it. **See Figure 3-37.**

Although it is possible to modulate the amplitude of the carrier signal as a means of encoding data, this is subject to noise and reliability problems. Instead, frequency modulation is a more robust method of encoding data because it alters the carrier signal between two distinct frequencies. There are many other modulation schemes, such as phase shift modulation, that provide even better performance (speed and reliability) but are more complex to implement.

One of the main problems with all of the modulation schemes is that there are many sources of natural and human-made interference and radio noise. Many different radio frequencies are in constant or intermittent use that can interfere with communications. Spread spectrum techniques use multiple carrier frequencies to mitigate these problems. For example, frequency-hopping spread spectrum (FHSS) signaling rapidly switches between several different frequencies in a pseudorandom order known to the sender and receiver.

The bandwidth of wireless signaling is limited by the frequency of the carrier wave. Larger bandwidth requires very high frequencies, which require more power and expensive components. Also, walls and steel structures have an attenuating effect on signals and can restrict the physical range of most wireless technologies to 300′ or less unless repeaters are used to boost the signals.

In 1969, the Electronic Industries Alliance created the standard RS-232-C. Years later, the RS (recommended standard) was renamed EIA-232. Since 1997, the standard has been called TIA-232-F, though it is still common to see the EIA and RS prefixes used.

Radio Frequency Encoding

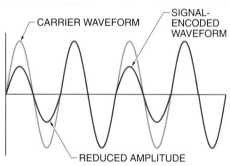

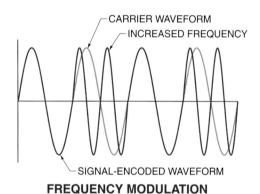

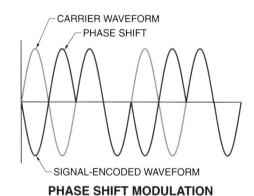

Figure 3-37. Modulating amplitude, frequency, or phase shift are three common methods of encoding digital data onto a radio frequency carrier waveform.

Powerline

A common drawback to communications wiring is the need for dedicated conductor, either copper or fiber optic cable. Also, wireless may not be feasible due to interference or bandwidth issues. Powerline technology, however, provides both power and data for each node by using the existing power wiring that is already installed in nearly every part of a building. *Powerline signaling* is a communications technology that encodes data onto the alternating current signals in existing power wiring.

Powerline signaling is similar to the way wireless technologies encode data onto radio frequency carrier waves. In fact, powerline signaling typically uses the same encoding techniques as radio frequency signaling, such as frequency or phase-shift modulation and frequency-hopping spread spectrum. Although there is a lot of development in the area of high bandwidth powerline carrier applications, automation systems mostly use relatively low-speed technology with data rates at or below 38 kbit/s.

Typically, powerline carrier transceivers incorporate both a power supply and data coupling circuit into the same AC connection. **See Figure 3-38.** The coupling isolates the electrical components from noise, transients, and other harmful effects, while passing modulated carrier wave signals through. Although modulation can be achieved through conventional discrete analog circuitry, most powerline carrier hardware uses digital signal processors because of the processors' flexibility and reliability.

COMMON BUILDING AUTOMATION MAC LAYERS

There are many different types of MAC layer technologies commonly used in building automation applications. Many of these are proprietary, meaning that they were developed by and possibly only used by a single company and its products. However, some technologies are published by the manufacturer as an open technology. There are also MAC layer technologies that are defined by national and international standards.

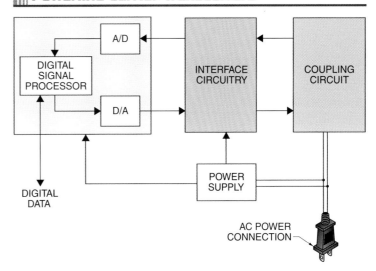

Figure 3-38. Powerline carrier transceivers include a coupling circuit that combines encoded digital data onto AC power lines.

Standards-based technologies are increasingly popular, especially since they are more likely to interoperate successfully, support system growth and long-term maintainability, and enable the availability of third-party devices and tools. The larger marketplace also drives pricing downward and expands the value of features or services available for the same cost.

The MAC layers commonly used for building automation systems include Ethernet, ARCNET, MS/TP, and wireless technologies. These technologies are families of frame-based protocols that define the physical layer characteristics and signaling, as well as the data link and media access schemes.

Ethernet

Ethernet uses a contention-based scheme for controlling access to the medium that is called carrier-sense multiple access with collision detection (CSMA/CD). Nodes listen to the line for activity and begin transmitting if no traffic is sensed. When a collision is detected, each node stops transmitting and retries again after a short wait. Each node retries after a randomly determined delay, reducing the chance that they will collide again. If a collision occurs again, the node keeps attempting for up to 16 times, which may cause another collision. After the maximum number of tries, the node stops the attempts, so the message never reaches its destination. Therefore, some messages never reach their destination due to repeated collisions.

Ethernet data is transmitted at 10 Mbit/s, 100 Mbit/s, 1 Gbit/s, or 10 Gbit/s on a variety of twisted-pair, coaxial, and fiber optic cabling. Ethernet is most commonly used in a star topology based on hubs or switches. Many modern hubs and switches can automatically sense the speed and adjust accordingly.

ARCNET

ARCNET uses token passing for media access, which makes it a deterministic network. The ARCNET technology is scalable to arbitrary speeds as high as 10 Mbit/s, though in practice, automation systems only use 2.5 Mbit/s or 156 kbit/s types. ARCNET may be transmitted on a variety of twisted-pair, coaxial, or fiber optic cabling and allows bus or star topologies, though this may depend on the media. At 2.5 Mbit/s, ARCNET requires the use of transformer-coupled transceivers, and distances of individual segments are limited. By lowering the speed to 156 kbit/s and using TIA-485 direct-coupled transceivers, it is possible to use longer segments up to 1000 m. Therefore, even though it is slower, the 156 kbit/s version is more commonly implemented.

Master-Slave/Token-Passing (MS/TP)

The master-slave/token-passing (MS/TP) MAC layer is defined in the BACnet protocol standard. MS/TP is used when there is only one or a limited number of master nodes that participate in token passing and share the medium with slave nodes. Slave nodes do not participate in token passing and only answer when invited to transmit. The token recovery scheme allows new nodes to join the token passing circle and subcircles to form if the network is broken. Like ARCNET, MS/TP is a deterministic network.

MS/TP is based on TIA-485 signaling at 9600 bit/s, 19.2 kbit/s, 38.4 kbit/s, or 76.8 kbit/s. The standard specifies the use of shielded, twisted-pair cable with specific electrical characteristics. In an MS/TP daisy-chain bus topology, network termination should be installed at the two nodes at the ends of the segment.

Wireless Networks

Wireless technologies include electromagnetic signals in the infrared, radio, and microwave frequencies, which are becoming practical options for building network infrastructure. Wireless building automation nodes are particularly useful in applications where physical wiring is not practical. They are also easier and faster to install and can be relocated if necessary. However, wireless systems may be less reliable due to range limitations or signal interference.

Wireless technologies often rely on mesh networks to route messages throughout the network. This allows wireless nodes to operate with less power, since each node needs only the signal range to reach its nearest neighbor. Although standards such as IEEE 802.15 and 802.11 cover wireless networking in various forms, there are no standards yet for applying these techniques as MAC layers for automation networking. An emerging consortium called the ZigBee Alliance is beginning to apply their adaptation of the 802.15.4 protocol to automation applications.

Summary

- Machines that are involved in network communication are known as nodes.
- Before any two nodes can effectively communicate with each other, they must agree on a protocol to govern the manner in which they exchange information.
- Each signaling method has characteristics that make it more appropriate for certain situations.

- All networks send individual messages in discrete units known as frames or packets.
- Information is conveyed within the boundaries of a frame using logical groupings called fields.
- The maximum length of node addresses places an upper limit on the number of nodes that may exist in a network.
- Data that exceeds the data payload per frame limit is broken into segments, which are reassembled at the destination back into one large data unit.
- If more than one node transmits at the same time, the messages can collide.
- A master node controls the message traffic, and slave nodes can transmit only when granted permission by the master.
- The OSI Model divides the very complex problem of node-to-node communication into several smaller pieces.
- Messages flow down a protocol stack and each layer may add data to the message according to its specific protocol.
- The physical and data link layer protocols form a special combination known as the MAC layer.
- A LAN utilizes a particular MAC layer type.
- In most collapsed architectures, some of the functionality of the missing layers may be provided by protocols in the other layers.
- Network architecture involves the physical design of the communication network, including the network devices and how they connect segments together for forming more complex networks.
- Network devices can be used to extend segments, connect segments together, and transition between different types of physical media, for example from twisted-pair cable to optical fiber.
- When two or more application protocols are being used to perform similar functions, a gateway can be used to integrate them together into one communication system.
- The signaling method affects the methods, locations, and shape of network connections.
- The four primary physical topologies used in network technologies are bus, star, ring, and mesh. Networks may also use a mixture of these topologies together, known as free topology.
- The most common physical media type for computer-based networks, including building automation networks, are copper conductors, particularly twisted-pair cable.
- Fiber optic cable allows for very high speed and is immune to electrical noise, transients, grounding, and lightning.
- Wireless signaling uses radio waves as the medium for conveying signals.
- Powerline technology provides both power and data for each node by using the existing power wiring that is already installed in a building.
- The MAC layers commonly used for building automation systems include Ethernet, ARCNET, and MS/TP, which are frame-based protocols that define the physical layer characteristics and signaling, as well as the data link and media access schemes.

Definitions

- A *node* is a computer-based device that communicates with other similar devices on a shared network.
- A *protocol* is a set of codes, message structures, signals, and procedures implemented in hardware and software that permits the exchange of information between nodes.
- A *signal* is the conveyance of information.
- *Signaling* is the use of electrical, optical, and radio frequency changes in order to convey data between two or more nodes.
- A *transceiver* is a hardware component that provides the means for nodes to send and receive messages over a network.
- *Bandwidth* is the maximum rate at which bits can be conveyed by a signaling method over a certain media type.
- A *media type* is the specification of the characteristics and/or arrangement of the physical conductors or electromagnetic frequencies used for digital communication.
- *Throughput* is the actual rate at which bits are transmitted over a certain media at a specific time.
- *Latency* is the time delay involved in the transmission of data on a network.
- *Simplex communication* is a system where data signals can flow in only one direction.
- *Half-duplex communication* is a system where data signals can flow in both directions, but only one direction at a time.
- *Full-duplex communication* is a system where data signals can flow in both directions simultaneously.
- A *packet* is a collection of data message information to be conveyed.
- A *frame* is a packet surrounded by additional data to facilitate its successful transmission and reception by delineating the start and end (or length) of the packet.
- An *octet* is a sequence of 8 bits.
- A *broadcast* is the transmission of a message intended for all nodes on the network.
- A *multicast* is the transmission of a message intended for multiple nodes, which are all assigned to the same multicast group.
- A *port* is a virtual data connection used by nodes to exchange data directly with certain application programs on other nodes.
- *Segmentation* is a protocol mechanism that controls the orderly transmission of large data in small pieces.
- A *collision* is the interaction of two messages on the same network media, which can cause data corruption and errors.
- The *Open Systems Interconnection (OSI) Model* is a standard description of the various layers of data communication commonly used in computer-based networks.
- A *protocol stack* is a combination of OSI layers and the specific protocols that perform the functions in each layer.
- The *physical layer* is the OSI Model layer that provides for signaling (the transmission of a stream of bits) over a communication channel.

- The *data link layer* is the OSI Model layer that provides the rules for accessing the communication medium, uniquely identifying (addressing) each node, and detecting errors produced by electrical noise or other problems.
- The *MAC layer* is a sublayer of the OSI Model that combines functions of the physical and data link layers to provide a complete interface to the communications medium.
- A *MAC address* is a node's address that is based on the addressing scheme of the associated data link layer protocol.
- A *local area network (LAN)* is the infrastructure for data communication within a limited geographic region, such as a building or a portion of a building.
- The *network layer* is the OSI Model layer that provides for the interconnection of multiple LAN types (MAC layers) into a single internetwork.
- An *address space* is the logical collection of all possible LAN addresses for a given MAC layer type.
- *Routing* is the process of determining the path between LANs that is required to deliver a message.
- The *transport layer* is the OSI Model layer that manages the end-to-end delivery of messages across multiple LAN types.
- The *session layer* is the OSI Model layer that provides mechanisms to manage a long series of messages that constitute a dialog.
- The *presentation layer* is the OSI Model layer that provides transformation of the syntax of the data exchanged between application layer entities.
- The *application layer* is the OSI Model layer that provides communication services between application programs.
- A *collapsed architecture* is a protocol stack that does not include layers that are not needed for the application.
- A *protocol data unit* is the portion of the frame containing fields belonging to a certain OSI Model layer and the layers above it.
- A *segment* is a portion of a network where all of the nodes share common wiring.
- A *logical segment* is a combination of multiple segments that are joined together with network devices that do not change the fundamental behavior of the LAN.
- A *repeater* is a network device that amplifies and repeats the electrical signals, providing a simple way to extend the length of a segment.
- A *hub* is a network device that repeats messages from one port onto all of its other ports.
- A *bridge* is a network device that joins two LANs at the data link and physical layers.
- A *switch* is a network device that can forward messages selectively to one of its other ports based on the destination address.
- A *router* is a network device that joins two or more LANs together at the network layer and manages the transmission of messages between them.
- An *internetwork* is a network that involves the interaction between LANs through routers at the network layer.
- A *firewall* is a router-type device that allows or blocks the passage of packets depending on a set of rules for restricting access.

- A *gateway* is a network device that translates transmitted information between different application protocols.
- *Mapping* is the process of making an association between comparable concepts in a gateway.
- *Topology* is the shape of the wiring structure of a communications network.
- A *bus topology* is a linear arrangement of networked nodes with two specific endpoints.
- A *daisy chain* is a wiring implementation of bus topology that connects each node to its neighbor on either side.
- A *star topology* is a radial arrangement of networked nodes.
- A *ring topology* is a closed-loop arrangement of networked nodes.
- A *mesh topology* is an interconnected arrangement of networked nodes.
- A *free topology* is an arrangement of nodes that does not require a specific structure and may include any combination of buses, stars, rings, and meshes.
- A *terminator* is a resistor-capacitor circuit connected at one or more points on a communication network to absorb signals, avoiding reflections.
- A *network tool* is a software application that runs on a computer connected to a network and is used to make changes to the operation of the nodes on a network.
- *Twisted-pair cable* is a multiple-conductor cable in which pairs of individually insulated conductors are twisted together.
- *Coaxial cable* is a two-conductor cable in which one conductor runs along the central axis of the cable and the second conductor is formed by a braided wrap.
- *Fiber optics* is a form of signaling based on light pulses to convey signals.
- *Radio frequency signaling* is a communications technology that encodes data onto carrier waves in the radio frequency range.
- *Powerline signaling* is a communications technology that encodes data onto the alternating current signals in existing power wiring.

Review Questions

1. Explain the difference between bandwidth and throughput.
2. What are the common fields included in message frames and what is their approximate order?
3. Why does the size of an address field affect the number of nodes allowed in a network?
4. Why is it important to control how nodes access the network media for sending messages?
5. How does the protocol stack affect the composition of the message frames that are transmitted on the network medium?
6. Why do some protocols use collapsed architectures?
7. How does a router filter network traffic?

8. How does a gateway translate between two different protocols?

9. What are the advantages and disadvantages of free topology?

10. What types of information does a signaling standard, such as TIA-232, typically include?

Chapter Four

LonWorks System Overview

Building automation networks based on LonWorks technology employ intelligent nodes connected to a communications network. The defining characteristics of the LonWorks technology include its conversion of physical and logical control information into data structures known as network variables. The network variables allow this control information to be shared among nodes and network management tools connected to a common network. LonWorks technology is based primarily on the LonTalk protocol, an open standard, and the Neuron® chip microprocessor specifically developed to manage LonTalk communications and node applications.

Chapter Objectives

- *Identify the standards and standards development organizations involved in maintaining LonWorks technologies.*
- *Describe the basic structure and distribution of information within a LonWorks network.*
- *Compare the roles of network variables, function blocks, bindings, and configuration properties.*
- *Identify the basic requirements of the LonWorks system architecture.*
- *Describe the role of the LonWorks Network Services (LNS®) platform in maintaining interoperability with different network management tools.*

LONWORKS DEVELOPMENT

In the mid-1980s, the then-CEO and co-founder of Apple Computer, A. C. "Mike" Markkula, challenged a handful of research engineers to develop an inexpensive computer small enough to put intelligence into everyday devices and share information over a variety of media. In 1988, the initial design was completed for a postage-stamp-sized microcontroller called the Neuron® chip. **See Figure 4-1.** In 1988, that small research group became Echelon Corporation. Today, the Neuron chip is the core of most LonWorks nodes.

Neuron® Chip

Echelon Corporation

Figure 4-1. LonWorks technology is primarily built around the implementation of the Neuron chip in control devices.

LonWorks technology was also developed to utilize this chip to create control networks. *LonWorks technology* is a platform developed by Echelon Corporation designed for networked control applications. The terminology of the LonWorks technology platform is built on the concept of a local operating network (LON®). A *local operating network (LON)* is a network of intelligent devices sharing information using the standard communication protocol LonTalk®. *LonTalk* is the open protocol standard used in LonWorks control networks.

The Neuron chip is licensed to and manufactured by Toshiba and Cypress Semiconductor. The processors are sold primarily through the manufacturer's sales channels, and Echelon receives a small fee for each Neuron processor produced. However, the LonTalk protocol specification is openly published and available for manufacturers to develop control devices that are interoperable in LonWorks networks. LonTalk can also be implemented on microprocessors other than the Neuron processor if desired.

Echelon Corporation supports the LonWorks technology platform with a wide variety of infrastructure products, development tools, and integration tools. Echelon estimates that thousands of manufacturers worldwide have implemented the LonWorks technology in over 100 million nodes so far.

LonMark International

Interoperability standards for LonWorks nodes are created and supported by LonMark® International. LonMark is a nonprofit, industry-supported organization consisting of manufacturers, systems integrators, application developers, and end users involved with LonWorks technology. LonMark is dedicated to creating and evolving standards that promote device interoperability across multiple industries. Control devices are evaluated by LonMark engineers and certified for compliance if all interoperability guidelines are met.

Industry-sponsored task groups within LonMark International develop standard definitions and functionalities for commonly used control variables and applications within specific industries. **See Figure 4-2.** All of the LonMark specifications are openly published and freely available.

LonWorks Standards

The LonTalk protocol was first introduced in 1990, and in October 1999, it was approved by the American National Standards Institute (ANSI) as ANSI/CEA 709.1. Supporting standards for twisted-pair, powerline, and IP communication for LonWorks networks were standardized as ANSI/CEA 709.2, ANSI/CEA 709.3, and ANSI/CEA 709.4, respectively. This series of communication standards could potentially be used for other applications. Additionally, the LonTalk protocol is included in the BACnet standard as one of the approved BACnet data link/physical layers for building automation networks.

LonMark Task Groups

Task Groups	Standard Functional Profiles	LonMark HVAC Functional Profiles
Automated Food Service Equipment/Catering	Access/Intrusion/Monitoring	8020_11: Fan Coil Controller
Building Automation Systems	Fire and Smoke	8030_11: Rooftop Unit Controller
Connectivity, Routers, and Gateways	HVAC	8040_10: Chiller
Elevator/Escalating/Moving Walks	Industrial	8051_10: Heat Pump
Fire/Smoke	Input/Output	8060_10: Thermostat
Home and Whitegoods	Lighting	8080_10: Unit Ventilator
HVAC	Management	8090_10: Space Comfort
Industrial	Motor Control	8010_11: VAV Controller
Lighting	Refrigeration	8110_11: Damper Actuator
Network Tools	Sensors	8120_10: Pump Controller
Refrigeration	Vertical/Conveyor (Elevators/Lifts)	8131_10: Valve Positioner
Safety	Whitegoods (Home Appliances)	8301_10: Boiler Controller
Security/Access/Notifiers		8500_20: Space Comfort Controller
Sunblinds/Weather	Standard Network Variables	8610_10: Discharge Air Controller
System Integration		
Transportation	Standard Configuration Properties	
Utilities		
	Device Interoperability Guidelines	

Figure 4-2. LonMark International is a group of manufacturers, integrators, and others involved in the implementation of LonWorks networks. Members participate in task groups that work on standardizing the aspects of the LonWorks technology.

The LonWorks control platform has also been standardized as the EN 14908 series of standards in Europe and the GB/Z 20177 series in China. In 2008, the LonWorks series of standards was formally approved by the International Organization for Standardization (ISO) as ISO/IEC 14908, *Open Data Communication in Building Automation, Controls and Building Management – Control Network Protocol*.

Additional standards have been created to support the LonWorks technology. For example, standards for twisted-pair cabling and powerline media infrastructure were designed for carrying LonTalk messages, though they could potentially be used in other ways.

LonWorks also incorporates other technology standards into its implementation. For example, the IP tunneling standard ANSI/CEA-852 is a recognized communication protocol, which allows control data and network management commands to be routed over Internet connections. The LonTalk packet is encapsulated within an IP packet according to the IP standard.

Standard machine-to-machine (M2M) communication protocols allow computer programs to effectively share information across different platforms. For example, Extensible Markup Language (XML) and Simple Object Access Protocol (SOAP) are used to request data from remote applications over the Internet. These protocols are supported by many LonWorks Internet servers, including the Echelon i.LON®100 and Tridium's Niagra Framework® network management platform.

LONWORKS TECHNOLOGY

The defining characteristics of LonWorks technology are the way in which information and control functions are organized and the methods used to share the information across a communication network. A typical node that implements the LonTalk protocol performs local control functions and shares control information with other LonWorks nodes in a peer-to-peer way. Control values are structured by individual nodes as network variables and transmitted onto the network in message packets. **See Figure 4-3.** These messages are independent from the signal type used to collect the data. Any node that is bound to the sending node picks up the sender's message packet and utilizes the network variable information to perform its application.

LonWorks Technology

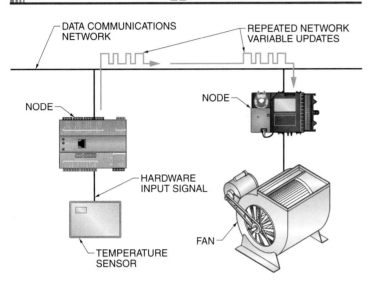

Figure 4-3. The basic concept of LonWorks is the structuring of control information as network variables, which are shared with other nodes on a communications network.

An example is a node that simply measures space temperature in an electrical distribution room. An analog temperature sensor wired to the node provides the hardware input signal, which may be based on resistance, voltage, or current changes. The node converts the hardware input signal to an output network variable that represents room temperature and transmits that value onto the network. The network variable can be transmitted repeatedly, either when it changes value or at a certain time interval.

A downstream node that actuates a fan serving the electrical room is bound to the upstream temperature sensor node. The exhaust fan node picks up the network message packet and delivers the temperature value to an input network variable value within its own application program. **See Figure 4-4.** The application program uses the temperature data input, along with its control configuration, to decide whether to start or stop the exhaust fan. The status of the fan may also be shared as an output network variable while the node sends the appropriate signal to start or stop the fan. Other nodes, such as smoke detectors and schedulers, may communicate additional network variable information that the exhaust fan node listens for.

With this control scheme, LonWorks technology provides many advantages. It allows the integration of all building systems onto one common network without the need for gateways or proprietary translation devices. Control intelligence resides in individual sensor and actuator nodes, which lowers installation and modification costs when compared to systems with legacy input/output devices hardwired to central controllers. This decentralization also avoids a single point of failure and facilitates making changes to the network without affecting unassociated nodes.

The event-driven update algorithms optimize network bandwidth, improving communication reliability. Also, since the LonWorks technology includes an open protocol, a variety of network management tools can be used with its network management platform.

Information Architecture

Interoperability first requires a standard way to structure information. Control information of both inputs and outputs is organized into network variables. Control logic is organized into function blocks whose behavior is adjusted with configuration properties. LonMark International specifies standard versions of all of these information components that are vital to interoperability. When arranged together to form a complete control network, the result is a network program.

Network Variables. LonWorks nodes share information by sending and receiving control data as network variables. A *network variable* is a basic unit of shared control information that conforms to a certain data type. **See Figure 4-5.** Many network variables represent a simple numerical value with a unit, such as temperature in degrees Celsius (°C) or pressure in pounds per square inch (psi). Other network variables can represent text strings, the node status, alarm data, enumerations, or structured setpoints. The LonMark interoperability guidelines allow for two classes of network variables: standard and user-defined.

Standard network variable types (SNVTs) are a key component of interoperability. A *standard network variable type (SNVT)* is a LonWorks network variable with a format, structure, and intended use that is defined and publicly documented for intended use by any node manufacturer. SNVTs are created and maintained by LonMark task groups. Currently, nearly 200 SNVTs are defined in LonMark documentation. **See Appendix.** The list of SNVTs is constantly growing according to industry needs.

> LonWorks technology can be scaled to very large networks. For example, the Italian electric utility Enel has deployed electric meters based on LonWorks in over 27 million homes and businesses across Italy, allowing the distribution of electricity to be managed remotely.

LonWorks Control Decisions

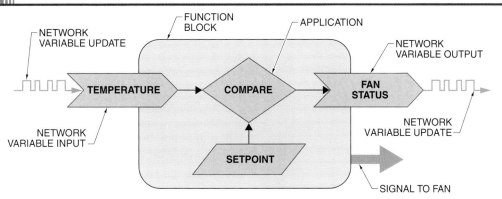

Figure 4-4. Control decisions are made by the function blocks that reside within nodes.

Network Variables

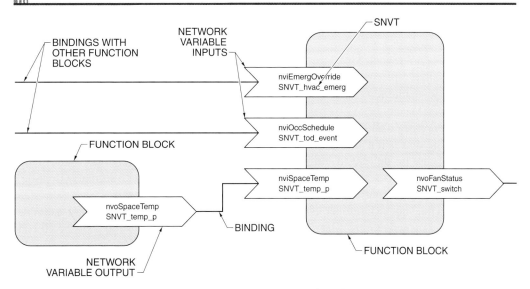

Figure 4-5. Network variables are the inputs to and outputs from function blocks.

Field Terminology

Integrators and installers working with LonWorks networks have developed a shorthand pronunciation for some of the otherwise awkward LonTalk terminology and acronyms. The acronym "SNVT" is often pronounced as "sniv-it" by LonWorks technicians. Likewise, the configuration property acronyms "SCPT" and "UCPT" are pronounced as "skip-it" and "you-keep-it," respectively.

Air-handling unit (AHU) controllers may include functional profiles for several types of function blocks, including ones representing sensors connected to the controller.

When there is no defined SNVT to represent certain desirable pieces of information, LonWorks node manufacturers can define custom network variables. A *user-defined network variable type (UNVT)* is a LonWorks network variable that is not standardized by LonMark and may be unique to individual node manufacturers. This feature allows flexibility in designing new control schemes, but the control data may not be interoperable with nodes from other manufacturers. Manufacturers of nodes with UNVTs should include a set of custom resource files documenting the UNVT format and structure for use with the network management tool.

Function Blocks. A *function block* is a software object within a node that represents a certain control task. Function blocks can represent a variety of control applications, such as simple analog sensors or complex package chiller controllers. **See Figure 4-6.** Nodes may include several function blocks, including multiple instances of the same functionality. For example, an environmental sensor unit may include multiple analog input function blocks, one each for the temperature, pressure, wind speed, and other sensors.

A function block receives information about the control system via inputs, performs some logic or algorithmic processing task, and generates one or more outputs. Inputs and outputs are either physical signals, such as variable voltage from a sensor or out to an actuator, or network variables that are shared between function blocks. Examples of function blocks include lighting scene controllers, air-handling unit (AHU) controllers, damper actuators, occupancy sensors, and fire alarm pull station initiators.

A function block is the implementation of a functional profile within a node. A *functional profile* is a specification of a particular control function. Standard functional profiles include mandatory and optional network variable types and configuration properties. Functional profile specifications are documented and freely published on LonMark's web site. Fundamental network variables and configuration properties are mandatory and must be included by all manufacturers of a specific node usage in order to be certified by LonMark. All certified products must contain at least one of the over 70 LonMark standard functional profiles.

Nodes implementing standard functional profiles allow a high degree of interchangeability between node vendors. These nodes can often be replaced with similarly LonMark compliant nodes, even those from a different manufacturer. **See Figure 4-7.** For example, if a node operating a rooftop package air conditioning unit fails, a replacement controller from a different manufacturer can be used if the new node contains the same standard functional profile. A replacement node that implements the same standard functional profile as a failed node must include the same mandatory application-specific network variables and configuration properties, regardless of the manufacturer.

Bindings. Two or more network variables connected together form a binding. A *binding* is a connection between a network variable input and a network variable output for the one-way sharing of dynamic control data. Bindings define the flow of information within and between nodes. Bindings are configured with connection properties that define message-timing parameters, address modes, and the message service type to be used for the network variable updates. In logic drawings of a control system, bindings are typically represented as a line connecting two network variables.

Configuration Properties. Each function block includes a set of configuration properties. A *configuration property* is an adjustable value that affects node, function block, or network variable behavior. Configuration property values affect how a node manages its own application and how network variable message traffic is sent. Examples include alarm limits, setpoints, default values, time delays, and network variable update intervals. Configuration properties are typically stored in the node's nonvolatile memory and are adjusted using network management tools. **See Figure 4-8.**

LonMark has defined over 150 standard configuration property types (SCPTs). **See Appendix.** These configuration properties are openly documented and available for any manufacturer to use within their node application. The use of SCPTs allows integrators to adjust configuration properties and configure nodes from different manufacturers using a single network management tool that includes the LonMark resource files.

As with UNVTs, node manufacturers can create custom user-defined configuration property types (UCPTs). However, access and adjustment of UCPTs may require the use of a manufacturer-specific configuration tool. Many manufacturers who utilize UCPTs provide software in the form of software plug-ins that can be run from within the network management tool. When nodes use UCPTs, plug-ins add the manufacturer's custom resource files to the network management tool that document their type, format, and enumeration structures.

Function Blocks

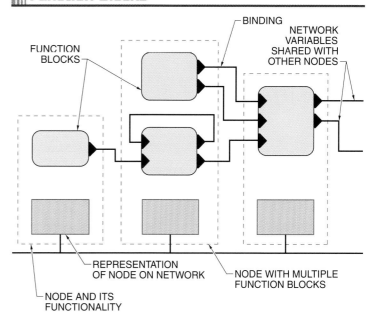

Figure 4-6. Each node includes instances of one or more function blocks, which represent a particular control functionality of the node.

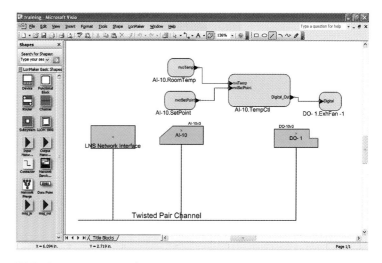

While the arrangement of graphical elements in the network program is up to the integrator, function blocks are typically placed above or near the node with which they are associated.

Compliance with LonMark Interoperability

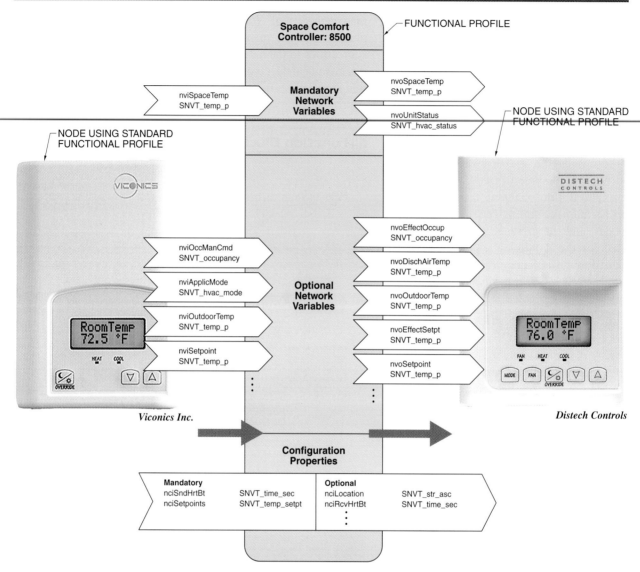

Figure 4-7. Nodes implementing the same LonMark standard functional profiles can be used in place of one another.

> On a node, additional configuration properties are available that affect network behavior and network variable traffic by defining how the output network variables are processed and transmitted across the network.

Network Programs. A *network program* is a collection of all node configurations and binding definitions within the control network. This is sometimes alternatively called a network image or network design. The network program is created and maintained with a network management tool. The network program includes the information on the organization of the network infrastructure, external interface files, function blocks in use, configuration property values, and the bindings between the network variables residing in function blocks. **See Figure 4-9.**

Chapter 4 — LonWorks System Overview

Configuration Properties

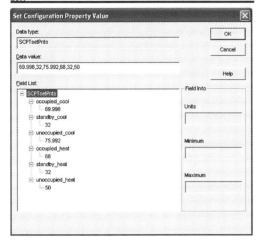

Figure 4-8. Configuration properties are set within network management tools to adjust node or function block behavior.

In LonWorks Network Services (LNS®)-based network architectures, the network management tool needs to be connected only when implementing adds, moves, or changes to the network program. After the tool is finished, it can be removed without affecting control system operations.

System Architecture

Components of a LonWorks system architecture fall into one of three basic categories: network infrastructure, nodes, and network tools.

Network Infrastructure. The LonTalk protocol is media independent and can be transmitted across a variety of media infrastructure, including twisted-pair cable, powerline, radio frequency, and fiber optic channels. A *channel* is one or more contiguous network segments that do not span routers. The LonWorks documentation specifies the requirements for the allowable types of network infrastructure. Up to 255 channels can be included in a single LonWorks network domain.

The most commonly used LonWorks communication channel is TP/FT-10, which uses relatively inexpensive twisted-pair cabling and allows flexibility in network topology. **See Figure 4-10.** LonTalk messages can also be transmitted using the IP-852 channel over IP networks, which are common in many buildings for computer workstation networks, and offer global connectivity. A LonWorks network can even include multiple channel types integrated together on one network through the use of LonTalk routers.

Network Program

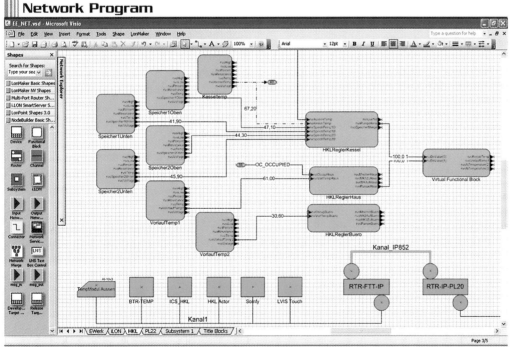

Ernst Eder

Figure 4-9. The network program encompasses all of the interconnected nodes and function blocks, along with the configuration property settings.

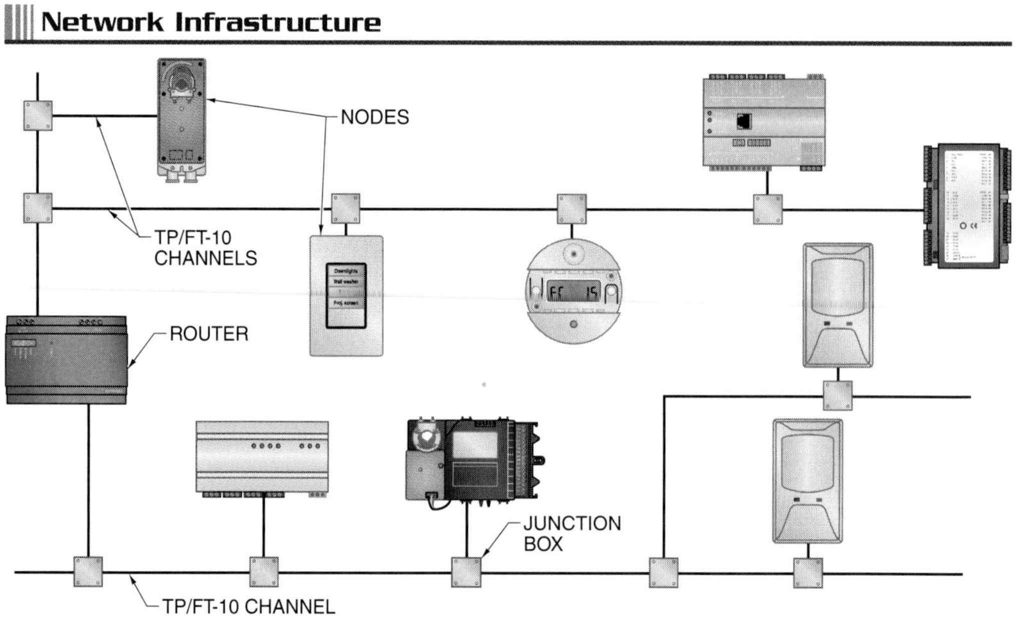

Network Infrastructure

Figure 4-10. LonWorks network infrastructure builds a communications network for the sharing of network variables. TP/FT-10 channels are particularly flexible in network structure.

LonTalk routers are used to connect channels together and isolate network traffic. Besides interfacing between different channel types, routers are also used to create logical divisions within the network. For example, nodes with a common application, such as HVAC control, are commonly arranged together on an HVAC channel, which is kept separate from other nodes, such as those on a lighting channel. Routers are then often used to connect these channels together, typically via a common backbone channel, and organize traffic between them. This helps manage network traffic between the nodes that commonly share information and the rest of the network.

Physical layer repeaters are also infrastructure components on LonWorks networks. These devices are used to extend the length of channel segments and the number of nodes that can be installed on the channel.

Nodes. LonWorks nodes are the control devices that implement the LonTalk protocol on microcontrollers, such as the Neuron chip, and perform the various control functions as required by sequence of operations descriptions. Node functionality is defined by the function blocks included in its application program.

Nodes that comply with LonMark interoperability guidelines include application-specific functionality that is explicitly defined and standardized. These node types may include thermostats, HVAC equipment controllers, occupancy sensors, lighting panels, and security access controls.

For customized applications that have no LonMark standard functional specification, freely programmable nodes are available and can be included in the control network. Also, system-wide controllers are available that perform global network functions such as scheduling, data logging, and alarm notification.

Network Tools. LonWorks network tools include software to create, manage, monitor, supervise, troubleshoot, and maintain the network program. **See Figure 4-11.** Network management tools commission nodes, create bindings, and configure node behavior. They are also used to maintain the network and perform such tasks as replacing nodes, updating node applications, relocating nodes to new channels, and backing up the network program.

The end user's view of the network is provided by a human-machine interface (HMI). HMIs provide the end user with the ability to monitor and control network functions, usually in a graphical format. An HMI can be a simple web page or a complex software application with dynamic graphics of equipment conditions, trend log graphs, and alarm logging. The current trend is toward web-browser-based user interfaces, which require a LonWorks Internet server device installed on the network.

LonWorks diagnostic tools are used to troubleshoot, verify, and document the health and status of nodes and channels. These are essential applications for network integrators because they ensure the high performance of the control network. Protocol analyzers report channel statistics such as bandwidth consumption, error rates, and response times. Protocol analyzers include a packet log that includes a real-time display of channel activity and detailed packet analysis. They may also include access to individual node error statistics and documentation of channel and node health.

Network Management Platforms

A *network management platform* is a software operating system that performs network management services and connectivity tasks utilized by network management tools. LonWorks networks typically employ the LonWorks Network Services (LNS) network management system that is developed and maintained by Echelon Corporation. LNS is an open platform that supports third-party node configuration tools, performs services used to manage nodes, provides network connectivity for tools over any network media, and includes a single network database that stores the complete network program. Additionally, LNS supports client/server architectures where multiple client tools can access the network database kept on a single server computer. **See Figure 4-12.**

LonWorks Network Tools

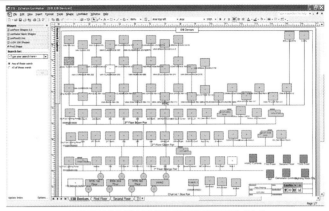
NETWORK MANAGEMENT TOOLS

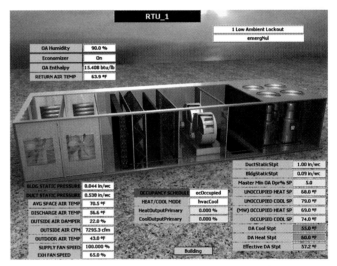
HUMAN-MACHINE INTERFACE (HMI)

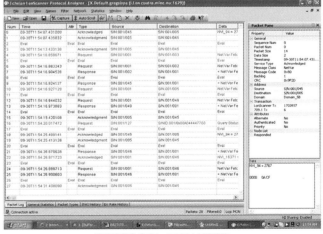
PROTOCOL ANALYZER

Figure 4-11. Network tools in the LonWorks environment include network management tools, human-machine interfaces (HMIs), and protocol analyzers.

Network Management Platform

Figure 4-12. The LonWorks Network Service (LNS) platform provides a single database of network information that can be accessed by a variety of network tools on computers connected in a variety of ways.

Echelon's LonMaker integration tool is an example of a network management application based on LNS. However, many third-party network management tools that are based on LNS are available from other control system manufacturers. Each of these tools relies on the LNS operating system services and can open and manage an LNS network database created by any other network management tool based on LNS. Additionally, these tools provide full support of LonMark standard network variable types and resource files.

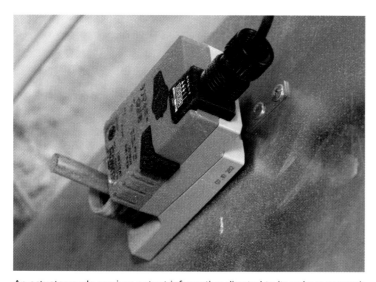

An actuator node receives output information directed to its unique network address.

Network Addressing

The sharing of control information over a communication network requires each node to have a unique logical address. This address allows LonTalk packets sent by nodes to be selectively forwarded or isolated by routers over the network channel, providing efficient management of network traffic and allowing packets to be transmitted across many different media types.

The top level of this logical address is known as the domain. **See Figure 4-13.** LonWorks nodes can only directly communicate with members of the same network domain. A domain ID is typically assigned using a network management tool and can be 1, 3, 5, or 7 bytes in length.

The next level of addressing is the subnet, which is a particular channel. Assigned by the network management tool, this may include one or more media segments, but does not span routers. A network domain may contain up to 255 subnets.

The lowest level of addressing is the node identification number. It is also assigned by the network management tool. There can be up to 127 nodes per subnet. Therefore, the maximum number of nodes in a LonWorks network domain is 32,385 (255 subnets/domain × 127 nodes/subnet = 32,385 nodes/domain).

Chapter 4 — LonWorks System Overview

LonWorks Network Addressing

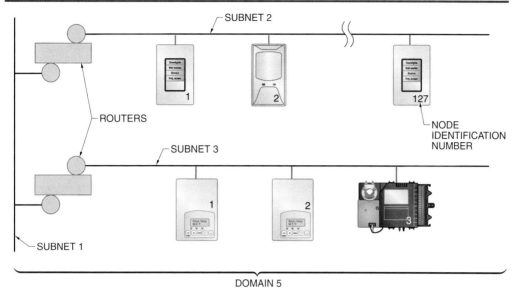

Figure 4-13. Nodes within a LonWorks network are addressed by the domain, subnet, and node number.

Summary

- The Neuron chip is the core of most LonWorks nodes.
- The LonTalk protocol specification is openly published and available for manufacturers to develop control devices that are interoperable in LonWorks networks.
- Interoperability standards for LonWorks are created and supported by LonMark International.
- LonWorks Technologies have been approved by the American National Standards Institute (ANSI) as the ANSI/CEA 709 series of standards. The International Organization of Standardization (ISO) has approved the technologies as the ISO/IEC 14908 series of standards.
- Control values are structured by individual nodes as network variables and transmitted onto the network within data message packets.
- Any receiving node that is bound to the sending node picks up the sender's message packet and utilizes the network variable information to perform its application.
- Network variables are transmitted repeatedly, either when they change values or at certain time intervals.
- A node application uses network variable input values, along with its control configuration, to determine the values of its hardware and network variable outputs.
- The LonMark interoperability guidelines allow for two classes of network variables: standard and user-defined.
- User-defined network variables allow node manufacturers flexibility in designing new control schemes, but the control data may not be interoperable with nodes from other manufacturers.
- A function block receives information about the control system via hardware and software inputs, performs some logic or algorithmic processing task, and generates one or more hardware or software outputs.

- A function block is the implementation of a functional profile within a node.
- Two or more network variables connected together form a binding.
- An input network variable can only be bound to output network variables.
- Configuration property values affect how a node manages its own application and how often network variable message traffic is sent.
- The network program includes the information on the organization of the network infrastructure, physical details of all the nodes on the network, external interface files, function blocks in use, configuration property values, and the bindings between the network variables residing in function blocks.
- The LonTalk protocol is media independent and can be transmitted across a variety of media infrastructure, including twisted-pair cable, powerline, radio frequency, and fiber optic channels.
- LonWorks network tools include software to create, manage, and maintain the network program.
- LonWorks Network Services (LNS) is an open platform that supports node configuration tools from multiple vendors, performs services used to manage nodes, provides network connectivity for tools over any network media, and includes a single network database that stores the network program and node information.
- LonWorks network logical addressing includes domain, subnet, and individual node numbers.

Definitions

- *LonWorks technology* is a platform developed by Echelon Corporation designed for networked control applications.
- A *local operating network (LON)* is a network of intelligent devices sharing information using the standard communication protocol LonTalk®.
- *LonTalk* is the open protocol standard used in LonWorks control networks.
- A *network variable* is a basic unit of shared control information that conforms to a certain data type.
- A *standard network variable type (SNVT)* is a LonWorks network variable with a format, structure, and intended use that is defined and publicly documented for use by any node manufacturer.
- A *user-defined network variable type (UNVT)* is a LonWorks network variable that is not standardized by LonMark and may be unique to individual node manufacturers.
- A *function block* is a software object within a node that represents a certain control task.
- A *functional profile* is a specification of a particular control function.
- A *binding* is a connection between a network variable input and a network variable output for the one-way sharing of dynamic control data.
- A *configuration property* is an adjustable value that affects node, function block, or network variable behavior.
- A *network program* is a collection of all node configurations and binding definitions within the control network.
- A *channel* is one or more contiguous network segments that do not span routers.
- A *network management platform* is a software operating system that performs network management services and connectivity tasks utilized by network management tools.

Review Questions

1. What is the difference between the terms "LonWorks" and "LonTalk?"

2. What is LonMark International?

3. Briefly explain the typical operation of a LonWorks network.

4. What are some of the advantages of the LonWorks system?

5. What are the disadvantages of implementing user-defined network variables?

6. What is the difference between a function block and a functional profile?

7. How are standard functional profiles important for interoperability?

8. How are routers typically used in LonWorks network infrastructure?

9. What types of network tools are commonly used on a LonWorks network?

10. What levels are involved in LonWorks node addressing?

Chapter Five

LonWorks Network Architecture and Infrastructure

LonWorks allows a large degree of flexibility in media types, network topologies, client/server arrangements, and node selection. Many nodes and infrastructure components are readily available for constructing a LonWorks control network while following relatively few requirements and guidelines. This results in a control network that is particularly simple to manage, update, and expand if necessary.

Chapter Objectives

- Describe the composition and function of the LonWorks Network Services (LNS) platform.
- Differentiate between local, full, and lightweight clients.
- Compare the characteristics of the channel types supported by LonWorks networks.
- Identify the purpose of network interfaces.
- Describe some of the considerations for installing twisted-pair cabling.

LONWORKS NETWORK ARCHITECTURES

The first stage in planning a LonWorks system is the selection of the network architecture. The customer requirements, control specifications, physical limitations, media types, available building infrastructure, network management tool options, and future expansion possibilities are deciding factors that affect network architectures. The choice of network architecture defines how tools connect, manage, monitor, and control nodes on the LonWorks network.

LonWorks networks are most often applied in a flat network architecture, where any node on the network can communicate directly with any other node on the network in a peer-to-peer way without the need for gateways or translators. In a flat architecture, nodes make control decisions based on their internal application program and network variables. Network management tools or human-machine interfaces (HMIs) can be connected or disconnected as needed to commission, modify, or monitor the network or its nodes without affecting the operation of the system.

Also, network management tools in a flat architecture can access any network information from any connection point on the network. With IP connectivity, the potential size of the network is effectively extended to anywhere in the world.

> LNS uses Microsoft® standards for Windows-component-based software, including COM components and ActiveX controls. This helps developers write LNS application programs using any programming language that supports these technologies.

LonWorks Network Services (LNS) Platform

LonWorks Network Services (LNS) is a network management system that provides an open platform for flat network architectures, as well as full support of LonMark interoperability standards. LNS services allow multiple applications to perform network management, monitoring, and control of LonWorks networks. For each network design, LNS includes a common database for storing the parameters of a network design, which include the physical node configurations and logical control functions. **See Figure 5-1.** Many controls vendors utilize the LNS platform as the basis for their network management tools.

The LNS platform is an object-oriented programming model that facilitates rapid network development and simplifies host-processing requirements. The LNS platform supports configuration of node behavior with device-specific software provided by node manufacturers.

A computer houses the LNS platform software. This computer is typically attached to a backbone channel through a network interface, but may be located anywhere on the LonWorks network, or even installed remotely. The database can be backed up and restored as needed.

The LNS model includes four major components: LNS databases and the LNS Server, the LNS Object Server, Network Service Devices (NSDs), and Network Interfaces.

LNS Databases and LNS Server. LonWorks network information is stored in two types of databases. A single LNS global database includes a collection of LNS network databases that are managed by the LNS Server application. The global database lists the locations of the individual network databases, since the network databases are not necessarily stored on the same computer as the global database and the LNS Server.

An LNS network database includes data about a single domain's infrastructure components, nodes, and network program configurations. This includes node-specific data such as Neuron IDs, Program IDs, channel location, and LonMark self-documentation. The complete network program is stored within the network database, including network variable binding definitions and configuration property values.

The LNS Server is an application that runs on the computer hosting the global database. This enables client applications, both local (on the same computer) and remote, to access the information in the databases.

LNS Object Server. Network management tools that are based on LNS utilize the LNS Object Server application to access and update the network databases while performing node installation, configuration, monitoring/control, diagnostics, and maintenance tasks. The LNS Object Server enables multiple applications to access and update the database. Each LNS application can open up to 100 network databases.

The LNS Object Server uses a hierarchical structure that includes a set of objects, properties, methods, and events associated with physical nodes and logical objects that together represent a collection of LonWorks network objects defined in the LNS global database. **See Figure 5-2.** Each network object contains a set of subsystem objects that include a collection of node application and router objects residing in that subsystem. For example, a network design for a five-story office building could be organized into five different subsystems by logically grouping nodes for each floor. As a network grows, new subsystems can be added to represent new control systems, such as chiller plants, a boiler room, or irrigation controls.

Object types defined in the LNS Object Server include properties and methods. Properties include configuration information such as a node's Neuron ID. Methods perform actions related to an object, such as requesting a test. A set of events allows LNS applications to be notified when certain operations are performed or when configurations have been modified.

The LNS Object Server manages multiple network resources, including up to 1000 channels, 1000 routers, 32,385 nodes, 62 or 4096 network variables (for a Neuron chip-based or host-based node, respectively), and an unlimited number of network variable selectors (used to identify network variables that are members of a binding).

Network Service Devices (NSD). A *Network Service Device (NSD)* is a software object used by LNS to communicate with nodes and routers during network management tasks. Unique Network Service Devices are assigned to each network. For example, a single network management tool could open instances of two separate network designs on the same LNS Server computer.

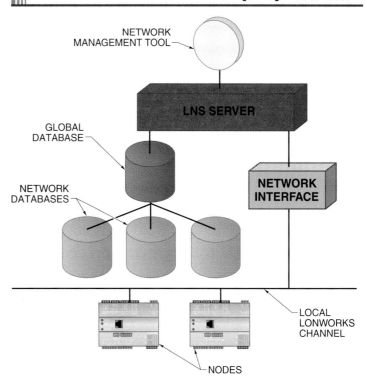

Figure 5-1. The LonWorks Network Services (LNS) platform manages the information that defines the structure and inter-node relationships of a LonWorks network.

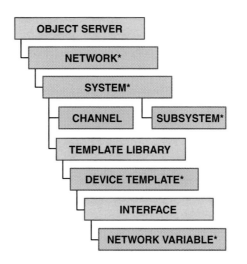

* OBJECT MAY BE ONE OF A COLLECTION OF SIMILAR OBJECTS

Figure 5-2. The Object Server defines a hierarchy of objects, properties, methods, and events that represent the attributes of the LonWorks system.

BUILDING AUTOMATION System Integration with Open Protocols

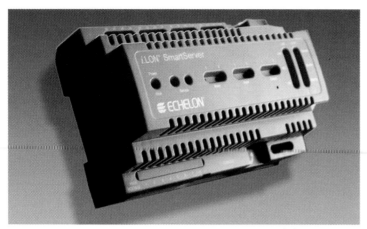

Echelon Corporation
Internet server devices are network interfaces that connect computers to the LonWorks LAN over IP networks.

Network Interfaces. Network management and user-interface tools access a LonWorks network control system through a network interface. A *network interface* is a hardware device and/or software driver installed in the network management tool computer for communicating on the network media. With the appropriate firmware, this is also known as a network services interface (NSI). Network interfaces coordinate with the LNS Server to communicate LNS message transactions to the nodes during network management and monitoring/control actions.

Several form factors and media/channel options for network interfaces are available, including many from Echelon and others. **See Figure 5-3.** The choice of a network interface depends upon the system architecture, channel, and host computer resources.

Temporary or portable network management tools, such as protocol analyzers or network management tools on laptop computers, typically use a USB or PCMCIA (Personal Computer Memory Card International Association) interface because they are easily connected and disconnected. Local desktop computers, since they are relatively permanent, typically use a PCLTA (Personal Computer LonTalk Adaptor) card installed in the computer's expansion slot.

Network interfaces can connect network management tools over IP networks. This requires an Internet server device attached to the network that employs a remote network interface (RNI). **See Figure 5-4.** This type of remote management can provide quick responses to control issues and customer needs. IP connectivity through Internet server devices on the network also provides opportunities for machine-to-machine (M2M) applications through web services using standard Extensible Markup Language (XML) and Simple Object Access Protocol (SOAP).

Client/Server Architectures

The LNS platform architecture provides connectivity options that allow network management tools access to the LNS Server and database from any connection point on the LonWorks network, or remotely over IP networks. These network management tools are client applications that are attached by the LNS Server to the network database through a network interface and update the databases as adds, moves, and other changes take place.

Selected Network Interfaces

Computer Interface	Device	TP/FT-10	TP/XF-1250	TP/XF-78	PL-20	TP-RS485-39
PCI	PCLTA-20			•		
PCI	PCLTA-21	•	•	•		•
PCMCIA	PCC-10	•	•	•		
TIA-232	SLTA-10	•	•	•		•
USB	U-10 or U-20	•			•	
Ethernet	i.LON 10 or i.LON 100	•			•	
IP-852	i.LON 600	•	•			
PC/104	PC/104					•

Figure 5-3. A variety of network interfaces are available from Echelon and other manufacturers. They depend on the available computer interface and the type of channels used in the LonWorks network.

IP Network Interfaces

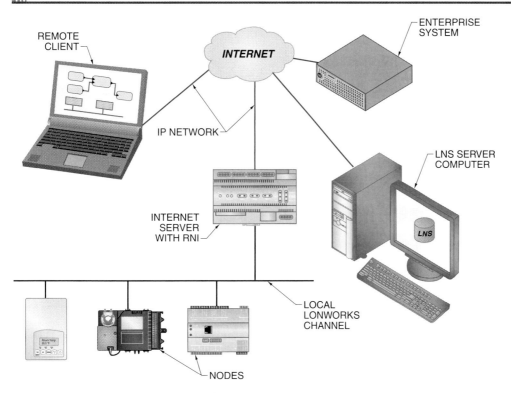

Figure 5-4. Internet server devices interface between a LonWorks network and IP networks, which can allow both on-site and remote systems to interface with the control network.

Multiple clients can access the LNS database at the same time. For example, teams of installers can commission nodes on different floors simultaneously using LNS network management tools on laptop computers. Each network management tool becomes a client to the LNS Server application and accesses the LNS database, implementing changes to the network database without synchronization issues.

Four different client/server configurations supported by LNS are local clients, full clients, lightweight clients, and independent clients.

Local Clients. A *local client* is an LNS network management tool that resides on the same computer as the LNS Server and database. **See Figure 5-5.** The local client application connects to the LonWorks network through either an NSI installed in the host computer or a remote network interface over an IP network if the LNS database and server are hosted on an off-site computer.

Local clients are typical of small network architectures and often consist of a laptop computer running a single network management tool application. An unlimited number of local client applications may access the network. However, the use of multiple network management tools, each hosting its own network database, may cause database synchronization issues.

Full Clients. A *full client* is a non-local LNS network management tool that communicates with the LNS Server and database computer through the LonWorks network. A full client can connect over any LonTalk channel type, including LonWorks over IP-852. **See Figure 5-6.** For network management and programming, full clients must be attached to the LNS Server in order to communicate changes to the LNS database. However, full clients can perform monitoring and control of network variable values even if the LNS Server is not available.

Local Clients

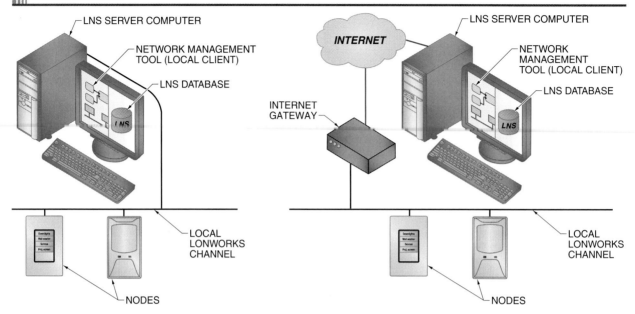

Figure 5-5. A local client is a network management tool that resides on the same computer as the LNS database, even if the computer is connected to the LonWorks network remotely.

Full Clients

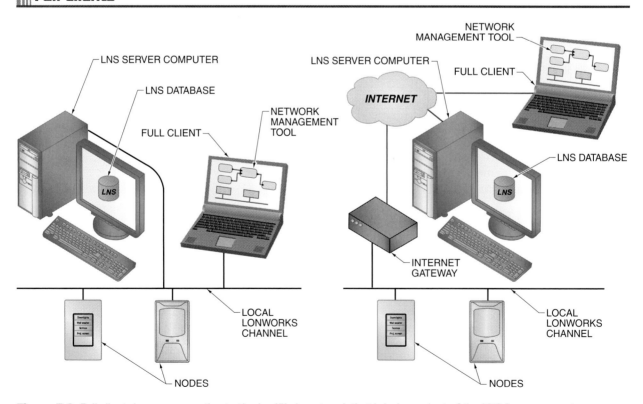

Figure 5-6. Full clients have a connection to the LonWorks network that is independent of the LNS Server computer.

Full clients are commonly used for large network architectures where multiple network management tools coexist among thousands of nodes, such as high-rise office buildings with numerous HVAC controllers, lighting controls, and access card readers. As remote full clients, maintenance technicians can replace nodes using portable laptop computers, facility managers can monitor comfort levels on HMIs, and building engineers can adjust equipment setpoints and modify node behaviors using computer workstations. Up to 10 remote full clients per network are supported by the LNS Object Server.

Lightweight Clients. LNS Server computers that include TCP/IP connectivity can support lightweight LNS clients. **See Figure 5-7.** A *lightweight client* is an LNS network management tool that communicates with the LNS Server computer directly over a TCP/IP network. Since they are connected over an IP network, lightweight client applications typically require knowledge of the LNS Server computer's IP address, as well as port assignments. The LNS Server listens for remote lightweight clients through TCP port 2540. All client requests for monitoring/control data and network management are routed through the LNS Server computer.

Lightweight clients are often used when an LNS Server computer is permanently hosted locally at the network location, but network management or monitoring/control tools are used remotely. The disadvantage of lightweight clients is that all network transactions must be routed through the LNS Server. Therefore, if the LNS Server is down, remote network tools cannot access network control data. As with full clients, up to 10 remote lightweight clients may be attached to the LNS Server.

Independent Clients. An *independent client* is an LNS network application that accesses the network only to perform monitoring and control of network variables. **See Figure 5-8.** Used primarily for HMI applications, independent clients must be aware of node addressing and network variable configurations, but do not perform network management tasks. Therefore, a server-independent client does not require access to the LNS Server or database.

> Full clients are also known as fullweight clients. They can be connected to the LNS Server computer in a variety of ways, including directly through the local LonWorks network, remotely over an IP network, and even remotely over the telephone network via modems.

Lightweight Clients

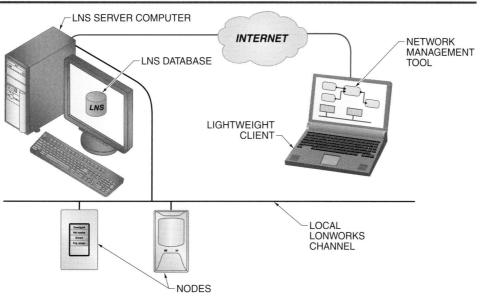

Figure 5-7. Lightweight clients communicate with LonWorks nodes only through the LNS Server computer.

Independent Clients

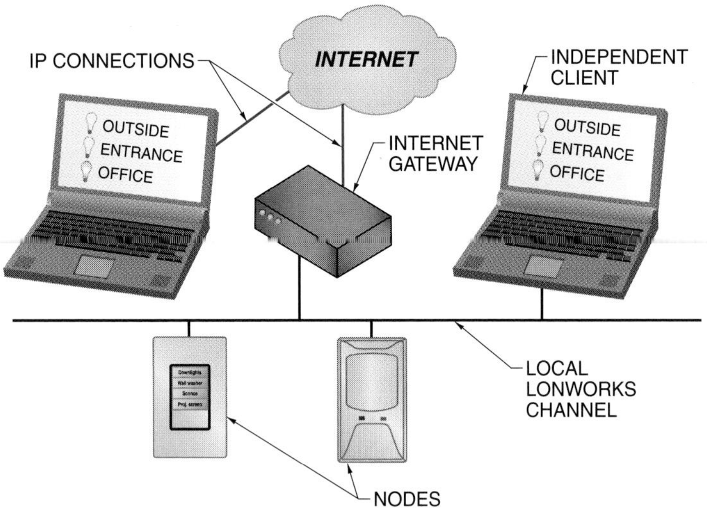

Figure 5-8. Independent clients are connected to the LonWorks network either directly or over IP, but do not require access to the LNS database or LNS Server.

Client Functionality

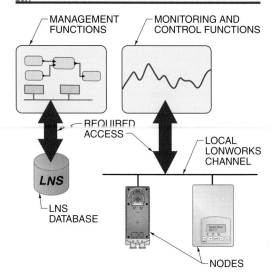

Figure 5-9. A client's functionality depends on its access to the LNS database and/or the control network.

Client Functionality

Access to the LNS database and the network affects a client's ability to perform network management and monitoring/control functions, respectively. **See Figure 5-9.** If the LNS Server computer is not present, the network may operate normally as programmed, but a client cannot make any changes to its operation, which requires access to the LNS database. Conversely, a client's monitoring/control functions are independent of the LNS Server computer, but rely on a connection to the network.

Therefore, a client's connection configuration determines its functionality. Local clients, since they reside on the same computer as the LNS Server, likely always have access to the database and retain management functionality. If the computer is not attached to the network, however, the changes cannot be loaded onto the nodes and network operation cannot be monitored. In fact, a local client that is unattached to a network is used as a method of programming a new network while off-site.

Full clients do not rely on the LNS Server computer for their connection to the network, so if it is not present, management functions are lost, but it can still monitor and send network variable updates into the network.

Lightweight clients lose all functionality if the LNS Server computer is disabled or not present. Therefore, a lightweight client configuration may be used for noncritical or temporary access into the network.

LONWORKS NETWORK INFRASTRUCTURE

The elements of a LonWorks network infrastructure include channels, network management tool interfaces, LonTalk routers, and Internet servers. As with LonWorks nodes, there are several choices among vendors of infrastructure components. It is important to understand the characteristics of each component to ensure high performance of the networked control system.

Channel Types

The construction of a LonWorks network typically begins with planning and installing the physical network channels. A channel is the physical network media that enables communications between nodes at the same physical layer. LonWorks nodes communicate over network channels through a hardware transceiver within the node.

Large networks that require many channels should include a backbone channel. **See Figure 5-10.** Backbone channels provide the connection points for LonTalk routers, which create the connection points for channels containing nodes. Backbone channels also allow for future expansion of the network and provide dedicated bandwidth for network management tools, such as HMIs.

LonWorks allows several different types of channels. Selection of an appropriate channel type involves balancing the channel features and capabilities with their costs, limitations, and feasibility within the building to be automated. **See Figure 5-11.** The most common media used for LonWorks channels is twisted-pair cabling, though other media types are supported. Nodes require transceivers that are specific to the channel type to which they are connected.

TP/FT-10. On twisted-pair cabling media, the most widely used channel is TP/FT-10, which is defined in the standard ISO/IEC 14908-2. The TP/FT-10 channel is proven in robustness even in noisy electrical environments. Several standard twisted-pair wire types can be used, including CAT5 and NEMA Level IV cable. The two-wire network connections are not sensitive to polarity and do not require shielded cable under normal conditions. A TP/FT-10 channel has a maximum traffic capacity of 78 kbit/s, or approximately 144 LonTalk packets per second.

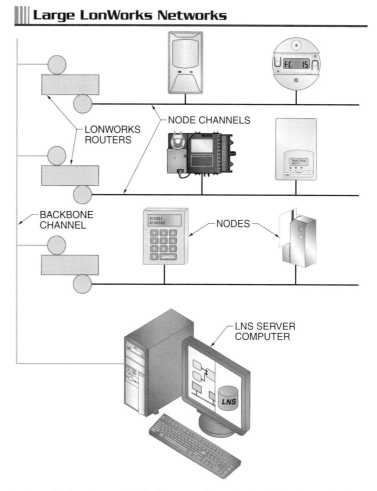

Figure 5-10. Large LonWorks networks typically divide the nodes into related groups and connect them together on separate node channels, which are then connected to a backbone channel.

LonWorks Channel Types

Channel Type	Typical Channel Use	Bandwidth*	Cabling	Topology Support	Required Terminators	Maximum Segment Length†	Maximum Nodes per Segment	Maximum Stub Length†
TP/FT-10	Node	78	CAT5 or Level IV	Free	1	500	64 (127 with link power nodes)	3
				Bus	2	2700		
TP/XF-1250	Backbone	1250	CAT5 or Level IV	Bus	2	130	64	0.3
TP/XF-78	Node	78	CAT5 or Level IV	Bus	2	2000	64	3
PL-20x	Node	3.6 (A band) or 5.4 (C band)	Existing power conductors	Free	Varies	Several kilometers	Hundreds	N/A
IP-852	Backbone	10,000 (10BASE-T) or 100,000 (100BASE-T)	CAT5, coaxial, RF, or other Ethernet media	Free	Determined by Ethernet	Unlimited	40	N/A
TP-RS485-39	Node	39	Level IV	Bus	2	1200	32	0
FO-20S FO-20L	Backbone	1250	Fiber optic	Bus	None	15,000	64 (FO-20S) or 512 (FO-20L)	N/A

* in kbit/s
† in m. For bus topology, distance is maximum bus length. May depend on cable type.

Figure 5-11. LonWorks allows for a variety of channel types, each with different signaling and cabling characteristics.

> ### Channel Designations
>
> Echelon and LonMark typically use consistent designations for channel types. Channels using twisted-pair cabling are preceded by a "TP" designation, while other letters are used to describe some defining aspect of the channel, such as "FT" for free topology or "XF" for transformer-coupled.
>
> However, there are many non-standard variations of channel designations used in the field or by other manufacturers that may cause confusion. Letter designations are often partially dropped from the standard designations. For example, "TP/XF-1250" may also be designated as "TP-1250" or "XF-1250." Alternatively, transceiver designations may be used to refer to a channel type, such as an "FTT-10A channel."
>
> Most non-standard designations are still understandable as to which channel is being referred to because there is a well-defined list of supported channel types. However, integrators should be aware of these inconsistencies throughout the industry. To avoid possible confusion, it is recommended to use the proper designations in all documentation.

This channel uses an FTT-10A transceiver that is manufactured by Echelon Corporation. A maximum of 64 FTT-10A transceiver loads are allowed per TP/FT-10 segment.

The TP/FT-10 channel can be installed in a free topology (hence the "FT" designation) configuration, but it allows for greater segment lengths in bus topology. Depending on the cable type, TP/FT-10 channels installed in bus topology can include wiring lengths up to 600 m, while in free topology the maximum wire length is 450 m.

Twisted-pair cabling is the most common type of media implemented for LonWorks networks. It is used for several channel types.

This type of channel can also be used for link power, which allows nodes to be powered from a regulated DC power supply applied on the twisted-pair channel media. **See Figure 5-12.** Power is transmitted on the same pair of conductors as the data. For example, many lighting control nodes use link-power transceivers to provide power for LonWorks light switches. Power is supplied to the channel from a link power supply of regulated 42.4 VDC. Up to 127 link-power transceivers may be installed on a twisted-pair segment.

While installing up to 128 link power nodes on a TP/FT-10 segment is electrically feasible, a segment is logically limited to 127 nodes. This also accounts for nodes that may draw slightly more current than a single link power unit.

Link-power nodes can be safely installed with regular TP/FT-10 nodes on the same channel segment as long as the TP/FT-10 nodes include blocking capacitors at the transceiver connections. Blocking capacitors provide DC voltage isolation for the FTT-10A transceiver in the event of a DC fault on the link power channel segment. **See Figure 5-13.**

TP/XF-1250. The TP/XF-1250 channel is a high-speed channel for twisted-pair cabling media that is typically used to create a high-speed backbone. The TP/XF-1250 channel runs at a high bitrate, 1250 kbit/s, which can accommodate up to 700 LonTalk packets per second. The TP/XF-1250 channel provides high bandwidth capacity to accommodate large user-interface applications where hundreds of data point values must be updated quickly. TP/XF-1250 channels must be installed in a bus topology and segment length is limited to 130 m. Also, no more than 8 nodes can be installed within any 16 m of cabling.

TP/XF-78. Like the TP/XF-1250 channel, the TP/XF-78 channel is a transformer-coupled channel for bus topology segments. Transceiver modules for both channel types include a transformer to isolate the module from the twisted-pair communication bus. However, this TP/XF-78 channel allows for significantly longer segments (up to 1400 m), though at the lower bandwidth of 78 kbit/s.

PL-20x. PL-20x is a family of powerline channels, allowing communication over existing line-voltage power circuits (if all nodes are on the same side of a transformer). These channels should not be confused with the link power channel. Powerline channels use the existing electrical infrastructure of a building as its physical network. The data signaling can coexist with the AC sine wave or be used with unpowered media. Link power nodes, in contrast, require separate network cabling, which provides both data communication and operating power through a pair of conductors.

Installation is simple and fast since the network communication channel is the existing power wiring of a building. Performance, while slow at about 5 kbit/s, is reliable over great distances and can accommodate hundreds of nodes per segment.

IP-852. IP-852 is a channel for routing LonTalk packets over IP media. It is based on ANSI/CEA-852, which is an Internet standard for tunneling control data over IP infrastructures. Network management tools that support IP-852 can deliver high throughput of up to 6000 LonTalk packets per second to the LNS Server.

LonWorks/IP-852 routers are commonly used to create IP backbones for buildings within a campus by leveraging IP networks already in place. LNS network management tools and an LNS Server computer can also become a member of a IP-852 channel, providing high-performance remote-network management.

TP-RS485-39. TP-RS485-39 is a standard channel type for nodes that use TIA-485 (formerly known as RS-485) transceivers. The channel media is twisted-pair cable installed in a bus topology only and has a recommended maximum segment length of 1200 m.

While the TIA-485 standard allows for bit rates up to 35 Mbit/s, the transmission speed of the LonMark TP-RS485-39 is 39 kbit/s. Up to 32 nodes may be installed on each TP-RS485-39 channel segment.

Link-Power Networks

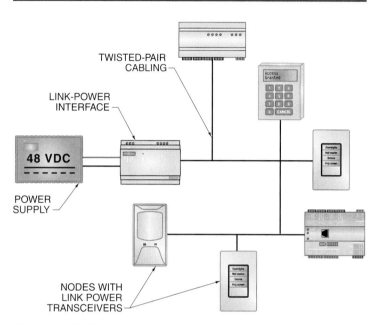

Figure 5-12. Link power networks combine data communication signal and node power supply on the same set of twisted-pair conductors.

Blocking Capacitors

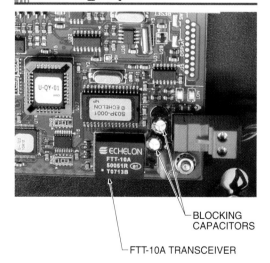

Figure 5-13. Blocking capacitors are required to protect FTT-10A transceivers when installed on link-power segments.

FO-20S and FO-20L. Fiber optic media channels are commonly used to create high-speed backbones. Performance is similar to the TP/XF-1250 channel, providing 1250 kbit/s throughput. Fiber optic media is immune to high electrical noise environments and segments can be very long, up to many kilometers. Nodes, typically routers for backbone installations, are daisy-chained on FO-20 channels. The two types of fiber optic channels, FO-20S and FO-20L, can support 64 and 512 nodes per channel, respectively.

> Network media may be dedicated or shared. Dedicated media is used only by a single network segment. Shared media is shared with other network segments or functions. For example, powerline and radio frequency channels are shared media. Shared media requires special attention to network traffic issues.

LonTalk Routers

Like other data communication networks, LonWorks systems use routers to create independent channels and filter network traffic. This helps use network bandwidth more efficiently. Some routers also connect channels of different media types, such as twisted-pair, fiber optic, radio frequency, or powerline channels. For example, a LonTalk router may include one TP/XF-1250 port for a network backbone and multiple TP/FT-10 ports for node channels. The use of LonWorks routers as network infrastructure components allows up to 127 nodes to operate on each of up to 255 separate channels within a single network domain.

LonTalk Repeaters

Physical layer repeaters are often confused with LonTalk routers, which can be configured to act as a repeater. While both act as network repeaters, there is a major difference. LonTalk routers, since they include LonTalk processors, error-check each message, and only forward packets that include a valid CRC (cyclical redundancy check). This prevents corrupted packets and noise from entering other LonTalk channels.

Physical layer repeaters are less expensive solutions for extending channel segments and increasing the number of nodes on a single channel. However, physical layer repeaters forward all traffic regardless of validity, which can propagate noise and corrupted packets through the network.

INFRASTRUCTURE PLANNING

The network design must be mapped to the building plan to indicate the physical locations of all network devices and nodes, and allow for the wiring routes between them. **See Figure 5-14.** The physical network design must be feasible and conform to the applicable building codes. For example, nodes must be located in accessible spaces, and wiring must not be required to pass through unbreachable walls. Also, channel-segment lengths must be calculated to determine if media-segment limitations could be exceeded according to planned cable type, topology, and transceiver type.

Twisted-Pair Cabling Installation

Installers should verify that the designed topology for backbone and node channels is supported by the device transceivers. For example, TP/XF-1250 channels can only be installed in bus topology and TP/FT-10 channels can utilize either bus or free topology. Also, transceivers must be compatible with the channel on which the node is installed.

In most building automation applications, installers should use unshielded twisted-pair cabling that has been tested and approved by the Echelon wiring guide. Noise immunity for twisted-pair cabling is exceptionally good, and improperly installed shielded cable can negatively impact network performance. Mixing cable from different vendors on the same segment is not recommended because it can change the electrical characteristic of the entire segment, which may cause communication issues. If it is necessary to change cable types, a physical layer repeater, or a router configured as a repeater, should be installed at the connection point.

Infrastructure Planning

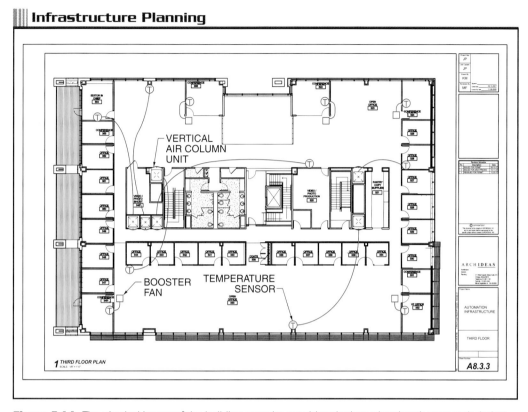

Figure 5-14. The physical layout of the building must be considered when planning the network design.

Detailed information about installing twisted-pair cabling, including cabling vendors, cable characteristics, and limitations, are listed in the Echelon document *Junction Box and Wiring Guideline for Twisted Pair LonWorks Networks*.

Segment Length Limitations. The maximum segment length depends on the twisted-pair cabling type. For example, a TP/FT-10 channel in bus topology using NEMA Level IV cable has a limit of 1400 m, while CAT5 cable has a limit of 900 m. Repeaters may be used to extend the length of a channel, but only one per channel. There are additional limits on maximum node-to-node distance.

Stubs are short lengths of cable that connect nodes to the channel. Typically, a stub runs from a T-tap in a junction box along the channel bus to the node. The maximum stub length depends on the channel type and topology. For example, on a TP/XF-1250 channel, stub lengths should not exceed 0.3 m, while a TP/FT-10 channel installed in bus topology may utilize a stub up to 3 m long.

Segment Termination. Terminators are resistor-capacitor (RC) circuits at certain points in the channel to absorb signal reflections. The required values of the resistors and capacitors depend upon segment channel and topology. **See Figure 5-15.** The location of terminations also varies. TP/XF-1250 and TP/FT-10 bus segments are terminated once at each endpoint, while TP/FT-10 free topology segments can be terminated at any single location along the segment. Terminators are available as separately installed devices in all of the four configurations and RC values.

High Electrical Noise Environments. It is recommended to avoid installing unshielded twisted-pair cabling near sources of electromagnetic interference (EMI), such as variable-speed motor drives or conductors carrying high-current triac-controlled loads. EMI can induce noise on nearby conductors, which can significantly interfere with data communications.

Segment Terminators

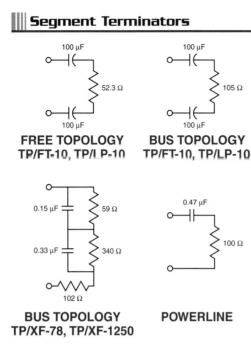

Figure 5-15. The required resistor-capacitor termination circuits vary depending on the channel type and topology.

If electrically noisy environments cannot be avoided, shielded cable may be used as long as the shield is terminated properly to ground. For all twisted-pair channel types, the shield conductor is terminated using an RC circuit. **See Figure 5-16.** The shield should be grounded at a minimum of one point on the segment. However, grounding the shield at every node helps suppress 50/60 Hz standing waves that can result in signal distortion.

Shielded-Cable Grounding

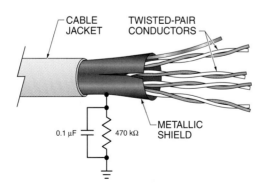

Figure 5-16. When a shielded cable is used, the metallic shield conductor should be grounded through a particular resistor-capacitor circuit.

Lightning Protection. For twisted-pair cabling installed outdoors, shielded cable and lightning protection should be used. **See Figure 5-17.** The shield at each building entry point should be connected to ground via a lightning/surge arrestor to conduct surges to ground. Each network data conductor should also be protected with lightning/surge arrestors designed for data lines. Low-capacitance, gas-discharge-type capacitors are recommended for data-line arrestors. Metal oxide varistors (MOVs) and transient voltage suppressors (TVSs) should not be used because they may corrupt network data packets, causing communication failures.

Lightning Protection

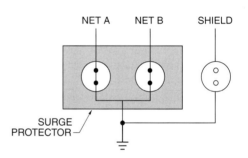

Figure 5-17. When twisted-pair cabling is used outdoors, it should be grounded for lightning protection.

Node Installation

Distributed control systems, such as LonWorks, have reduced installation costs when compared to centralized systems, where all control wiring is home run to master control panels. This is because less conduit and wiring is required when nodes can be daisy-chained and each node is installed adjacent to the equipment to be controlled or monitored.

Additionally, good node installation practices include grouping nodes on channels by usage, application, or location. Network bandwidth is optimized when node channels primarily include nodes that regularly communicate with each other. For example, the traffic between lighting control nodes installed on a lighting channel is isolated and not forwarded to the HVAC channel unless a specific network variable binding is created. This requires less forwarding by the LonTalk routers to send messages to other channels, which improves the performance and responsiveness of the network.

Common Node Power Supplies

Most manufacturers recommend an independent power supply for each node, but it is possible to share a common power supply among multiple network nodes. However, there are special considerations for using a common power supply. First, power conductors must be sized properly to accommodate voltage drops over the segment distances.

Also, nodes that utilize half-wave rectified power cannot be mixed with nodes designed to utilize full-wave power, or serious damage can occur. **See Figure 5-18.** Half-wave rectified power supplies typically have one terminal of the AC transformer grounded. Full-wave power supplies have a ground connection in the rectification circuit. If the full-wave power supply is used in a system designed for half-wave rectification, the two ground connections create a short circuit across the transformer (through a diode). This can either trip an overcurrent protection device, cause physical and permanent damage to the diode or transformer, or both.

Installers should always check with the node, sensor, and power supply manufacturers to confirm the type of rectified power before using secondary power supplies.

Power Supplies

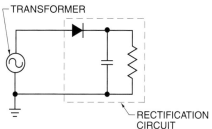
HALF-WAVE POWER SUPPLY

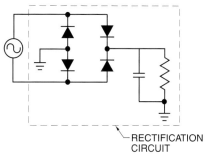
FULL-WAVE POWER SUPPLY

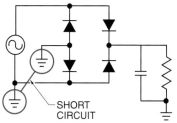
FULL-WAVE POWER SUPPLY CONNECTED TO HALF-WAVE POWER SUPPLY

Figure 5-18. Mixing half-wave and full-wave rectified power supplies in the same system can cause a short circuit that permanently damages the electronic components.

Summary

- The choice of network architecture defines how tools connect, manage, monitor, and control nodes on the LonWorks network.

- LonWorks networks are most often applied in a flat network architecture.

- LonWorks Network Services (LNS) is a network operating system that provides an open platform for flat network architectures, as well as full support of LonMark interoperability standards.

- A computer houses the LNS platform software.

- An LNS network database includes data about infrastructure components, nodes, and network-program configurations within a specific network domain.
- The LNS Server is an application that runs on the computer hosting the global database and enables client applications to access the information in the databases.
- Network management and user-interface tools access a LonWorks network control system through a network interface.
- The LNS platform architecture provides connectivity options that allow network management tools access to the LNS Server and database from any connection point on the LonWorks network, or remotely over IP networks.
- Four different client/server configurations supported by LNS are local clients, full clients, lightweight clients, and independent clients.
- Access to the LNS database and the network affects a client's ability to perform network management and monitoring/control functions, respectively.
- LonWorks nodes communicate over network channels through a hardware transceiver within the node.
- Selection of an appropriate channel type involves balancing the channel features and capabilities with their costs, limitations, and feasibility within the building to be automated.
- On twisted-pair cabling media, the most widely used channel is TP/FT-10.
- Detailed information about installing twisted-pair cabling, including cabling vendors, cable characteristics, and limitations, are listed in the document *Junction Box and Wiring Guideline for Twisted Pair LonWorks Networks*.
- If electrically noisy environments cannot be avoided, shielded cable may be used as long as the shield is terminated properly to ground.
- Good node installation practices include grouping nodes on channels by usage, application, or location.

Definitions

- A *Network Service Device (NSD)* is a software object used by LNS to communicate with nodes and routers during network management tasks.
- A *network interface* is a hardware device and/or software driver installed in the network management tool computer for communicating on the network media.
- A *local client* is an LNS network management tool that resides on the same computer as the LNS Server and database.
- A *full client* is a non-local LNS network management tool that communicates with the LNS Server and database computer through the LonWorks network.
- A *lightweight client* is an LNS network management tool that communicates with the LNS Server computer directly over a TCP/IP network.
- An *independent client* is an LNS network application that accesses the network only to perform monitoring and control of network variables.

Review Questions

1. What are the advantages of using LonWorks in a flat architecture?

2. How is information about LonWorks networks stored in LNS databases?

3. What is the role of a network interface?

4. What is the difference between a full client and a lightweight client?

5. What types of applications are independent clients?

6. How does a client's connection to the LNS Server computer affect its functionality?

7. What is the primary difference between LonTalk routers and physical layer repeaters?

8. What is involved in planning the physical installation of system infrastructure?

9. What factors influence the installation location of nodes?

10. Why must half-wave rectified powered and full-wave rectified powered nodes not be mixed on a common node power supply?

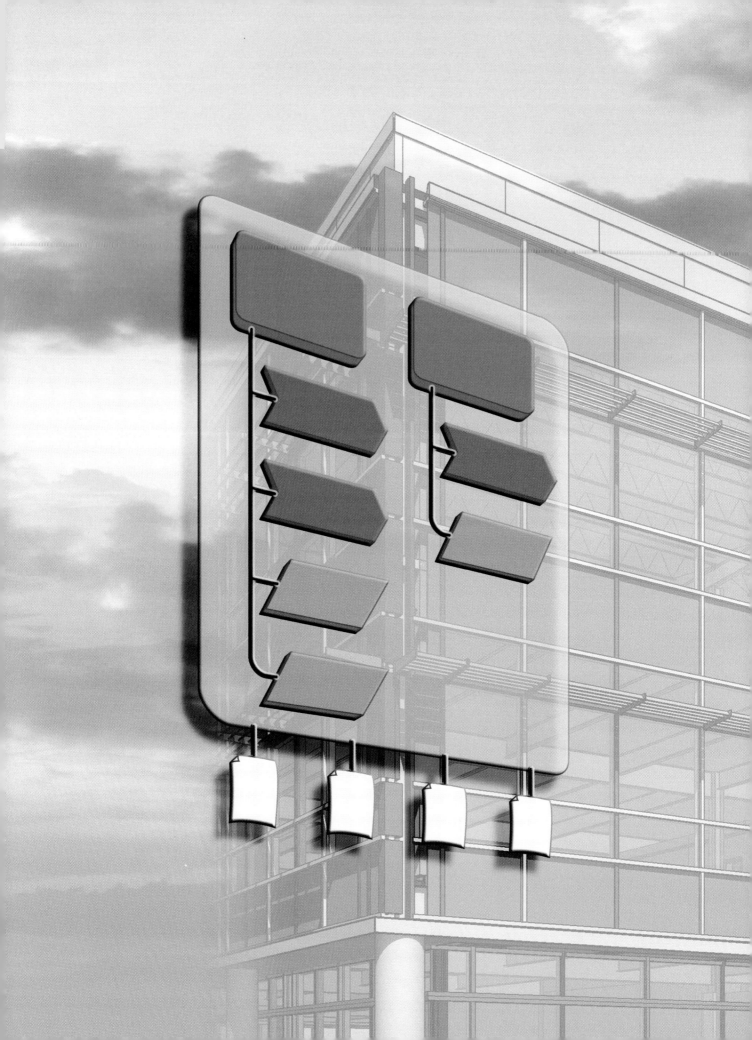

Chapter Six

LonWorks Nodes

All LonWorks nodes contain a core set of hardware and software components. In order to be LonMark-certified, which ensures interoperability with any other certified LonWorks node, certain components must be present and documented in resource files. Standardization of LonWorks components allows nodes from multiple manufacturers to operate together seamlessly in a LonWorks control network.

Chapter Objectives

- Identify the role of LonMark International in maintaining system interoperability.
- Describe the hardware components required for LonWorks nodes.
- Compare the logical structures in which information and functions are stored in node software.
- Differentiate between the types of LonWorks nodes.

LONMARK CERTIFICATION

It is recommended to use nodes certified by LonMark International in LonWorks systems. Uncertified nodes may operate adequately, but do not ensure interoperability, especially if nodes from more than one manufacturer are used together on the network. LonMark-certified nodes are tested to conform to the specifications in the LonMark Interoperability Guidelines, which provide a high degree of interchangeability between different node vendors. **See Figure 6-1.**

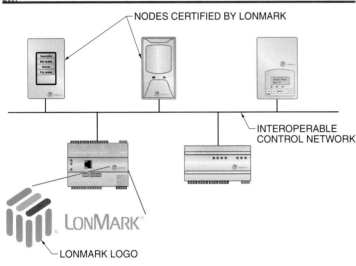

Figure 6-1. Nodes certified by LonMark can be used together in a control network, even if they are from different manufacturers.

There may be minor differences in the inclusion of optional and manufacturer-defined software components, but all of the required software components are included, regardless of manufacturer. Therefore, interoperable LonWorks networks may include nodes from several different manufacturers.

LONWORKS NODE HARDWARE COMPONENTS

The minimum hardware components required for a node to send, receive, and act on control information are a microprocessor and a transceiver. For practical purposes, most nodes also include a number of other components, such as additional memory, power supplies, and input/output electronics, depending upon the node's application and manufacturer requirements. **See Figure 6-2.**

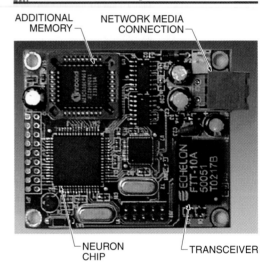

Figure 6-2. A LonWorks node must include a microprocessor and transceiver as hardware components. Additional components, such as external memory, are also commonly included.

Microprocessors

All LonWorks nodes include microprocessors, which implement the LonTalk protocol and the node's application program. The LonTalk protocol is openly published and freely licensed for use on any manufacturer's microcontroller. Most LonWorks node manufacturers utilize the Neuron chip to run the node's application program. Some, though, utilize supplementary or replacement microprocessors to take over some or all of the processing requirements.

Neuron Chip. The Neuron chip is a single microcontroller designed by Echelon and manufactured by Toshiba and Cypress Semiconductor. The Neuron chip is capable of running all seven layers of the OSI Model for data communications. The Neuron chip contains three microprocessors, each handling different parts of the OSI Model. **See Figure 6-3.**

Neuron Chip

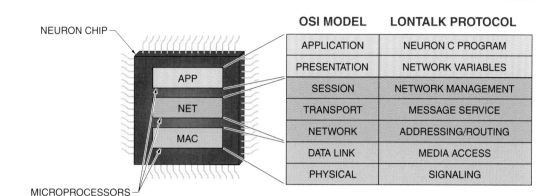

Figure 6-3. The Neuron chip includes three processors that implement the seven layers of the LonTalk protocol.

The media access control (MAC) processor supports the physical layer in the transmission of raw bits of information over the physical network media, such as twisted-pair conductors. The MAC processor also supports the data link layer by organizing and framing the raw bits into data frames (LonTalk packets). The MAC processor acknowledges received messages when required and provides message time slots for priority message packets.

The network (NET) processor manages the node's network layer, which includes a table of the logical addresses of individual or groups of node destinations. The NET processor implements services at the transport layer, such as different message service types, and the session layer, such as the processing of network management commands from network management tools, request/response messaging (polling), and security authentication.

The application (APP) processor runs a node's application program, which is created by the node manufacturer. The node program is written in Neuron C, which is an extension of the ANSI C programming language that is designed specifically for control applications performed by LonTalk microcontrollers. Application development tools, such as Echelon's NodeBuilder, provide access to Neuron chip firmware objects that define input/output functions (physical layer), network variables (presentation layer), and configuration properties (application layer).

Memory space limitations restrict Neuron chip nodes from running large application programs and from monitoring or controlling large numbers of physical input/output points. For example, the Toshiba TMPN3120 Neuron chip includes 16 kB of onboard ROM (read only memory), 2 kB of RAM (random access memory), 2 kB of EEPROM (electrically erasable programmable read only memory), and 11 input/output pins, and it is limited to a maximum of 62 network variables.

Host Microprocessors. To address manufacturers' requirements for additional resources for complex tasks, Echelon provides Neuron chip firmware to use with other microprocessors. The *Microprocessor Interface Program (MIP)* is software that transforms the Neuron chip into a co-processor and moves the upper two layers of the LonTalk protocol, the presentation and application layers, off the Neuron chip and onto a more sophisticated external processor. A *host-based node* is a LonWorks node that employs the Microprocessor Interface Program (MIP) to use a supplemental microprocessor. Host-based nodes expand the resources of the LonTalk protocol while retaining the networking capabilities of the Neuron chip. Typical host-based applications include network management and monitoring/control applications where a computer acts as a node on the network. A host-based node can include up to 4096 network variables.

Non-Neuron-Chip Microcontrollers. The LonTalk protocol can also be fully implemented on a totally non-Neuron-chip solution. For example, an independent company, Adept Systems, has demonstrated an implementation of the LonTalk protocol on a Motorola MC68360 microcontroller. This satisfied a requirement by the Electrical Industry Association (EIA) for LonTalk to become an American national standard. Adept Systems provides a family of LonTalk/IP routers that use non-Neuron-chip microcontrollers.

Neuron IDs. When each Neuron chip is manufactured, it is assigned a Neuron ID. A *Neuron ID* is a 48-bit serial number that is unique to a single Neuron chip. The Neuron ID is used to define a unique physical address for each network node, much like a MAC address. The Neuron ID is stored on the chip in nonvolatile memory and cannot be modified. The Neuron ID is required by network management tools when the node is added onto the network during the commissioning process.

Neuron IDs are also provided by Echelon to manufacturers of LonTalk microcontrollers or developers providing software implementations of the LonTalk protocol. Vendors who have signed a Protocol Patent License Agreement pay a royalty for each Neuron ID.

Transceivers

All nodes include a hardware transceiver that transmits and receives network information, allowing the nodes to communicate over network media. **See Figure 6-4.** The Neuron chip's communication port is software configurable for speeds of 600 bit/s to 1.25 Mbit/s. The external transceiver is connected to the communication terminals and electrically isolates the Neuron's integrated circuit from the network media, allowing the node application program to be media-independent.

Microcontrollers are integrated circuits (chips) that include not only a microprocessor, but also memory, input/output functions, and possibly other electronics. A microcontroller is often known as a "computer on a chip."

Transceivers

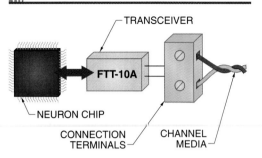

Figure 6-4. Transceivers communicate data from the microprocessor onto the channel media.

The transceiver type must match the network channel type. Therefore, transceivers are available for twisted-pair, powerline, and fiber optic channels. **See Figure 6-5.** The most common transceiver is the Echelon FTT-10A, which communicates over twisted-pair cabling at a speed of 78 kbit/s.

LonWorks Transceivers	
Channel Type	Transceiver
TP/FT-10	FTT-10A, LPT-11
TP/XF-1250	TPT/XF-1250
TP/XF-78	TPT/XF-78
PL-20x	PL-3120, PL-3150
IP-852	Various
TP-RS485-39	TP-RS485-39
FO-20	Various

Figure 6-5. Transceivers must be matched to the channel media on which the node or network device will be installed.

External Memory

The model 3150 Neuron chip provides an external memory bus that allows up to an additional 58 kB of memory. This can be divided into two additional storage areas: one for hosting the Neuron firmware program (up to 16 kB) and another for application code space (up to 42 kB). The external memory may be in various combinations of ROM, EEPROM, NVRAM, flash memory, and static RAM.

Input/Output Connections

Neuron-chip-based nodes include built-in communication and control functions that reduce development time necessary to create an intelligent LonWorks node. Neuron chips include 11 programmable input/output pins that can be configured to monitor or control a variety of physical and data signals while requiring a minimum of external electronics. Input/output flexibility is further enhanced by the 34 control function objects included with the Neuron firmware.

Physical input/output types and quantities can vary greatly between manufacturers depending upon the node application. For example, a typical air-handling unit controller includes five sensor inputs wired to temperature and pressure sensors. The inputs may be configured for digital or analog sensors, including current, voltage, or resistive signal types. Six configurable physical outputs then provide ON/OFF or analog control of equipment actuators.

Power Supplies

LonWorks nodes operate on a variety of power-supply configurations, depending on transceiver type, usage, and manufacturer requirements. For example, nodes that utilize powerline or link power transceivers draw supply power from the network media. Nodes that include other twisted-pair transceivers are normally powered from external low-voltage transformers, DC power supplies, or line-voltage input power.

Most LonWorks nodes are powered by an external 24 VAC or DC power supply. Internal electronics further condition and distribute the power to the node components, such as the microprocessors, interfaces, transceiver, and displays. The Neuron chip requires a 5 VDC power supply. The Neuron chip includes a low-voltage detection (LVD) circuit that resets the chip if input voltage falls below 4.15 VDC.

A common external power supply may be shared between multiple nodes, as long as each node uses the same type of rectification (converting AC power to DC). Nodes that use a full-wave power rectification circuit should never share the same input power supply with nodes that use half-wave rectified power conversion circuits, which can cause permanent damage to electronic circuits. Also, a disadvantage of having a common power supply is that it risks causing major problems from a single source of failure.

Service Pins

A *service pin* is a node pushbutton that causes the node to transmit an identifying message onto the network. **See Figure 6-6.** Pressing the service pin causes the node to broadcast its unique Neuron ID number on the network. This is used to identify the individual node to a network management tool, which receives the broadcast and logs the node's Neuron ID, Program ID, and channel location.

Service Pin and Service LED

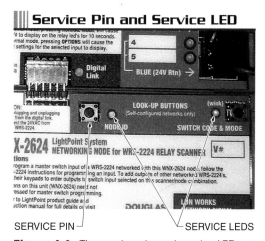

Figure 6-6. The service pin and service LED are hardware components used for basic interfacing with a node or network device.

Service LEDs

The service LED is an optional feature of Neuron-chip-based nodes. A *service LED* is an LED that physically indicates the node's operating status. **See Figure 6-7.** For example, if the node's service LED is blinking at half-second intervals, then the node includes an application program, but has not yet been commissioned into the network. If the service LED is ON continuously, the node has no application program loaded. If the service

LED turns ON briefly after power-up and then turns OFF, it has been commissioned into a network. Node manufacturers can also provide additional service LED behaviors according to their application requirements.

Service LED Behaviors

Figure 6-7. The behavior of the service LED indicates the state of the node.

A node's Program ID identifies its general type and application, such as a commercial lighting controller.

LONWORKS NODE SOFTWARE COMPONENTS

LonWorks nodes include a collection of software components that share and manage control information. The node is programmed to perform one or more specific control tasks, and the software components provide the integrator with access into its building blocks. LonWorks software components include application programs, function blocks, network variables, configuration properties, and resource files. **See Figure 6-8.**

LonWorks' interoperability relies on the standardization of these software components whose explicit definitions are freely available from LonMark for use by any node manufacturer. The implementation of standard software components means that nodes are interchangeable, regardless of manufacturer.

When the standard software components are inadequate for a certain control task, LonWorks provides for non-standard, user-defined (manufacturer-defined) software components for custom applications. This allows for significant flexibility in developing control actions, but sacrifices interoperability. A node that uses a manufacturer-defined software component is not easily interfaced with or replaced by a node from another manufacturer. It may also require the use of proprietary configuration and installation tools. However, the LonWorks platform is updated periodically with new standard software components, which provides a way for popular user-defined components to eventually become standardized.

Application Programs

Software residing on Neuron-chip-based nodes is divided between system-image, application-image, and network-image categories. The system image software includes the Neuron chip firmware, which consists of the LonTalk protocol stack, the Neuron C runtime library, and a task scheduler. The system image is kept in ROM. The application image software contains the compiled program code created by the developer and is loaded into EEPROM. In addition to manufacturer-defined control algorithms, the application image contains definitions of network variables, optional self-documentation strings, buffer sizes, Program ID, and transceiver type.

The application program code is written in the Neuron C programming language using development software from multiple vendors. The network image software contains node address information and network variable binding definitions and is typically stored in node EEPROM. Network management tools update the network image over the network media.

Program ID. LonWorks nodes are assigned an identification number by the application development tool to specify the node's intended use and type. The *Program ID* is a 64-bit number that provides specific details and unique identification for a LonWorks node. **See Figure 6-9.** This number is organized into six fields and typically presented in hexadecimal format.

The first field is a 4-bit device format number that indicates conformance to LonMark guidelines. An "8" indicates that the node has been reviewed by LonMark and has been certified as meeting interoperability standards. A "9" indicates that the manufacturer has created a LonMark-compliant node, meaning that it also meets interoperability standards but is not certified. This is commonly used for prototype nodes that will later go through the certification process. All other formats are either reserved by Echelon for future use or used by network interfaces or legacy devices.

The second field is a 20-bit number that identifies each LonWorks node manufacturer. These member IDs are assigned by LonMark. For example, Hubbell has a LonMark member ID of 199 (0x000C7). The LonMark International web site includes a list of the more than 500 members, along with their designated member ID number.

The third field is the 16-bit device class and is associated with a LonMark standard functional profile (SFP) included in the node application. There are currently hundreds of LonMark device classes, which are collected into 21 groups for different industries or applications, such as HVAC, lighting, and energy management. Within each group are standard functional profiles for specific control tasks within that industry. For example, a lighting scene controller profile is "32.51" (0x2033). If a node implements multiple functional profiles, one is designated as the primary functional profile and its number is used in the Program ID.

LonWorks Node Software Components

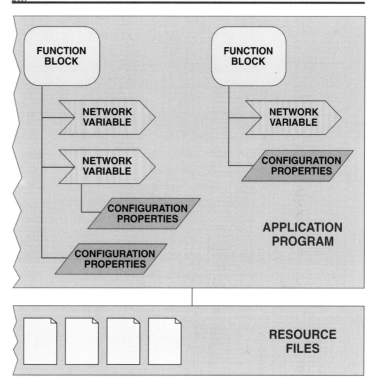

Figure 6-8. Node software components include the application program, function blocks, network variables, configuration properties, and resource files.

Program ID

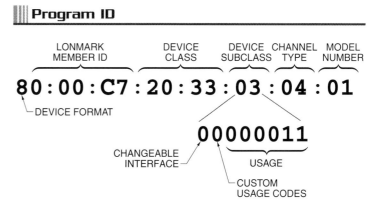

Figure 6-9. The Program ID specifies information about a node, including LonMark conformance, manufacturer, class, usage, transceiver type, and application version number.

The fourth field is an 8-bit device subclass designation, which indicates the node usage. The most significant bit indicates whether the node has a changeable interface, which allows for changeable or dynamic network variables. Such network variables are used when the node developer cannot know the correct type of the network variable in advance, such as for a generic sensor device. The next bit indicates whether the usage field (the following six bits) indicates a standard or functional profile-specific usage. If this is set to 1, the program uses a custom set of usage values in the functional profile definition. The final six bits of the subclass field is a usage ID, which indicates the environment in which the node is intended to be used. **See Figure 6-10.** For example, usage code 000101 (5) indicates residential usage.

Channel Type Codes	
Code*	Channel Type
01	TP/XF-78
03	TP/XF-1250
04	TP/FT-10
05	TP-RS485-39
09	PL-10
0F	PL-20A
10	PL-20C
11	PL-20N
12	PL-30
18	FO-20S
98	FO-20L
9A	IP-852

* in hexadecimal format

Figure 6-11. A node's channel type is documented in its Program ID with a special code.

Device Subclass Usage Codes	
Code*	Usage
000000	Network Management
000001	Connectivity
000010	Industrial
000011	Commercial
000100	Industrial/Commercial
000101	Residential
000110	Residential/Commercial
000111	Transportation
001000	Telecommunications
001001	Medical
001010	General
001011	Utility

* in binary format

Figure 6-10. Standard device subclass usage codes in the Program ID indicate the environment in which a node is intended to be used.

The fifth field is an 8-bit field that specifies the communication channel type supported by the node. The channel type is determined by the hardware transceiver supplied with the node. **See Figure 6-11.** Checking the channel ID from a node's Program ID verifies that it should only be installed on specific channel.

The sixth field is an 8-bit manufacturer-defined model number. The use of this field is entirely up to the manufacturer, and it is typically used to indicate an application version number.

Function Blocks

The control intelligence within a LonWorks node is organized into function blocks. A function block represents a certain control task. A function block performs a task by receiving control information inputs, processing the data, and sending out new control information as outputs. Inputs and outputs may be from the network, hardware attached to the node, or from other function blocks on the node. A function block can be as simple as an analog sensor or as complex as a chiller controller. A node may include one or more function blocks, representing each control task that the node can perform.

Function blocks are implementations of functional profiles, either LonMark standard functional profiles or manufacturer-defined functional profiles. Functional profiles detail a specific control task, including the network variables and configuration properties required to execute and configure that task. **See Figure 6-12.** Functional profiles are independent from the control device, so they can be implemented in a standard way in any type of node requiring that control task. Therefore, nodes that utilize the same standard functional profiles are highly interchangeable.

LonMark has approved over 70 standard functional profiles for control tasks in a variety of applications and building systems. Profiles are created by LonMark members belonging to industry-specific task groups and many profiles are under development. To become standard functional profiles, they undergo an extensive review and approval process by the associated task group. The LonMark Interoperability Guidelines allow for the potential creation of many standard functional profiles by providing hundreds of device class definitions.

In order to be certified by LonMark, a node must contain at least one LonMark functional profile. Each LonMark functional profile is assigned an identification number and specifications for each profile are freely available on the LonMark web site. Nodes certified by LonMark may also include non-standard, user-defined functional profiles (UFPs) to perform control tasks that are not defined within the LonMark standards.

Many nodes include a generic function block called the Virtual Function Block, which provides a holding place for network variables and configuration properties that are not associated with a specific function block. These may include data and settings that affect the node globally.

LonMark-compliant nodes that include more than one function block profile must also include a standard management function block called a Node Object. **See Figure 6-13.** The Node Object is used to monitor the status of multiple function blocks within a node and report node state, mode, and error statistics. Node manufacturers can optionally include custom self-test diagnostics for troubleshooting node hardware. The Node Object can also be used to disable/enable and override node function blocks.

Network variables and configuration properties included within a function block provide the interface to application specific network data and tools.

Network Variables

LonWorks nodes send and receive control data as network variables. Network variables provide a standard way for sharing information among nodes of different types and from different manufacturers. For example, an interoperable air temperature sensor can transmit the outdoor temperature to multiple thermostats, which then use the shared information to control their portion of the system.

Functional Profile

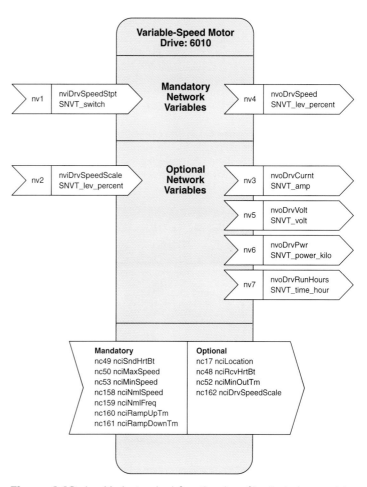

Figure 6-12. LonMark standard functional profiles include mandatory and/or optional network variables and configuration properties.

All standard software components are documented in detail and freely available for download from the LonMark web site at www.lonmark.org. Also available are the *LonMark Layer 1-6 Interoperability Guidelines* and *LonMark Application Layer Interoperability Guidelines* documents.

Node Object Function Block

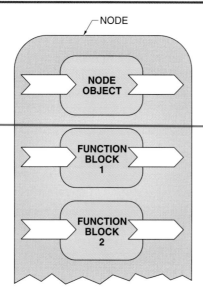

Figure 6-13. The Node Object is a standard functional profile that manages, monitors, and reports the status of a node. It is included on any node that contains more than one functional profile.

Many network variables are based on the values of input devices, such as sensors, that measure certain building or system conditions.

Network variables can contain relatively simple values, such as temperature, pressure, or voltage, or complex data such as alarms, modes, and status information. Network variable types define the content and structure of the data. Industry task groups within LonMark have created a set of nearly 200 standard network variables types (SNVTs). SNVTs are a key component of interoperability because their format, structure, and usage are publicly documented and freely available for use by any node manufacturer. LonWorks node manufacturers can optionally define non-standard user-defined network variable types (UNVTs) within their node application program.

A Neuron-chip-based node can support up to 62 network variables, while a host-based node that uses an external processor to run the node application can support up to 4096 network variables.

The three basic configurations of network variables are numeric, enumerated, and structured types. When a manufacturer creates a node application program, the appropriate network variables are selected according to node usage. Optionally, node developers can provide changeable network variables in their node application program, which promotes interoperability between different manufacturers.

Numeric Network Variables. Numeric network variables are used to convey quantitative data such as temperature, pressure, and flow. All network variable values are shared in SI (metric) units. Network management tools and user interfaces convert these values to other units and formats, such as U.S. customary units, as required. Numeric network variable types include specific ranges and resolutions, depending on the type. **See Figure 6-14.** For example, the variable SNVT_temp_p, which is intended for HVAC applications, uses two bytes to represent temperature in degrees Celsius. It has a range of –273.17°C to 327.66°C and a resolution of 0.01°C. Alternatively, SNVT_temp_f was designed for temperature applications requiring a high-precision floating-point number. Four bytes are used to represent temperature in degrees Celsius, resulting in a range of -3.40282×10^{38}°C to 3.40282×10^{38}°C.

Enumerated Network Variables. Enumerated network variables include a single numeric value to represent a specific control mode, state, or value. For example, SNVT_date_day uses one byte to represent each day of the week; a value of 0 indicates Sunday, 1 indicates Monday, 2 indicates Tuesday, and so on. **See Figure 6-15.** The lists of enumerations are defined for standard network variable types and included in resource files. **See Appendix.**

Numeric Network Variable

SNVT_temp_p

SNVT Index	Measurement	Type Category	Type Size
105	Temperature	Signed Long	2 bytes
Valid Type Range	**Type Resolution**	**Units**	**Invalid Value**
−273.17 to 327.66	0.01	Degrees Celsius	32,767

Figure 6-14. A numeric network variable, such as temperature, allows for a range of values and units of measure.

Enumerated Network Variable

SNVT_date_day

SNVT Index	Measurement	Type Category	Type Size
11	days_of_week_t	Enumeration	1 byte
Valid Type Range	**Type Resolution**	**Units**	**Invalid Value**
days_of_week_t	1	N/A	DAY_NULL

days_of_week_t

Value	Identifier	Notes
−1	DAY_NUL	Invalid Value
0	DAY_SUN	Sunday
1	DAY_MON	Monday
2	DAY_TUE	Tuesday
3	DAY_WED	Wednesday
4	DAY_THU	Thursday
5	DAY_FRI	Friday
6	DAY_SAT	Saturday

Figure 6-15. Enumerated network variables use a single number to represent control modes, states, and levels. The associated list of enumerations may be found in separate documentation.

Structured Network Variables. Structured network variables include multiple fields and each may contain either numeric values or enumerations with specific meanings and formats. For example, the commonly used variable SNVT_switch includes two fields. **See Figure 6-16.** One holds an analog numeric value and the other an enumerated state. The analog value may represent the position of a light dimmer and the state may represent a pushbutton to enable/disable the dimmer. Therefore, SNVT_switch is 100.0 1 for fully ON and 0.0 0 for fully OFF. Some LonWorks variable-speed motor drives use the SNVT_switch value field to control the speed of the motor and the state field to enable/disable the drive.

> A set of custom resource files that document UNVT formats and structures must be developed by the node manufacturer and loaded onto the computer running the network management tool. These are required in order to display node data.

Structured Network Variable

SNVT_switch

SNVT Index	Measurement	Type Category	Type Size
11	Switch	Structure	2 bytes

Field	Measurement	Field Type Category	Type Size
value	Value	Unsigned Short	1 byte

Valid Type Range	Type Resolution	Units	Invalid Value
0 to 100	0.5	Percent of Full Scale	—

Field	Measurement	Field Type Category	Type Size
state	State*	Signed Short	1 byte

Valid Type Range	Type Resolution	Units	Invalid Value
0 or 1	1	N/A	–1

* This field can be –1 (NULL), 0 (OFF), or 1 (ON).

Figure 6-16. Structured network variables include multiple fields, which may be any combination of numeric and enumerated variables.

Configuration Properties

Configuration properties affect node behavior, such as alarm limits, setpoints, default values, and time delays. Configuration properties can apply to one or more network variables, specific functions within a node, or the entire node application. LonMark has defined over 300 standard configuration property types (SCPTs). The implementation of SCPTs allows integrators to configure nodes from different manufacturers using a single network management tool.

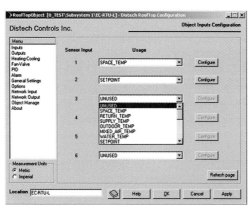

Distech Controls

Plug-ins provide user-friendly software interfaces for adjusting node behavior via configuration properties from within the network management tool.

Each SCPT uses a specific measurement type and format. Like network variable type definitions, SCPT data may be a numeric value, enumeration, or group of structured fields. Network management tools are used to access and modify SCPT values, which are stored in a node's nonvolatile memory. This is done during commissioning or network changes, but otherwise the values are typically constant. Alternatively, configuration properties may be implemented as network configuration inputs (nci), which allow modifications to node behavior by writing network variable values to configuration properties. However, since configuration properties are stored on a node's EEPROM, which can be written to a limited number of times, network configuration inputs must be used carefully.

Node manufacturers can create nonstandard user-defined configuration property types (UCPTs) for custom applications. However, adjustment of UCPTs may require a manufacturer-specific configuration tool or plug-in. A *plug-in* is a third-party software add-on that works within a network management tool that is based on LNS to provide a user-friendly interface for adjusting configuration properties. **See Figure 6-17.**

Manufacturers can also include safe limits, documentation, and monitoring applications within the plug-in.

External Interface Files (XIF)

LonMark requires descriptive information about a node's control functions that is openly exposed to network management tools. When a node application program is developed, an external interface file is generated and stored on the node that describes the inputs and outputs related to the control algorithm that the node performs. An *external interface file (XIF)* is a file on a LonWorks node that documents its Program ID, network variables, configuration properties, function blocks, and number of address table entries. **See Figure 6-18.** Node control algorithms and operational characteristics are not included in the XIF.

Network management tools require the information in the XIF for programming a control network. This information can be read from the node during commissioning, but can also be downloaded from the manufacturer's or LonMark's web site. LonMark maintains a database of over 600 certified devices and the XIF for each is available for free download. Downloading an XIF and importing the information into a network management tool allows an integrator to program a control network with that node before it is physically available.

The XIF is a plain-text file that can be viewed with a standard text editor, but it should never be modified. Importing an altered XIF into the network management tool may result in unpredictable and potentially damaging behavior. However, for performance optimization, some network management platforms, including LNS, convert the XIF from a plain-text file format to an optimized binary file (.XFO or .XFB extension) format.

The XIF format has undergone several revisions. The current version 4.0 XIF format adds manufacturer-defined default configuration values, which are used to define default node behavior and was not supported in earlier XIF formats.

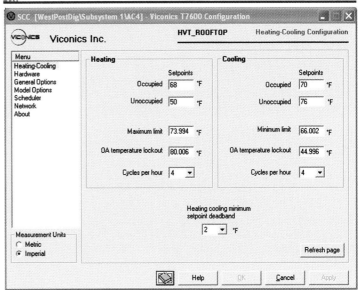

Figure 6-17. Configuration property plug-ins are used to adjust node configuration properties.

Network Management Tool Resource Files

The *LonMark Application Layer Interoperability Guidelines* specify a set of standard and user-defined resource files that describe a node's network variables, configuration properties, and functional profiles. Resource files are used by network management tools to manage node information and present control data in meaningful structures, units, and engineering formats. Resource files can apply to a single node or multiple nodes, depending on the manufacturer or usage. Four file types are included in the resource file set.

- Type files (.typ extension) include definitions of network variables, configuration properties, and enumerations.
- Functional profile template files (.fpt extension) include definitions of functional profiles.
- Format files (.fmt extension) include formulas for converting raw control values into engineering units such as United States customary or SI (metric) formats.
- Language files (such as the .enu extension for U.S. English) provide translation for data descriptions according to geographic locations.

External Interface File (XIF)

```
File: Dimmer.XIF generated by LONNCC32 Version 4.03.09, XIF Version 4.400
Copyright (c) ABC Corp
All Rights Reserved.  Run on Wed Aug 09 13:46:17
                                    ┌─PROGRAM ID                    ┌─MAXIMUM NUMBER OF
                                    │                               │ NETWORK VARIABLES
80:02:D8:1E:14:0A:11:00
2 15 0 5 0 3 3 3 3 3 3 11 9 2 4 0 0 4 11 1 1 11 5 0 0 0 0 0 0 0 0 2 15 0 0
0 0 0 2 1 0 ─── NUMBER OF STATIC NETWORK VARIABLES
15 5 14 13 28 532 0 15 5 3 174 4 10000000
1 9 1 2 0 8 3 15 200 0
3987 0 1 1 0 14 1 0 16 0 0 0
73 16 101 175 335 0 0 0 5 0 63 166 119 103
*    └─CHANNEL TYPE            ┌─SELF-DOCUMENTATION STRING
"&3.4@0NodeObject,3040LampActuator  (INCLUDES LIST OF FUNCTIONAL PROFILES)
              ┌─NETWORK VARIABLE
VAR nviRequest 0 0 0 0
0 1 63 0 0 1 0 1 0 1 0 0 0
"@0|1
       └─NETWORK VARIABLE DIRECTION (0 = INPUT, 1 = OUTPUT)
92 * 2
2 0 0 0 0  ─── SNVT INDEX
1 0 0 1 0
VAR nvoStatus 1 0 0 0
0 1 63 1 0 1 0 1 0 1 0 1 0
"@0|2
93 * 26
2 0 0 0 0
3 0 1 0 0
3 1 1 0 0
```

```
VAR nvoFileDirectory 2 0 0 0
0 1 63 1 0 1 0 1 0 1 1 0 0
"@0|8
114 * 1
2 0 0 0 0
VAR nviLampValue 3 0 0 0
0 1 63 0 0 1 0 1 0 1 0 0 0
"@1|1;Lamp Value
95 * 2         └─SELF-DOCUMENTATION STRING
1 0 0 0 0
1 0 0 1 0
VAR nvoLampValueFb 4 0 0 0
0 1 63 1 0 1 0 1 0 1 0 0 0
"@1|2;Lamp Value Feedback
95 * 2
1 0 0 0 0
1 0 0 1 0
```

Figure 6-18. An external interface file (XIF) contains information about the software components contained within it, including the Program ID, function blocks, and network variables.

Resource file sets are stored on the computer hosting the network management tool, typically in a separate folder labeled by manufacturer. A resource file catalog stored with these folders allows the network management tool to access the files. A resource file catalog browser updates the catalog as new resource files are installed. Node manufacturers providing plug-ins typically include the installation of resource files and update the resource catalog when the plug-in is installed.

LONWORKS NODE TYPES

During network design planning, the required physical and logical control points are identified. Physical control points include temperatures, pressures, actuator positions, and equipment activations/deactivations. Logical control points include setpoints, alarm limits, and schedule outputs. This point list, along with the sequences of operation, influences the selection of control devices. The nodes must include the necessary functions and input/output capabilities.

The four types of LonWorks control nodes are application-specific, application-generic, system-wide controllers, and freely programmable.

Application-Specific Nodes

An *application-specific node* is a control device that performs dedicated control functions defined by the application program loaded by the manufacturer. **See Figure 6-19.** While the node behavior can be modified by adjusting configuration properties, the node program is fixed and can only be changed by the node manufacturer. Examples of application-specific nodes include space comfort controllers, occupancy sensors, lighting controls, and chiller controllers.

Nodes that conform to LonMark interoperability standards include at least one standard application-specific functional profile, which must be appropriate for the necessary control task.

Application-Specific Node

Distech Controls

Figure 6-19. Application-specific nodes are available for commonly used control functions in a wide variety of building automation applications, such as controlling VAV units.

Application-specific nodes are designed with all the necessary inputs, outputs, and control software to operate a certain type of equipment, such as the variable frequency drives on HVAC fan units.

The external interface file (XIF) is also known as the device interface file. This file documents software components of the node that are exposed over the LonWorks network, such as network variables. The control software algorithms that are part of the application program are not included in the XIF, only the inputs to and outputs from those algorithms.

> Device selection criteria typically include listed safety certifications (such as UL, CE, and CSE), an accessible service pin, self-test functions, and a service LED that displays device status.

Application-Generic Nodes

An *application-generic node* is a control device that converts traditional control signals into network variable data. For example, an outside air sensor provides a 4 mA to 20 mA output signal to an application-generic node, which converts the analog signal to a data value and sends it as a SNVT_temp_p network variable to other nodes. **See Figure 6-20.** Application-generic output nodes are also used to position legacy (non-intelligent) actuators, such as relays and analog actuators, according to the values of received network variable inputs.

Application-Generic Nodes

Figure 6-20. Application-generic nodes provide universal hardware inputs and/or outputs, which are configurable through plug-ins.

System-Wide Controllers

A *system-wide controller* is a control device that provides general-purpose network functions for multiple nodes. These include scheduling, data logging, alarm generation, web serving, protocol translation, and event notifications. Setup normally requires use of a network management tool and may include a plug-in configuration tool provided by the node manufacturer. Several system-wide controllers can be accessed from the Internet via a web browser interface. **See Figure 6-21.**

System-Wide Controller Interface

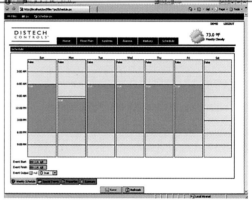

Distech Controls

Figure 6-21. System-wide controllers perform global control functions such as scheduling. This information is often available through a web browser interface.

Freely Programmable Nodes

A *freely programmable node* is a control device that is loaded with a custom control application. This type of node is used when no other node exists for a particular control task. Freely programmable nodes are popular for creating industrial process control applications and complex control sequences. Freely programmable nodes often include universal inputs and/or outputs that can be configured for common signal types, such as 4 mA to 20 mA.

Manufacturers of freely programmable nodes typically provide proprietary software for developing applications and loading them onto nodes. This may be a stand-alone program or a plug-in module for writing node application programs from within a network management tool environment. Specialized training may be required in the use of a manufacturer-specific programming language. **See Figure 6-22.** The custom application program can be backed up separately so that the program can be loaded onto replacement nodes if needed.

Freely-Programmable Node Programming

```
task

TOPOFTASK:

OutEnth = enthalpy(nviOSATemp, uiOutsideRH, ATMOSPHERIC_PRESSURE)
InEnth = enthalpy(nviReturnAirTemp, nviReturnAirRH, ATMOSPHERIC_PRESSURE)
nvoOSAEnthalpy = kjkg_to_btulb(OutEnth)
nvoReturnEnthalpy = kjkg_to_btulb(InEnth)

//Cooling setpoint reset schedule
      ramCoolingSetpt = linear_scale (c_to_f(nviReturnAirTemp),
f_to_c(70.0),f_to_c(74.0),f_to_c(60.0),f_to_c(50.0))

      if ((nvoOSAEnthalpy + 1.5) < nvoReturnEnthalpy) and (nvoMechCool =
False) then
            iconogo = True
      else
            if (InEnth < OutEnth) or (nvoMechCool = True) then
            iconogo = False
      end if
      end if
```

Figure 6-22. Freely programmable nodes use manufacturer-specific programming tools that may require writing customized application program code.

Summary

- It is recommended to use nodes certified by LonMark International in LonWorks systems.
- Nodes certified by LonMark are tested to conform to the specifications in the LonMark Interoperability Guidelines, which provides a high degree of interchangeability between different node vendors.
- The minimum hardware components required for a node to send, receive, and act on control information are a microprocessor and a transceiver.
- Most LonWorks node manufacturers utilize the Neuron chip to run the node's application program.
- The LonTalk protocol can also be implemented in nodes with non-Neuron chips.
- When each Neuron chip is manufactured, it is assigned a Neuron ID.
- All nodes include a hardware transceiver that provides the means for transmitting and receiving network information.
- Input/output connection types and quantities vary between manufacturers depending upon the node application.
- Most LonWorks nodes are powered by an external 24 VAC or DC power supply.
- The service LED is an optional feature of Neuron-chip-based nodes that is used to physically indicate the node's status.

- LonWorks' interoperability relies on the standardization of these software components whose explicit definitions are freely available from LonMark for use by any node manufacturer.
- When the standard software components are inadequate for a certain control task, LonWorks provides for non-standard, user-defined (manufacturer-defined) software components for custom applications.
- LonWorks nodes are assigned an identification number by the application development tool to specify the node's intended use and type.
- The control intelligence within a LonWorks node is organized into function blocks.
- Function blocks are implementations of functional profiles, either LonMark standard functional profiles or manufacturer-defined function profiles.
- LonWorks nodes send and receive control data as network variables.
- The three basic configurations of network variables are numeric, structured, and enumerated types.
- Configuration properties affect node behavior, such as alarm limits, setpoints, default values, and time delays.
- Network management tools require the information in the external interface file (XIF) for programming a control network. This information can be read from the node during commissioning, but can also be downloaded from LonMark's web site.
- Four types of resource files (type, functional profile template, format, and language files) are included in the resource file set stored on the computer hosting the network management tool.
- The four types of LonWorks control nodes are application-specific, application-generic, system-wide controllers, and freely programmable.

Definitions

- The *microprocessor interface program (MIP)* is software that transforms the Neuron chip into a co-processor and moves the upper two layers of the LonTalk protocol, the presentation and application layers, off the Neuron chip and onto a more sophisticated external processor.
- A *host-based node* is a LonWorks node that employs the Microprocessor Interface Program (MIP) to use a supplemental microprocessor.
- A *Neuron ID* is a 48-bit serial number that is unique to a single Neuron chip.
- A *service pin* is a node pushbutton that causes the node to transmit an identifying message onto the network.
- A *service LED* is an LED that indicates the node's operating status.
- The *Program ID* is a 64-bit number that provides specific details and unique identification for a LonWorks node.
- A *plug-in* is a third-party software add-on that works within a network management tool that is based on LNS to provide a user-friendly interface for adjusting configuration properties.
- An *external interface file (XIF)* is a file on a LonWorks node that documents its Program ID, network variables, configuration properties, function blocks, and number of address table entries.
- An *application-specific node* is a control device that performs dedicated control functions defined by the application program loaded by the manufacturer.

- An *application-generic node* is a control device that converts traditional control signals into network variable data.
- A *system-wide controller* is a control device that provides general-purpose network functions for multiple nodes.
- A *freely programmable node* is a control device that is loaded with a custom control application.

Review Questions

1. Why is it recommended to use nodes certified by LonMark in LonWorks systems?
2. In what three ways are microprocessors used to implement the LonTalk protocol in LonWorks nodes?
3. Why are multiple types of transceivers available for use in LonWorks nodes?
4. How does the service LED indicate the status of a node?
5. What is the relationship between standard and user-defined software components?
6. What is the Program ID and what information does it include?
7. What is the Node Object?
8. Explain the difference between the three types of network variables.
9. How can an external interface file (XIF) be obtained for a particular node?
10. What is the difference between an application-specific node and an application-generic node?

Chapter Seven

LonWorks Network Programming

Programming a LonWorks network defines the information that creates the relationships and interactions between nodes. These relationships and interactions determine how the system operates. Therefore, programming is the most influential task in creating or modifying an automation system. LonWorks network programming can be done with any of a number of LNS-based network management tools, but all of these tools create network programs based on the standardized LonTalk protocol and services. During the commissioning process, this network program information is loaded onto installed LonWorks nodes, which then become operable network nodes.

Chapter Objectives

- Describe the role of a network management tool in programming and commissioning a LonWorks network.
- Differentiate between ad hoc and engineered program design methods.
- Describe the underlying information that forms a network variable binding.
- Compare network variable binding arrangements and addressing modes.
- Describe the importance of configuration properties and how they are adjusted.
- List the steps involved in commissioning LonWorks nodes.
- Identify common node and network device commissioning considerations.

LONWORKS NETWORK PROGRAMMING

Programming a LonWorks network consists of creating a network design to meet the control specification and required sequences of operations. Programming tasks include identifying nodes, binding network variables, and adjusting configuration properties. A network management tool is required to create the network program, which contains all of the node configuration and communication information for a LonWorks network. **See Figure 7-1.** The network program may be created prior to node installation or while the network management tool is connected to an installed network. During commissioning, the network program is loaded onto the nodes.

Network Management Tools

While the LonTalk communication messages used to set up a LonWorks network are standardized, the software that controls this process is not. All LonWorks network programming and node commissioning requires a network management tool. Network management tools are used to engineer, install, program, commission, manage, and maintain the network design. A number of network management tools can be used to program, commission, and maintain a LonWorks network design, including Echelon's LonMaker® Integration Tool and Distech Controls' Lonwatcher. **See Figure 7-2.**

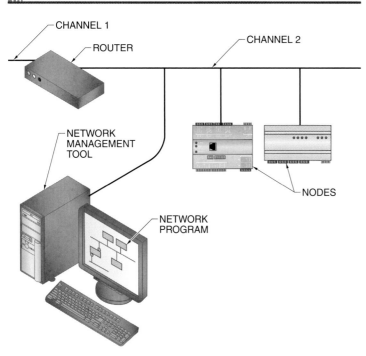

Figure 7-1. LonWorks network programming involves making the software bindings that represent the interactions between the physical nodes.

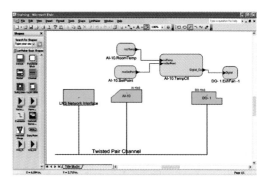

ECHELON LONMAKER INTEGRATION TOOL

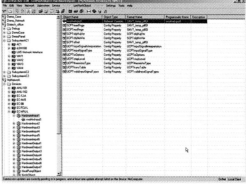

Distech Controls

DISTECH CONTROLS LONWATCHER

Figure 7-2. Network management tools may use a variety of interfaces for programming and configuring LonWorks networks.

> Some LonWorks network management tools reinforce the concept of object-oriented programming by using graphical icons to represent the software objects, including physical nodes, function blocks, and network variable bindings.

Echelon's LonMaker® Integration Tool is the most commonly used LonWorks network management tool. This software is based on Microsoft's Visio® program for drawing diagrams and flow charts. LonMaker uses a custom library of graphical symbols to represent infrastructure components, nodes, function blocks, and network variable connections. These elements are combined to show both the physical and logical relationships between LonWorks nodes and their functions. It then uses the relationships of these drawing elements to generate the network program.

There are also many other network management tools from other vendors, typically companies that also develop nodes certified by LonMark. These software tools use a variety of ways to show the network structure and functional relationships in a user interface. For example, Circon's Network Integrator tool uses a network tree structure and submenus to organize management and programming functions. Network management tool software is often chosen based on the interface that is the most user-friendly for the system integrator. There may be some minor differences in features, but typically the exact same network can be designed by any of a variety of network management tools.

Most, though not all, network management tools generate a network program based on standard LonWorks Network Service (LNS) tools.

Network Program Design

If the network management tool is based on the LNS platform, all network program information is stored within a single database. This information is accessed while the network is being designed and when changes are made. The necessary parameters for the network to operate autonomously are then loaded onto the individual nodes. This programming scheme allows for flexibility in the way nodes are configured with the network information. It may be done incrementally as the system is being installed, or it may be done all at once after installation with a predesigned network program.

Control System Programming

Some network management tools simplify the programming process by providing the integrator with graphical tools for establishing relationships and communication settings between nodes. This is a very intuitive and easy-to-use method of programming, and reinforces the object-oriented concept of the LonWorks information architecture.

Alternatively, control system programming may involve writing or modifying lines of code. For an integrator unfamiliar with this type of programming, adds, moves, and changes can be difficult to implement and prone to introducing errors that can cause serious disruptions in the control system. Also, many digital control systems utilize a proprietary programming language that requires special training. For example, the following code is an excerpt of a program controlling lighting.

```
...
if (nvi0ComericaLtg8 = ST_ON) or
(nvi1CBREMgmtLtg8 = ST_ON) or
(nvi2CBREOfficeLtg8 = ST_ON) or (nvi3HSLtg8 =
ST_ON) or (nvi4InglewoodLtg8 = ST_ON) then
    LtgGo = TRUE
    else
    LtgGo = FALSE
end if
if (ui0Ltg8 = ST_ON) then
    tmrLtg8 = 7200
end if
if (tmrLtg8 > 1) or (LtgGo = TRUE) or
(nvi5Ltg8 = ST_ON) then
    nvo5Ltg8 = 1029
    else
    if (tmrLtg8 = 0) and (LtgGo = FALSE) and
(nviLtg8 = ST_OFF) then
        nvo5Ltg8 = 65535
    end if
end if
...
```

This type of programming is highly structured with rigid rules on syntax, style, and sequence. However, it can also be very powerful, enabling a skilled programmer to customize the functionality of the system in a way that may not otherwise be possible with graphical interfaces.

Ad Hoc Program Design. When LonWorks networks use a flat network architecture, programming changes and additions can be made easily with previously commissioned and configured nodes. *Ad hoc program design* is a method of creating a new LonWorks network design while the network management tool is attached to the network. **See Figure 7-3.**

As each node is logically installed and connected to the network, its objects are added to the system program and interfaced with other commissioned nodes. Ad hoc program design combines the network programming and device commissioning processes and builds the network program in incremental steps. This design method is most commonly used for smaller network designs.

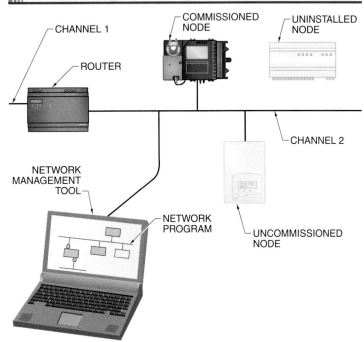

Figure 7-3. Ad hoc program design requires the network management tool to be connected to the nodes because they are commissioned as they are added to the network program.

An advantage of ad hoc program design is that it allows for some testing of network communication at each step, which can catch some node or programming problems early in the process. However, this method requires special considerations for network traffic and node availability. Network traffic temporarily increases as configuration changes are downloaded to nodes, which may then cause updates to other nodes and routers. Also, the node's application program may go OFFLINE briefly while updates are written to the node's memory. If the node is controlling actuators, this can cause unpredictable behavior. It is recommended to manually place equipment in a safe state before modifying node settings. It is also recommended to commission sensor devices before actuator devices.

Engineered Program Design. *Engineered program design* is a method of creating a new LonWorks program without the network management tool being attached to the network. **See Figure 7-4.** This method is used to create the network program while off-site. As the design is programmed, infrastructure, node, and network variable configurations are saved in a network database. When programming is complete, the network management tool is attached to the network and the entire network program loaded into the nodes and network devices during commissioning.

Engineered program design is popular for large networks where programming responsibilities are distributed to several network designers. As each part of the project is completed, it is merged into a master design. Network programming can also be performed in parallel with building construction or infrastructure installation, reducing the time spent to fully implement the control system.

The network management tool requires information about the function blocks and variables supported by the nodes that will compose the physical network. The process of commissioning a node transfers this information from the node's memory to the network management tool. However, since the nodes are not connected to the tool when using the engineered program design method, the information must be acquired in another way. The external interface file (XIF) includes complete descriptions of a LonWorks node's attributes and features. When loaded, the network management tool interprets the coded information in the file and makes all of the node's features available for programming. For most nodes and network devices, these files are readily available for download from manufacturer web sites. LonMark provides free XIF downloads for all nodes and network devices it certifies.

Engineered Program Design

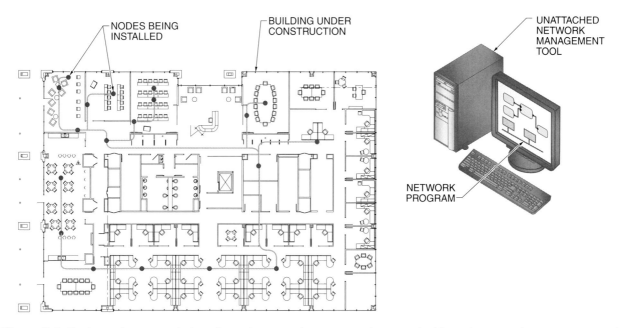

Figure 7-4. Engineered program design allows the network program to be created without the network management tool being attached, which is useful for planning an automation system while a building is under construction.

NETWORK VARIABLE BINDINGS

The primary task in programming a LonWorks network is making network variable bindings. A network variable binding is a connection between two network variables for the sharing of dynamic control data. Network variable bindings can be thought of as a virtual wire between nodes through which data flows. **See Figure 7-5.**

Since all LonWorks nodes are considered peers in the network, network variable bindings can be made between any two nodes, as long as they use the same network variable types. Network variable bindings between function blocks in two different nodes are the most common type, but bindings can also be made between two function blocks within the same node, or even between two variable objects in the same function block.

There are two general rules for network variable bindings. First, a network variable output can only be connected to a network variable input. A *network variable output (nvo)* is the representation of data that a node can send to one or more nodes on the network. A *network variable input (nvi)* is the representation of data that a node can receive for use within its application. Variable names typically include "nvo" or "nvi" as prefixes to distinguish their roles. For example, a variable representing a temperature value may be known as "nvoTemp" from the temperature sensor and "nviTemp" at the controller using the information.

Second, the network variables in a binding must be of the same type. Variable types include character strings, numbers, and enumerations (defining certain states or modes). For example, a network variable output representing a temperature reading cannot be connected to a network variable input representing a day of the week. However, sometimes a certain binding requires data to be shared among different types of network variables. In this case, translator function blocks may be available in the node to convert between certain variable types.

Network Variable Bindings

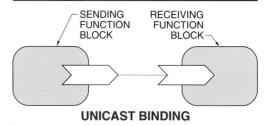

Unicast Bindings

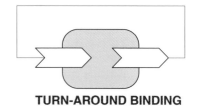

Figure 7-6. Unicast bindings include one sender and one receiver, though they may be within the same function block.

Figure 7-5. Like a virtual wire, network variable bindings allow control information to be shared between function blocks or nodes.

The process of creating a network variable binding varies for the different network management tools available, but for the most widely used network tool, the LonMaker Integration Tool, it is done graphically. The binding is represented by a line that connects a network variable output arrow with a network variable input arrow. This binding is created simply by clicking on these endpoints with a special line drawing tool.

Binding Arrangements

Network variable bindings can be arranged in a number of ways, depending on the need to share control data among multiple function blocks or nodes. The simplest type of binding is a unicast binding. A *unicast binding* is a single binding where one sending function block transmits network variable updates to one receiving function block. **See Figure 7-6.** A special type of unicast binding is a turn-around binding. A *turn-around binding* is a binding where a function block sends network variable updates to itself.

Multiple network variable connections involving a common network variable create a multicast binding. A *multicast binding* is a group of two or more bindings involving a common function block. Multicast bindings include fan-out and fan-in configurations. **See Figure 7-7.** A *fan-out binding* is a group of bindings where one function block sends network variable updates to two or more receiving function blocks. A *fan-in binding* is a group of bindings where two or more function blocks send network variable updates to one receiving function block.

Selector IDs

The binding of network variables with a network management tool generates Selector IDs, which uniquely identify each binding. A *Selector ID* is a unique identifying number that is shared by the network variables involved in a binding. Both output and input network variables within a binding share the same Selector ID. Each LonWorks node includes a configuration table of the network variable Selector IDs for its bindings to other nodes. **See Figure 7-8.** This table matches message packets received to local network variables. When a node receives a message packet containing a certain Selector ID, it knows which network variable input should process the information.

Multicast Bindings

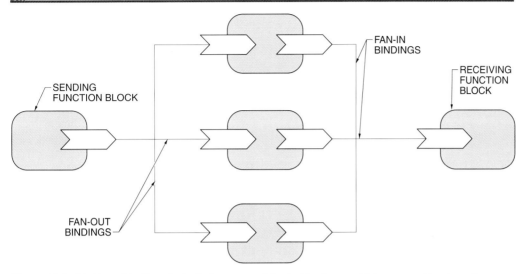

Figure 7-7. Multicast bindings include fan-out and fan-in bindings, which are used to share the same control information among several function blocks.

Multiple input network variables within the same node cannot share the same Selector ID, and this is problematic for certain types of connections, such as fan-in bindings. However, the use of alias network variables compensates for this constraint. An *alias network variable* is a software duplicate of a network variable that allows a different Selector ID to be used for the same network variable data. This allows for more complex bindings that otherwise could not be made, though it results in an additional message transaction. The alias feature must be supported by the node application program in order to be implemented.

For example, the node "LampNode" operates two separate indicator lamps that are assigned the names "nviLamp1" and "nviLamp2." **See Figure 7-9.** A control device known as "SwitchNode" includes a switch value represented by the network-variable output "nvoSW." When the switch wired to SwitchNode is closed, a network variable update, including a unique Selector ID, is sent to LampNode. LampNode uses its network variable configuration table to match the incoming message's Selector ID to the input network variable nviLamp1, which has been assigned the same Selector ID. SwitchNode then uses an alias network variable for nvoSW to communicate the same control data to LampNode's nviLamp2 with a new Selector ID.

Selector IDs

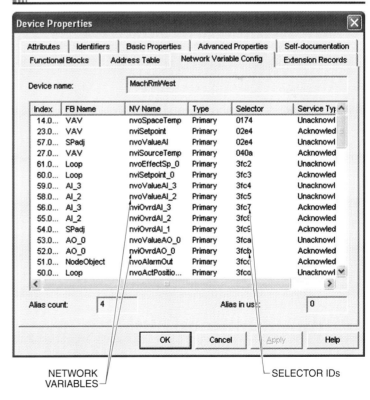

Figure 7-8. The network variable configuration table includes the Selector IDs that are associated with a node's network variables.

Alias Network Variables

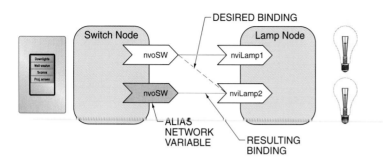

Figure 7-9. Alias network variables allow multicast bindings to avoid Selector ID conflicts.

Addressing Modes

LonWorks nodes identify each other with logical addresses based on their physical location within the network domain. The domain is composed of one or more subnets, which are logical collections of nodes on separate channels. **See Figure 7-10.** A LonWorks network domain can include up to 32,385 nodes, requiring a precise way to address a message packet to any one of them. Inside each message packet is a destination address that identifies a single node, groups of nodes, or all nodes in a subnet or network domain. Each node listens to every packet transmitted on its subnet to determine if a packet is intended for it.

Each LonWorks node includes an address table that specifies the destination address of nodes that are members of a network variable binding. **See Figure 7-11.** There can be up to 15 address table entries for a single node. When a network variable binding is created between nodes, entries in the sending node's address table provide a method of identifying the receiving node or groups of nodes. There are three addressing modes: subnet/node, group, and broadcast.

Address Tables

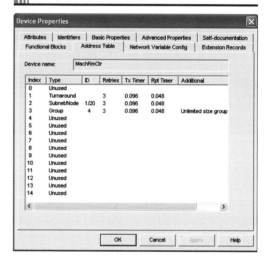

Figure 7-11. Each node includes an address table that specifies the destination address for the network variable updates it transmits onto the network.

LonWorks Network Domain

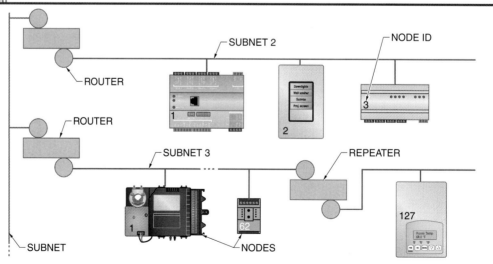

Figure 7-10. LonWorks network domains include one or more subnets that include one or more nodes. The numerical identification of the domain, subnet, and node are used to precisely address message packets to individual nodes.

Subnet/Node Addressing. *Subnet/node addressing* is a LonWorks addressing mode where a sending node identifies a single receiving node by its logical subnet/node address. Subnet/node addressing is typically applied in unicast connections. The sending node identifies in its address table the sole recipient node of the network variable update. The entry includes the receiving node's logical subnet/node address. No address table entry is required in the receiving node's address table.

From a sending node's point of view, fan-in bindings are no different from unicast bindings. Each sender in the fan-in connection uses subnet/node addressing, which adds an address table entry that identifies the receiving node.

Group Addressing. Fan-out bindings require group addressing. *Group addressing* is a LonWorks addressing mode where a sending node uses a Group ID to send message packets to a specific set of nodes. A *Group ID* is a unique number that identifies the set of nodes that all must receive a certain network variable update. The Group ID is used as the address in a message packet meant for these nodes. If a node on the network is a member of the group, it receives and processes the message packet.

The Group IDs of fan-out bindings that a particular node belongs to are stored in its address table. Each Group ID consumes an address table entry. Due to limited number of address table entries in its program, a node can belong to a maximum of 15 groups.

Broadcast Addressing. *Broadcast addressing* is a LonWorks addressing mode where all nodes on a subnet (for a subnet broadcast) or domain (for a domain broadcast) are identified as receivers of a network variable update. **See Figure 7-12.** When a node picks up a broadcast packet, it searches its network variable configuration table for a matching Selector ID. If no match is found, the message packet is discarded. For group connections, the broadcast mode can be used to conserve address table entries in receiving nodes. Only the sending node uses an address table entry for broadcast addressing. Routers isolate subnet broadcasts but forward domain broadcast messages to other subnets.

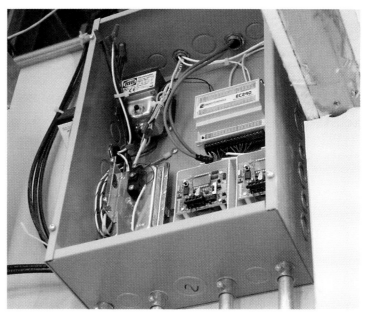

An enclosure may contain several devices and nodes for controlling or monitoring a single unit of building equipment.

Excessive use of broadcast addressing can cause a node's message receiving buffers to overflow, resulting in poor response times and communication failures. Additionally, some network management tools reuse Selector IDs and broadcast messages may cause a connection bleed. A *connection bleed* is a situation where a node that is not a member of a broadcast group erroneously responds to a network variable update.

Message Service Types

Control information sent on a LonWorks network is event driven. When sensor, actuator, and controller values change, nodes transmit network variable messages onto the network. This strategy results in efficient bandwidth utilization and a robust control system.

The LonTalk protocol supports several message service types that define how message packets are sent and delivered to network nodes and tools. Each type has unique characteristics that affect network performance, channel bandwidth, and control response times. Network management tools are used to assign message service types and/or modify parameters during programming.

Broadcast Addressing

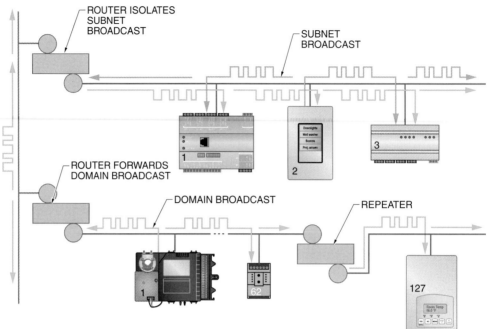

Figure 7-12. Subnet broadcasts are isolated to a certain subnet by routers, but domain broadcasts are forwarded by routers to all nodes in the network domain.

Acknowledged Message Service. The most widely used message service type is acknowledged message service. *Acknowledged (ACK) message service* is a LonWorks message service type where a sending node expects a response from the receiving node, confirming receipt of the message packet. The receiving node sends back an acknowledgment response packet indicating that the message packet was successfully received. **See Figure 7-13.** This process is comparable to certified mail.

If the acknowledgment is not received within a certain time interval, as measured by the transaction timer, the sending node resends the message packet. After a certain number of retries, which is configurable with the network management tool, the sending node logs a transaction timeout error in its error log.

While connection properties can be user-modified, most network management tools include binding algorithms that automatically define the most appropriate defaults, such as addressing modes, network variable connection properties, and timing parameters.

Acknowledged Message Services

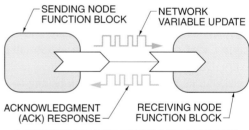

ACKNOWLEDGED MESSAGE SERVICE

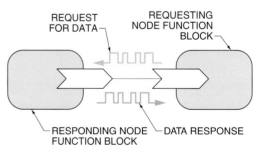

REQUEST/RESPONSE MESSAGE SERVICE

Figure 7-13. Acknowledged message service requires a return response from the receiving node, confirming that a network variable update was received.

By confirming the receipt of message packets, the acknowledged message service provides very reliable communication. However, this message service substantially increases network traffic, especially with fan-out addressing, since each receiving node must send an acknowledgment response packet back to the sender. The resulting sudden burst of network traffic increases the chances of message packet collisions, which further increases network traffic when failed message packets are retried. Therefore, a maximum of 64 node members are allowed in a group when using acknowledged message service. The broadcast addressing mode cannot utilize the acknowledged message service type.

Request/response (REQ-RESP) message service is a variation of acknowledged message service that is more generally known as polling. *Request/response (REQ-RESP) message service* is a LonWorks message service type where a sending node requests control data from a receiving node and requires a response message packet with the requested data. If the receiving node does not have the requested information, an error message is returned instead.

Since the request requires a data response, a lack of response indicates that the original message was likely not received. Similar to the acknowledged message service, the sending node resends the message until the receiving node responds or the transaction times out. In a LonWorks system, request/response message service is typically used by HMIs for gathering and displaying control data.

Unacknowledged Message Service. *Unacknowledged (UNACK) message service* is a LonWorks message service type where a sending node transmits message packets without expecting any subsequent action by the receiving node. **See Figure 7-14.** This service is appropriate in noncritical applications and when node applications are not sensitive to an occasional loss of data. Unacknowledged message service consumes the least amount of network bandwidth. The broadcast addressing mode can only utilize this message service type. There is no limitation to the number of nodes belonging to a group when using unacknowledged message service.

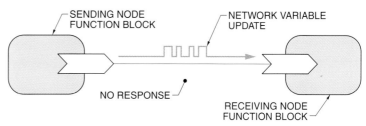

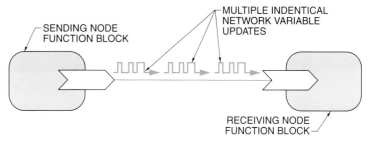

Figure 7-14. Unacknowledged message service is a one-way transaction where no response is expected from the receiving nodes.

Repeated message service is a variation of the unacknowledged message service. *Repeated (UNACK_RPT) message service* is a LonWorks message service type where a sending node transmits a series of identical message packets. The number of repeats and their intervals are configurable with a network management tool. Each node temporarily stores received messages in a transaction database. If message duplicates are received by the receiving node, it discards all but one. Duplicate message packets are identified by their identical transaction IDs.

When used for large group bindings, repeated message service contributes less to network traffic than acknowledged message service. For example, when using acknowledged message service, a group binding consisting of one sending node and ten receiving nodes results in at least 11 message packets on the network for each transaction. With repeated message service with three repeats, the same transaction results in only four total message packets while maintaining a high probability of successful message packet delivery.

Occupancy sensor nodes may use the authenticated message service for security reasons.

Authenticated Message Service. The *authenticated message service* is a LonWorks message service type where the receiving node determines if the sending node is authorized to communicate with it. This message service type is used in high-security applications.

The network management tool loads a unique 48-bit authentication key onto all nodes belonging to an authenticated binding. When a node receives an authenticated message packet, it transmits back to the sending node a 64-bit, randomly generated challenge number. **See Figure 7-15.** The sending node performs a mathematical calculation using the challenge number and authentication key and returns the result to the receiving node. If the result matches the number calculated by the receiving node, then the two nodes must have the same authentication key, which authenticates the sending node. The receiving node sends an acknowledgment to the sending node, regardless of the authentication result.

Authenticated message service can significantly increase network traffic if used excessively because each transaction results in four message packets. The challenge and reply packets are also larger than normal because they contain the 64-bit challenge and 48-bit result data.

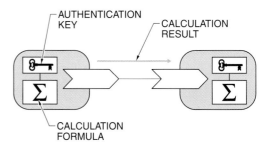

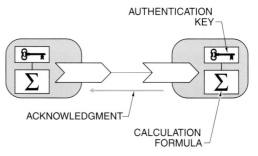

Figure 7-15. Authenticated message service provides a high-security method of sending network variable data by requiring nodes to prove that they share the same authentication key.

Channel Priority

The LonTalk protocol also supports an optional priority mechanism to improve response time for nodes sending critical network variable updates on channels with high network traffic. Nodes configured for priority are assigned unique time slots for transmissions. **See Figure 7-16.** This guarantees access to the network media when needed. Priority transactions are processed ahead of nonpriority transactions by routers.

Assigning priority for certain nodes creates deterministic network variable updates. However, overuse of the priority feature can adversely affect channel traffic as it may reduce the bandwidth capacity for nonpriority network traffic.

Configuration Properties

Network programming also includes the adjustment of individual node configuration properties, which change node, function block, or network variable behavior. Configuration property values affect the node's local behavior and may impact network performance. Node developers define the configuration properties within the node application program and may assign default values. New configuration properties cannot be added without modifying the node application program.

The node application program performs control functions based on network variable and configuration properties values. For example, a node application may include a PID (proportional-integral-derivative) function block that is used to control duct pressure. Setpoint and measurement network variables are received from interface and sensor nodes, and the PID function's network variable output is sent to the node that controls fan speed. The output value is calculated based on a formula that uses coefficient values that are set by the installer. These PID coefficient values, along with other settings (such as scan intervals, bias, and deadband), are set or adjusted as configuration properties.

A node's configuration properties are normally defined during its initial installation but may also be modified later with a network management tool. For a selected function block, a list of the corresponding configuration properties is displayed. **See Figure 7-17.** These are listed in a simple table interface that differentiates configuration properties by their abbreviated name, such as SCPTovrValue. Changes are made by selecting a configuration property and typing a new value. This is an effective method, but not very easy to use, especially since there is no information about what values are allowable for each configuration property or their meanings. Alternatively, configuration properties can be changed more easily with configuration plug-ins.

Channel Priority

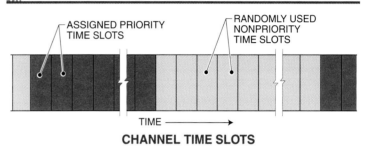
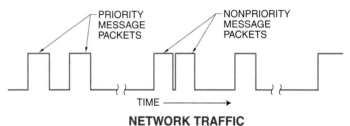

Figure 7-16. The use of a network channel can be divided into time slots for nodes to transmit message packets onto the network. Assigning a priority time slot to a node guarantees that it can communicate reliably without packet collisions.

Configuration Properties

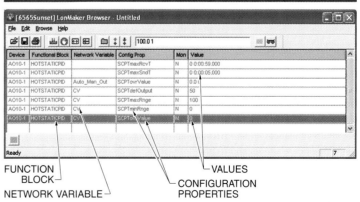

Figure 7-17. Configuration properties can be changed to customize node behavior and the control functions.

Plug-Ins. Network management tools that are based on LNS support plug-ins, which are third-party software add-ons that provide a user-friendly method of adjusting configuration properties. **See Figure 7-18.** A plug-in interface appears as a window over the network management tool screen. Configuration properties are shown with names that are more easily interpreted and may include units of measure. It groups configuration properties in logical order categories, separated by tabs, and provides convenient ways to change the values, such as check boxes or drop-down menus. This also helps limit user inputs to allowable and meaningful values.

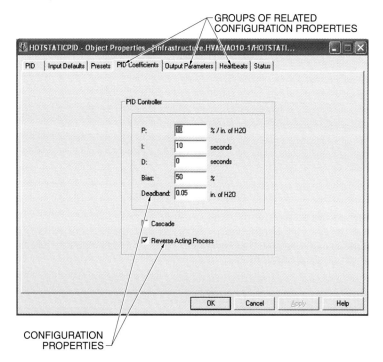

Figure 7-18. Plug-ins from node or network device manufacturers provide a user-friendly way to adjust configuration properties.

Plug-ins are developed and distributed by node manufacturers in order to interface with their products. Plug-ins are specific to hardware models, so a network program may use many different plug-ins. LonMark provides a listing on its web site of links for downloading the plug-in software files. Hundreds of plug-ins are listed and most are available to download for free. Plug-in downloads also include resource files that document manufacturer-defined variables (UNVTs) and configuration properties (UCPTs).

If a manufacturer does not offer a plug-in, the configuration of node behavior may require the use of proprietary tools. In supervisory system architectures such as the Tridium Niagra platform, third-party node configuration requires the use of manufacturer-specific drivers.

Heartbeats. Some of the most common and important configuration properties are related to heartbeats. A *heartbeat* is a LonWorks network variable update that continuously repeats at a configured time interval. Heartbeats can be used to create a fail-safe binding between critical nodes. A receiving node expects to receive regular network variable updates. However, a node or network failure may prevent updates from being received. In this case, the receiving node suspends its application program for this function block. The affected function block then enters a locked-out condition.

While locked-out, hardware outputs default to safe values until the updates are restored. This minimizes the possible system problems caused by erratic control information until the failure can be remedied. Heartbeat-related configuration properties may apply to individual network variables or globally to all network variables within a node or function block.

Heartbeats are defined by a group of configuration properties. The sending node is configured with SCPTmaxSendTime, which is the maximum time interval between the transmissions of network variable updates. **See Figure 7-19.** Updates can be sent at shorter intervals, but not at greater intervals than the SCPTmaxSendTime value, even if the variable has not changed since the last update.

The receiving node is configured with SCPTmaxRcvTime, which is the maximum time interval allowed between receiving network variable updates. **See Figure 7-20.** If a receiving node does not receive an update within the SCPTmaxRcvTime interval, it will enter its locked-out mode. It is recommended to set the SCPTmaxRcvTime interval at about four times the SCPTmaxSendTime interval.

It is common for occasional message packets to be lost, so this avoids nuisance lock-out conditions among receiving nodes. If the SCPTmaxRcvTime interval is set to 0 (zero) for a receiving node, it always operates utilizing the last value received, regardless of when an update was last received.

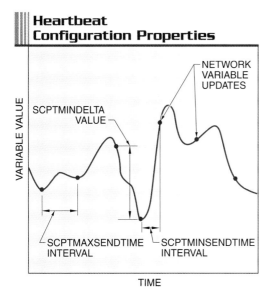

Figure 7-19. Heartbeat configuration properties define how often a network variable update is transmitted from a sending node.

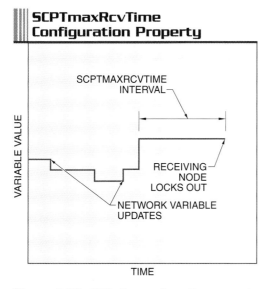

Figure 7-20. With the configuration property SCPTmaxRcvTime, a receiving node can be set to lock out if a heartbeat is not received from a sending node within a certain period.

Additional configuration properties affect the transmission of network variable updates according to node usage and application. A group of configuration properties with names that begin with "SCPTminDelta" specify the amount of change allowed in a measured variable before an update is transmitted. This ensures that the most accurate network variable values are available when the sensor measurements are changing rapidly. This type of parameter is commonly known as "send on delta." Standard send-on-delta configuration properties are available for angle, carbon dioxide, flow, level, relative humidity, and temperature variables (for example, SCPTminDeltaTemp).

SCPTminSendTime vs. SCPTminSndT

Some pairs of time-related configuration properties have very similar names and represent the same type of time information, but have different ways of representing the information. For example, SCPTminSendTime and SCPTminSndT both define the minimum interval of time between the sending of network variable updates. SCPTminSendTime represents the interval as a number of seconds, from 0 to 6553.5. Therefore, this range is limited to up to about 109 minutes. Heartbeat intervals typically fall well within this range. However, if a longer range or a different format is desirable, SCPTminSndT uses a more complex structure to define this interval in numbers of days, hours, minutes, seconds, and milliseconds.

Similar relationships exist between the following configuration properties:
- SCPTdriveTime and SCPTdriveT
- SCPTmaxDefrostTime and SCPTmaxDefrstTime
- SCPTmaxRcvTime and SCPTmaxRcvT
- SCPTmaxSendTime and SCPTmaxSndT

However, frequent network variable updates adversely affect network performance due to excessive traffic. The configuration property SCPTminSendTime specifies a minimum time between network variable updates. This parameter is commonly known as a throttle because it defines the maximum update rate, reducing network traffic created by fast changing sensors.

Digital inputs can cause a related problem when they change states. When a set of mechanical contacts is opened or closed, their momentum and elasticity cause bounce, which is a set of rapid pulses instead of a clean transition between zero to full current. The configuration property SCPTdebounce is set on the sending node to avoid transmitting network variable updates until the contacts settle into the new state. This parameter sets the minimum time interval before the new digital signal is updated. **See Figure 7-21.**

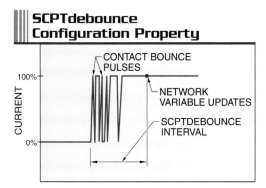

Figure 7-21. SCPTdebounce sets a delay period before a digital network variable is updated to avoid transmitting the pulses of a set of physical contacts that are changing state.

Node developers define the configuration properties within the node application program and assign default values. These values are included in XIF of version 4 or later.

DEVICE COMMISSIONING

Device commissioning is the process of assigned logical addresses and loading network program information onto installed nodes. Device commissioning involves matching the physical network that is installed in a building to the logical network that is programmed into the network management tool. The system is not capable of monitoring or controlling anything until the bindings, configuration properties, and messaging options are loaded onto the physical nodes and network devices. The network management tool is used to apply the network program to the installed and powered-up nodes.

The device commissioning process involves identifying nodes by their Neuron IDs, assigning subnet/node addresses, and loading network variable connections and configuration property values onto node memories.

Neuron ID Identification

The first step in the commissioning process requires each node to be identified by its unique 48-bit Neuron ID number. This correlates the physical device with the node and function objects in the network program. There are three ways to acquire a node's Neuron ID and add this information to the network program. **See Figure 7-22.**

Service Pins. The most common method for obtaining a node's Neuron ID is to physically push its service pin. When a node is commissioned, the network management tool prompts the installer to press the service pin. When this button is pressed, the node broadcasts its Neuron ID, Program ID, channel, and location onto the network. The network management tool must be in a commissioning mode at this time so that it is expecting to receive the node identity information. If the Program ID and channel from which the broadcast was received matches the network design, the network management tool completes the commissioning process. This method requires a person to be physically at each device in order to press the service pin. This requirement may be difficult to fulfill, depending on the installation location.

Neuron ID Labels. Many manufacturers provide peel-off labels on their nodes that include the Neuron ID number and an associated barcode. As a node is installed, a label is removed and attached to the network documentation. During commissioning, the Neuron ID is then entered into the network database either manually or with a barcode scanner.

Neuron ID Identification

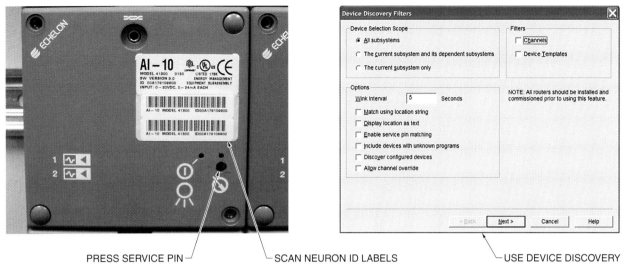

PRESS SERVICE PIN — SCAN NEURON ID LABELS — USE DEVICE DISCOVERY

Figure 7-22. For commissioning, the Neuron ID of a LonWorks node or network device can be obtained by any of three methods.

Device Discovery. Several network management tools include the ability to discover Neuron IDs from uncommissioned nodes. The network management tool broadcasts onto the network a message asking for unknown nodes. These nodes, if there are any, respond with messages that include their identity and location. They can then be matched to the nodes defined in the network design.

Device discovery does not require access to the device's service pin. In fact, in some circumstances, it may even be done from a remote location over an IP network. However, device discovery may not be a reliable means of Neuron ID identification for some networks, such as those with multiple instances of the same type of node.

Wink Command

Many network management tools and nodes support the Wink function for verifying communication with the node. Nodes respond to a Wink command with some physical reaction, usually by blinking an LED. **See Figure 7-23.** (The particular Wink physical behavior is defined by the manufacturer.) This command might be used for troubleshooting or confirming communication with a node without commissioning it.

Wink Command

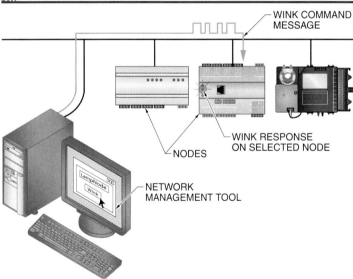

Figure 7-23. The Wink command confirms that the tool can communicate with a node by causing the node to exhibit a physical response, such as flashing an LED.

Subnet/Node Address Assignment

In the next step of commissioning, a unique address is assigned to the node by the network management tool. This domain/subnet/node address (often referred to more simply as the subnet/node address) is a logical address independent from a node's Neuron ID. A subnet/node address is determined by the ID of the subnet where the node is installed and a serial node number that is assigned in the order of commissioning. Subnet/node addressing enables routers to isolate or forward network packets according to destination subnet locations. To prevent duplicate addresses, many network management tools automatically assign a subnet/node address that cannot be modified.

Node Configuration

After a subnet/node address is assigned, the network management tool downloads network variable binding definitions and configuration property values to the node, configuring its interaction and behavior on the network. Network variable binding information includes Selector IDs for the node's address table and binding properties for addressing modes, message service types, and message timing. If the node is a member of group bindings, the Group IDs are also entered into the node's address table.

The source of configuration property values may include values previously entered into the network database while in the engineered design mode, default values from the external interface file, or current values in the node if it was configured with a third-party tool.

By default, most network management tools set the node into the OFFLINE state after commissioning. While OFFLINE, a node suspends its application program and applies its default values on hardware outputs. An OFFLINE node does not send message packets or respond to network variable updates. However, after an OFFLINE node is reset, such as from a power cycle, it will automatically become ONLINE and run its application.

Router Commissioning

LonTalk routers are commissioned in the same way as nodes but have special considerations. Commissioning of LonTalk routers includes assigning subnet/node addresses for each segment port and defining the router as configured, repeater, learning, or bridge type. On most network architectures, routers are set up as configured routers, with routing tables that are updated by the network management tool.

Routers within a LonWorks network affect the order in which nodes can be commissioned. Because the network management tool and node must communicate during commissioning, all configured routers between the two must already be commissioned and ONLINE. Therefore, a router must be commissioned before the network management tool can commission any nodes on the other side of the router. **See Figure 7-24.** Similarly, multiple routers in a hierarchical arrangement must be commissioned in a topological order, from the network management tool to the nodes.

Commissioning Modes

Since updating configuration information into nodes on an operating network may cause network traffic issues or brief node problems, it is sometimes desirable to carefully control this process. Some network management tools have the ability to address this issue by commissioning the network in either OFFNET or ONNET mode.

These commissioning modes are different from the two program design methods (ad hoc and engineered), though they can be easily confused. The programming methods are distinguished by whether the network management tool is attached to the network. In contrast, both OFFNET and ONNET modes can involve the network management tool being attached to the network while commissioning, but the distinction is how the tool updates network program changes onto the physical nodes.

OFFNET. *OFFNET* is a LonWorks commissioning mode in which any network configuration changes are applied only to the LNS database. If the network management tool is attached and OFFNET, it can communicate with nodes and obtain Neuron IDs, Program IDs, and channel locations during commissioning. However, it does not load the network program into the node. This information is verified and queued for later processing, when the network management tool is placed ONNET.

Router Commissioning

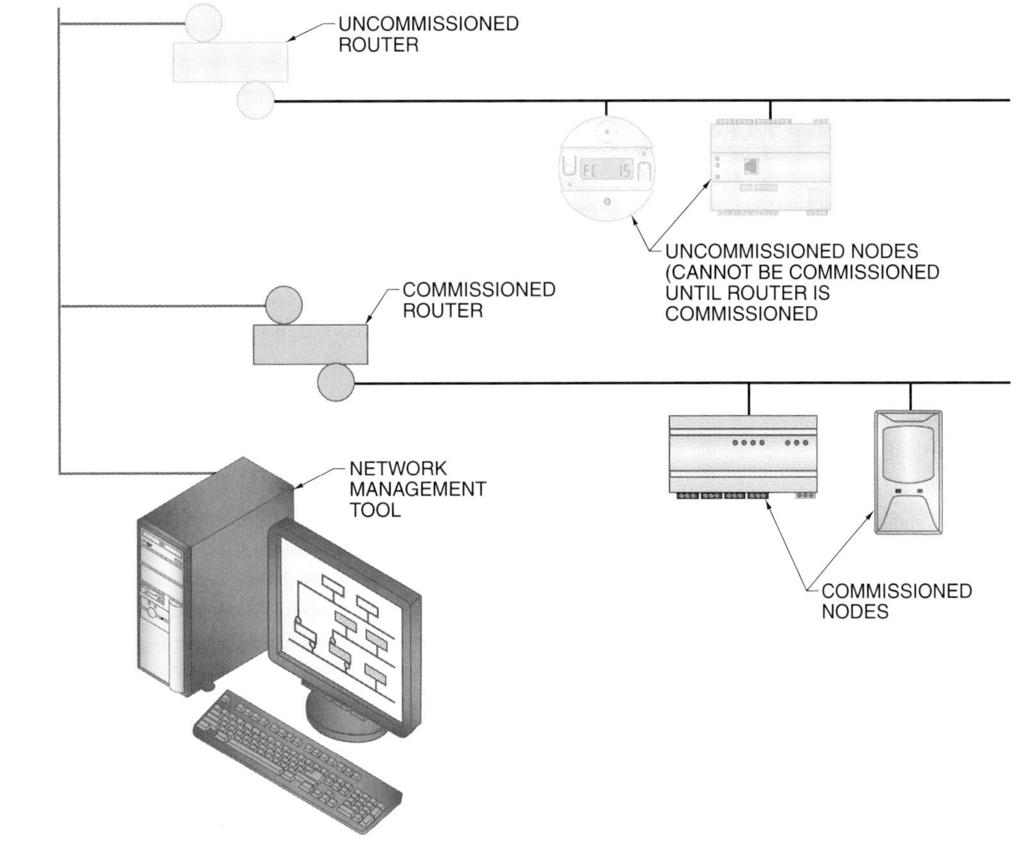

Figure 7-24. Routers must be commissioned and on-line in order for the tool to communicate with nodes on their other side.

Commissioning and network configuration changes are typically done OFFNET during critical or high network traffic periods. The information can then be saved to be loaded onto the nodes at a later time when the additional network traffic will not significantly impact the system operation.

ONNET. *ONNET* is a LonWorks commissioning mode in which network configuration changes are loaded onto the nodes at the time of commissioning. The ONNET mode can be used while programming, which loads all the design parameters to the nodes as they are commissioned. Or, the network management tool can be placed in ONNET mode to recommission and load at once all of the network parameters that were saved from programming in an OFFNET or engineered design mode.

Refer to Quick Quiz® on CD-ROM

Summary

- Programming a LonWorks network is the process of engineering the network design to meet the control specification and required sequences of operations.
- All LonWorks programming and device commissioning requires a network management tool.
- The most commonly used LonWorks network management tool is Echelon's LonMaker® Integration Tool.

- The ad hoc design method combines the system programming and device commissioning processes and builds the network program in incremental steps.

- The engineered design method is used to create the network program while unattached to the network.

- The primary tasks in programming a LonWorks network are making network variable bindings and adjusting configuration properties.

- Network variable bindings can be thought of as a virtual wire between nodes through which data flows.

- Unique Selector IDs are assigned to each binding.

- Each LonWorks node includes an address table that specifies the destination address of nodes that are members of network variable bindings.

- Inside each message packet is a destination address, which could be a single node, groups of nodes, or all nodes in a subnet or network domain.

- A Group ID is used as the address in a message packet sent to a set of certain nodes.

- Excessive use of broadcast addressing can cause a node's message receiving buffers to overflow, resulting in poor response times and communication failures.

- The most widely used message service type is acknowledged message service.

- Request/response (REQ-RESP) message service is a variation of acknowledged message service that is known more generally known as polling.

- Unacknowledged message service is appropriate in noncritical applications and when node applications are not sensitive to an occasional loss of data.

- Authenticated message service type is used in high-security applications.

- Configuration property values affect the node's local behavior and may impact network traffic.

- A node's configuration properties are normally defined during its initial installation but may also be modified later with a network management tool.

- Plug-ins are developed and distributed by node manufacturers in order to provide a user-friendly way to modify the configuration properties in their products from within a network management tool.

- Heartbeats are the repetition of network variable updates that can be used to create fail-safe bindings between critical nodes.

- Heartbeat intervals are defined by a group of configuration properties.

- Device commissioning involves matching the physical network installed in a building to the logical network programmed into the network management tool.

- The device commissioning process involves identifying nodes by their Neuron IDs, assigning a subnet/node address, and loading the network variable bindings and configuration property values into the node's memory.

- Because the network management tool and the node must communicate during commissioning, all configured routers between the two must already be commissioned and on-line.

- Updating configuration information into nodes on an operating network may cause network traffic issues or brief node problems, so some network management tools carefully control this process by commissioning nodes while in the OFFNET mode.

Chapter 7—LonWorks Network Programming

Definitions

- *Ad hoc program design* is a method of creating a new LonWorks network design while the network management tool is attached to the network.

- *Engineered program design* is a method of creating a new LonWorks program without the network management tool being attached to the network.

- A *network variable output (nvo)* is the representation of data that a node can send to one or more nodes on the network.

- A *network variable input (nvi)* is the representation of data that a node can receive for use within its application.

- A *unicast binding* is a single binding where one sending function block transmits network variable updates to one receiving function block.

- A *turn-around binding* is a binding where a function block sends network variable updates to itself.

- A *multicast binding* is a group of two or more bindings involving a common function block.

- A *fan-out binding* is a group of bindings where one function block sends network variable updates to two or more receiving function blocks.

- A *fan-in binding* is a group of bindings where two or more function blocks send network variable updates to one receiving function block.

- A *Selector ID* is a unique identifying number that is shared by the network variables involved in a binding.

- An *alias network variable* is a software duplicate of a network variable that allows a different Selector ID to be used for the same network variable data.

- *Subnet/node addressing* is a LonWorks addressing mode where a sending node identifies a single receiving node by its logical subnet/node address.

- *Group addressing* is a LonWorks addressing mode where a sending node uses a Group ID to send message packets to a certain set of nodes.

- A *Group ID* is a unique number that identifies the set of nodes that all must receive a certain network variable update.

- *Broadcast addressing* is a LonWorks addressing mode where all nodes on a subnet (for a subnet broadcast) or domain (for a domain broadcast) are identified as receivers of a network variable update.

- A *connection bleed* is a situation where a node that is not a member of a broadcast group erroneously responds to a network variable update.

- *Acknowledged (ACK) message service* is a LonWorks message service type where a sending node expects a response from the receiving node, confirming receipt of the message packet.

- *Request/response (REQ-RESP) message service* is a LonWorks message service type where a sending node requests control data from a receiving node and requires a response message packet with the requested information.

- *Unacknowledged (UNACK) message service* is a LonWorks message service type where a sending node transmits message packets without expecting any subsequent action by the receiving node.

- *Repeated (UNACK_RPT) message service* is a LonWorks message service type where a sending node transmits a series of identical message packets.

- The ***authenticated message service*** is a LonWorks message service type where the receiving node determines if the sending node is authorized to communicate with it.
- A ***heartbeat*** is a LonWorks network variable update that continuously repeats at a configured time interval.
- ***Device commissioning*** is the process of assigning logical addresses and loading network program information onto installed nodes.
- ***OFFNET*** is a LonWorks commissioning mode in which any network configuration changes are applied only to the LNS database.
- ***ONNET*** is a LonWorks commissioning mode in which network configuration changes are loaded onto the nodes at the time of commissioning.

Review Questions

1. What are the advantages of using the engineered design method?
2. Between what types of objects can network variable bindings be made?
3. What are the two general rules for making network variable bindings?
4. How does an alias network variable make certain multicast bindings possible?
5. What is the role of a router in broadcast addressing?
6. Why can unacknowledged message service be used to maintain reliable communication with large group bindings while minimizing network traffic?
7. How do plug-ins make it easier to adjust configuration properties?
8. How are heartbeats used to create fail-safe bindings between critical node functions?
9. What is the relationship between the commissioning of routers and the commissioning of nodes?
10. What is the primary advantage of commissioning network OFFNET?

Chapter Eight

LonWorks Network Testing

The process of creating a LonWorks network design includes verifying control functionality, node health, and channel performance. The control algorithms must perform as expected, the nodes must operate adequately, and the network channels must not be overloaded with excessive message traffic. Various network management tools can be used to investigate the operation of the control network and uncover problems with adverse effects. They may also be used to help calibrate the network operation to improve network performance.

Chapter Objectives

- Describe ways to verify correct control sequence operation.
- Describe ways to verify network infrastructure integrity.
- Identify the types of error information available from nodes.
- Identify the capabilities of protocol analyzers to investigate network performance.
- Compare the effects of configuration properties and other settings on network traffic.

TESTING AND VERIFYING THE NETWORK

Several tools are available that have features for testing and verifying network and node performance. Network management tools are used to verify control functionality and report node error statistics. Protocol analyzer tools provide channel traffic statistics, such as bandwidth consumption, error rates, and detailed packet information.

Echelon Corporation
Network interfaces provide a way for desktop or laptop computers to connect to the network for programming or troubleshooting.

LonWorks uses an object-oriented system to organize control information. Function blocks were originally known as "objects," which explains the origin of the terms for the Node Object, SNVT_obj_request, and SNVT_obj_status. These terms remained, even after most objects became known as "function blocks."

Control Sequence Operation

The operation of a control sequence can be observed by monitoring the values of network variables as the system runs. However, this is not a practical way to test the system's reaction to all possible scenarios, especially for those that are infrequent or unlikely. Therefore, control functionality verification typically requires node management. *Node management* (or device management) is the manual forcing of a node's function blocks into specific operating modes in order to send control data onto the network. This is a software feature of most network management tools, though it may be known by a different name. **See Figure 8-1.** Network designers use this feature to simulate control conditions and confirm that the control application operates acceptably within a range of various sensor values and control commands.

Node Object. LonMark compliant nodes that include more than one application-specific function block also include a Node Object function block. This provides a standard method for managing, testing, and reporting node-level and function-block-level status.

The Node Object includes a mandatory input network variable for network management tools to request a particular mode for one or all of the node's function blocks. SNVT_obj_request is a structured network variable that includes two fields: a numeric object_id for the index number of the function block to be managed, and an enumerated object_request that specifies the type of management command to be applied to the function block. **See Figure 8-2.** If the object_id is 0, the management command applies to all function blocks within the node. The object_request enumerations include 18 actions, such as requesting override, disable, manual control, and reset. **See Figure 8-3.** Three commands, RQ_NORMAL, RQ_UPDATE_STATUS, and RQ_REPORT_MASK, must be supported by the node. The other 15 commands are optional. The RQ_REPORT_MASK command is used to query which commands are supported by that node.

A mandatory output network variable reports function block status. SNVT_obj_status is a structured network variable that contains multiple fields with which a node manufacturer can report abnormal conditions such as mechanical fault, electrical fault, override condition, or self-test failure. **See Figure 8-4.**

Device Management

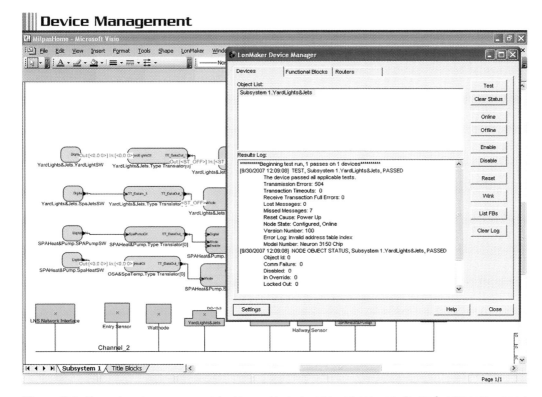

Figure 8-1. The network management tool is used to test node and network device functionality, report Neuron error statistics, and provide device status.

Node Object

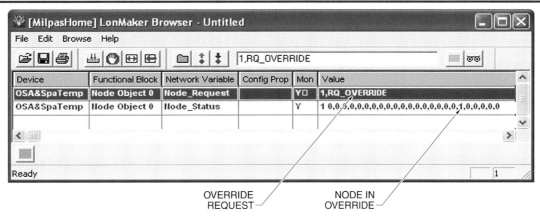

Figure 8-2. The Node Object provides a standard method to manage, test, and report function block status.

Override Values. Many node manufacturers support override capability for node hardware inputs and outputs through the Node Object. This allows user-inputted override values to be used in the network program in order to verify control functionality. Often manufacturers include the override feature in the node plug-in. **See Figure 8-5.**

A node that is SOFT OFFLINE has an application and network configuration, but the application program is at a standstill. The service LED is off. Function blocks within SOFT OFFLINE nodes use their default values. However, a SOFT OFFLINE node becomes ONLINE after a power cycle reset, allowing the function blocks to resume normal operation.

object_request Enumerations

Value	Identifier	Notes
−1	RQ_NUL	Invalid value
0	RQ_NORMAL	Enable object and remove override
1	RQ_DISABLED	Disable object
2	RQ_UPDATE_STATUS	Report object status
3	RQ_SELF_TEST	Perform object self-test
4	RQ_UPDATE_ALARM	Update alarm status
5	RQ_REPORT_MASK	Report status bit mask
6	RQ_OVERRIDE	Override object
7	RQ_ENABLE	Enable object
8	RQ_RMV_OVERRIDE	Remove object override
9	RQ_CLEAR_STATUS	Clear object status
10	RQ_CLEAR_ALARM	Clear object alarm
11	RQ_ALARM_NOTIFY_ENABLED	Enable alarm notification
12	RQ_ALARM_NOTIFY_DISABLED	Disable alarm notification
13	RQ_MANUAL_CTRL	Enable object for manual control
14	RQ_REMOTE_CTRL	Enable object for remote control
15	RQ_PROGRAM	Enable programming of special configuration properties
16	RQ_CLEAR_RESET	Clear reset-complete flag (reset_complete)
17	RQ_RESET	Execute reset-sequence of object

Figure 8-3. The SNVT_obj_request input variable is used to invoke one of 18 possible commands for managing function blocks.

Function Block Status

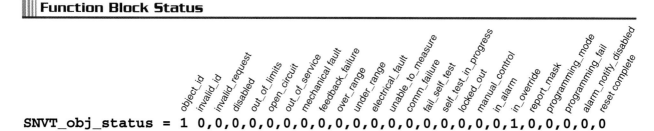

Figure 8-4. The variable SNVT_obj_status contains 26 structured data fields that indicate the status of function blocks within a node.

Default Values. Node management is also used to test fail-safe behavior. This verifies that a node or function block will default to a set of predefined "safe" values for outputs associated with hardware if some part of the control network becomes disabled. This is important because missing or inaccurate control information may otherwise cause actuators to inadvertently damage equipment or create a hazardous condition. Default network variable values place output hardware in safe positions or conditions until the problem can be corrected. There are two ways in which fail-safe default values can be tested.

The use of heartbeats is one method to induce function blocks or entire nodes to default safe values. Receiving nodes (or their function blocks) set hardware outputs to a safe default condition when network variable updates (heartbeats) are not received within a configured time interval. This heartbeat reaction can be tested by disabling upstream function blocks. The disabled function block suspends its local application and does not send network variable updates. **See Figure 8-6.**

When downstream function blocks that are bound to disabled function blocks do not receive any updates within the heartbeat-receive interval, they enter a locked-out condition and use their default network variable values. (The SNVT_obj_status network variable includes a 1-bit field for reporting the locked-out condition.) Once the network variable update is restored, downstream function blocks are typically released automatically from the locked-out condition and resume normal operation. (The lockout function and the automatic lockout-release upon subsequent receipt of heartbeats should be verified with the node manufacturer. Not all nodes support these features.)

Alternatively, node manufacturers can include a software feature to force a node to use its default values on command. When disabled and locked out, the function blocks apply default values to the associated outputs. Manually disabled function blocks remain disabled, even after resets such as power cycles. The condition must be manually released in order to return the function block to normal operation.

Network Variable Monitoring. During initial commissioning or after subsequent changes, it is also important to verify the correct operation of control sequences by monitoring the values of network variable updates. **See Figure 8-7.** When monitoring is enabled, a network management tool polls nodes (with REQ_RESP messages) for selected network variable values at a configurable polling interval, typically once per second. On network variable bindings, the value of an input network variable at downstream nodes should match the output network variable sent by upstream nodes. Input values that do not match output values on a network variable binding can be an indication of disabled function blocks or OFFLINE or locked-out nodes.

Manually writing to network variable values from the network management tool can also prove useful in verifying control algorithms. Many network tools include a browser plug-in that provides a table view of node network variables and configuration properties. The browser allows a manual value to be written to input network variables and configuration properties.

Override Values

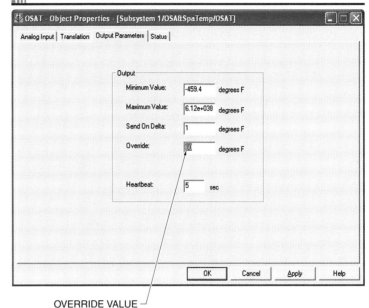

Figure 8-5. Plug-ins are often used to define and implement function block override values.

Heartbeats

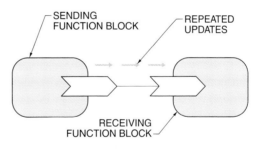

NORMAL OPERATION

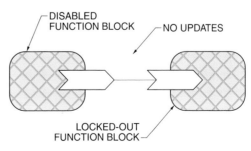

LOCKED-OUT CONDITION

Figure 8-6. When periodic network variable updates (heartbeats) are not received when expected, receiving function blocks use default values for the network variable inputs instead.

Network Variable Monitoring

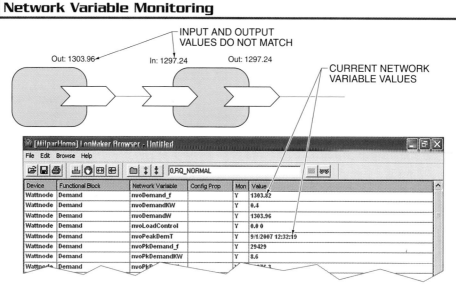

Figure 8-7. Network management tools can be used to monitor network variable values sent and received on bindings.

Node Error Statistics

Good design practice includes testing and verifying the health of each node in the network. Particularly important are the critical nodes, including sending nodes in large fan-out group bindings, nodes with frequent network variable updates, receiving nodes in large fan-in bindings, nodes performing life-safety functions, and nodes controlling equipment with high horsepower motors.

Each LonWorks node logs node error statistics and status information. Reports of the node error statistics should be reviewed during normal network operations and under peak conditions, which include traffic contributed by network tools such as HMIs. **See Figure 8-8.** Using the node management feature of the network management tool, integrators can access the error log for troubleshooting or node health verification.

The error log also reports node status, such as the cause of the last reset, current state, and microprocessor information. A node can be in one of three states: application-less, unconfigured, or configured. When provided by the manufacturer, a node's service LED indicates the current node state.

A node in the application-less state either has not been loaded with an application program or the application program has been lost. The node's service LED, if present, indicates this condition with a continuous ON illumination.

The unconfigured state is one where the node has been loaded with an application program but has not yet been assigned a network address and binding definitions. A Neuron-chip-based node does not run its application program while in this state. A host-based node runs its application program while in the unconfigured state in order to communicate with its host processor. Nodes based on other processors may respond differently, so this must be verified by the manufacturer's documentation. An unconfigured node typically flashes the service LED once per second.

A node in the configured state is running an application program, has an assigned network address, and may include binding and configuration property values. This is the normal state for a node and is indicated by an OFF service LED.

The SCPTminSendTime, SCPTminSndT, and SCPTminDelta configuration properties are also known as throttles because they control the maximum rate at which heartbeats are sent. A throttle configuration property is disabled by setting it to 0, allowing network variable updates to be sent more frequently.

Neuron Errors

Error Statistic	Description	Cause	Resolution
Transmission errors	The node received a packet containing an invalid CRC	Channel noise; improper termination; excessive traffic	Investigate physical wiring issues
Transaction timeouts	A receiving node did not send back an ACK after a number of retries	Receiving node is offline, not powered, or disconnected from network	Test receiving nodes
Receive transaction full	The node's receive transaction database overflowed, causing it to drop packets	Excess network traffic address to node; excess broadcast addressing causing each node to receive packets, regardless of network variable binding	Use group addressing or aliases
Lost messages	The node application buffer overflowed, causing it to drop packets	Node application was too busy to process packets; excessive channel traffic	Contact node manufacturer; reduce channel bandwidth consumption
Missed messages	The node network buffer overflowed, causing it to drop packets	Node network processor could not process the number of packets addressed to the node	Reduce channel bandwidth consumption

Figure 8-8. Errors reported in Neuron node error logs indicate possible problems with network communication or configuration.

The statistics of the function block test may include the results of the optional manufacturer-defined self-tests. **See Figure 8-9.** Self-tests look for various inconsistencies in the way the function block is configured, in both software and hardware. For example, a self-test routine may check that the software configuration for an analog input's signal type (voltage, current, or resistance) matches the hardware configuration of the node's physical jumpers for the signal type. If a problem is detected, manufacturer-specific error messages are displayed that describe the configuration issues.

Network Infrastructure Integrity

Verifying the integrity of the network infrastructure includes testing the network media, routers, and network interfaces.

Media Testing. Prior to node or power supply installation, the physical media should be verified. The total conductor resistance of copper media types should be measured with an ohmmeter and compared to the cable specification. On bus topologies, temporarily shorting the conductors at the end of a segment makes it easy to measure the total resistance. **See Figure 8-10.** For example, a Level IV twisted-pair cable has a resistance specification of 18 Ω per 1000′ of conductor length (0.018 Ω/ft). If the segment is 800′ in length, and there are two conductors in the segment, the total resistance should be close to 28.8 Ω (0.018 Ω/ft × 800 ft × 2 = 28.8 Ω).

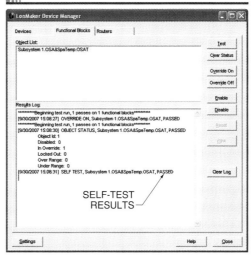

Figure 8-9. Function block test features include reports of function block status, mode, and self-tests.

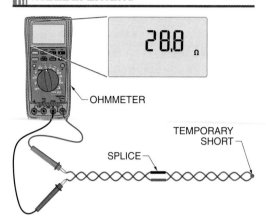

Figure 8-10. Measuring the electrical resistance of copper channel media, such as twisted-pair cabling, avoids communication problems due to wiring or connection issues.

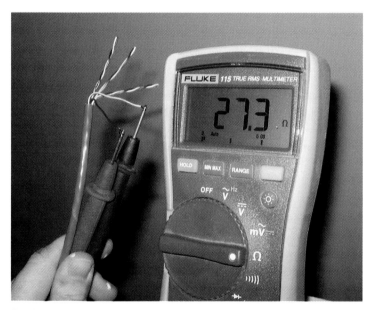

An ohmmeter, or a multimeter set to measure resistance, is used to check the electrical integrity of copper conductors.

Problems with the network infrastructure may require reterminating connections or pulling new cables, either of which can be a significant task. Problems due to excessive network traffic, however, are typically easier to remedy. Often, changing heartbeat or other configuration properties can notably improve the communication reliability. If this does not help, nodes may need to be moved to new channels.

A higher resistance typically indicates breaks in the conductors or poor connections. Any connectors in the segment should include vapor-tight connectors such as terminal blocks, crimped connectors, or soldered connections. Additional twisted-pair cable specifications are included in the Echelon twisted-pair wiring guidelines.

Network Device Testing. When verifying network infrastructure integrity, it is particularly important to test communication to and operation of network devices such as routers, repeaters, and network interfaces. Network devices can be verified if they have been installed, powered up, and commissioned.

The simplest way to confirm communication with a network device is with the Wink command, which is a common optional feature of both network management tools and network devices. This simple command causes a physical response by the network device that confirms media integrity and communication with the network management tool. On a bus topology segment, applying the Wink command to the farthest node or network device verifies the continuity of the entire channel segment. For free topology segments, each device on the channel must be Winked. The physical response to the Wink command is typically in the form of an LED blink, audible beep, or some other simple behavior. **See Figure 8-11.** The Wink command also verifies that the device object logically defined in the network design matches the physical node installed in the network. This is particularly useful when the network includes more than one of a particular type of node.

Network devices should be tested after commissioning to verify that they are ONLINE and configured. Since they typically also contain Neuron chips, they include a log of error statistics and status information that can be accessed with the network management tool and examined for network performance problems. For example, a router error log that reports high missed packet error statistics indicates excessive network traffic. **See Figure 8-12.** High transaction timeouts indicate unsuccessful polls when monitoring network variable values.

Network Device Testing

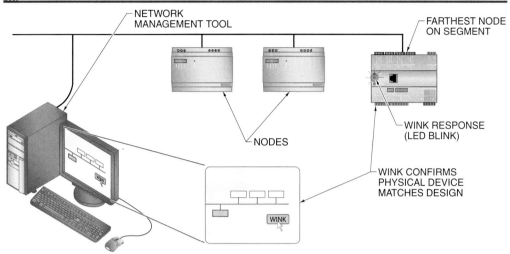

Figure 8-11. The Wink command confirms that the network management tool can communicate with nodes and physically verifies that the installed node matches the network design.

Protocol Analysis

Channel health is verified using a protocol analyzer tool. A *protocol analyzer* is software that provides detailed channel traffic statistics, packet logs, individual packet analysis, and other information about network traffic. **See Figure 8-13.** Several manufacturers offer protocol analyzers for LonTalk networks.

To collect channel traffic information, protocol analyzers require a hardware interface that must include a transceiver that matches the channel type. Some interfaces provide connections to multiple channel types, such as TP/FT-10, TP/XF-1250, and TP-RS485-39. In addition, various computer interface form factors are available, including USB, PCMCIA for laptop computers, PCI for desktop computers, and internally installed interfaces for LonTalk Internet servers and IP-852 routers.

Protocol analyzers can collect and report only message packets and traffic statistics on the channel segment to which they are connected. This is because the network's routers isolate much of the network traffic on the channel unless messages are destined for nodes on different channels. Therefore, for networks that include multiple configured routers, a protocol analyzer tool on a laptop computer makes it possible to move to each channel and document network traffic conditions.

Router Error Log

```
**********Beginning test run, 1 passes on 1 routers**********
[6/29/2009 19:26:55]   TEST, Infrastructure.RTR-1, PASSED
    >>> Router Near Side <<<
    The device passed all applicable tests.
    Transmission Errors: 0
    Transaction Timeouts:  0
    Receive Transaction Full Errors: 33
    Lost Messages: 0
    Missed Messages: 0
    Reset Cause: None
    Node State: Configured, Online
    Version Number: 125
    Error Log: No error.
    Model Number: Neuron 3150 Chip

    >>> Router Far Side <<<
    The device passed all applicable tests.
    Transmission Errors: 0
    Transaction Timeouts:  0
    Receive Transaction Full Errors: 0
    Lost Messages: 0
    Missed Messages: 0
    Reset Cause: None
    Node State: Configured, Online
    Version Number: 125
    Error Log: No error.
    Model Number: Neuron 3150 Chip
```

Figure 8-12. Neuron error logs for routers include information for each port.

The router port that is closest to the network interface device is known as the "near side" and the port farthest from the network interface is designated as the "far side." These designations are used to differentiate a router's two Neuron error logs.

Protocol Analyzer

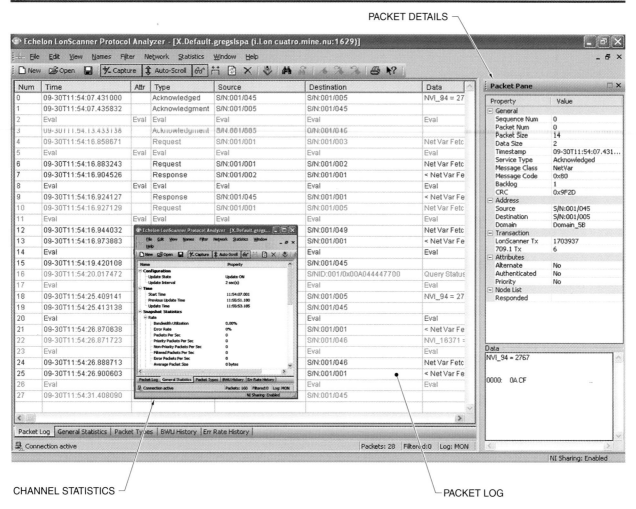

CHANNEL STATISTICS
PACKET DETAILS
PACKET LOG

Figure 8-13. Protocol analyzer tools provide detailed network traffic statistics, packet logs, and individual packet analysis.

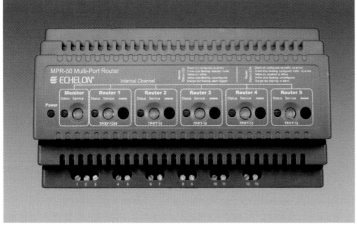

Echelon Corporation

Routers collect information on communication errors in error logs. Each port has its own error log.

Packet Statistics. Protocol analyzer reports include a packet log and statistics on bandwidth consumption percentage, error rate, maximum packets/second, and average packets/second. **See Figure 8-14.** The packet log is updated in real time, including a display and log of message packets as they are transmitted on the channel. Details of each packet can be individually examined, including sender/receiver address information, message service type, addressing mode, Selector ID, and network variable update values. **See Figure 8-15.**

Also included within each message packet is a date/time stamp for the moment at which the protocol analyzer recorded the packet transmission. The date/time stamp allows control response times to be measured. For example, consider a switch contact in a sending node that is bound to an actuator output in a receiving node. **See Figure 8-16.** When the contact closes, the network variable is updated by sending a message packet out onto the network. The receiving node picks up the message and actuates the output. A feedback network variable associated with the actuation is bound back to the sending node. This results in a second message packet when the action takes place. Subtracting the time recorded in the switch update packet from the time of the feedback packet provides a control response measurement.

Packet Error-Checking. Electrical interference signals, loose cable connections, and excessive network traffic can cause corrupted packets. A protocol analyzer is often used to report and log corrupted packets that occur on the network channel. LonTalk packets contain a cyclic redundancy check (CRC) calculated value used to validate complete packets. Receiving nodes only accept a message when it has the correct number of bits and is uncorrupted, as indicated by comparing the included CRC value with one calculated by the receiving node. The protocol analyzer tool also examines the packet CRC and reports messages that are too short, too long, or incomplete.

Packet Filtering. Packet filtering is a feature of a protocol analyzer that isolates and reports specific packet types and message transactions between selected nodes. This is useful during troubleshooting as a method for identifying nodes and message service type packets that are contributing to channel overload conditions.

OPTIMIZING NETWORK PERFORMANCE

Excessive channel traffic can adversely affect node performance and control response times. As bandwidth consumption approaches a saturation level, the occurrence of packet collisions increases, resulting in packet corruption and CRC mismatch. **See Figure 8-17.** Damaged packets are discarded by receiving nodes, causing transaction timeouts in sending nodes and poor control response times.

Packet Statistics

```
Rate
Bandwidth Utilization                    20.12%
Error Rate                               0%
Packets Per Sec                          44.88
Priority Packets Per Sec                 0
Non-Priority Packets Per Sec             44.88
Filtered Packets Per Sec                 9.97
Error Packets Per Sec                    0
Average Packet Size                      12.76 bytes
Maximum
Max Bandwidth Utilization                41.83%
Max Error Rate                           2.97%
Max Average Packet Size                  13.01 bytes
Max Packets Per Sec                      93.02
Max Priority Packets Per Sec             0
Max Non-Priority Packets Per Sec         93.02
Max Filtered Packets Per Sec             15.98
Max Error Packets Per Sec                1.50
Cumulative Statistics
Elapsed Time                             00:03:28.319
Average Bandwidth Utilization            20.75%
Average Error Rate                       0.10%
Average Packet Size                      12.76 bytes
Average Packets Per Sec                  46.27
Average Priority Packets Per Sec         0
```

Figure 8-14. Packet statistics reported by a protocol analyzer include information on throughput, errors, and priority packets that help analyze the health of the communication network.

Packet Details

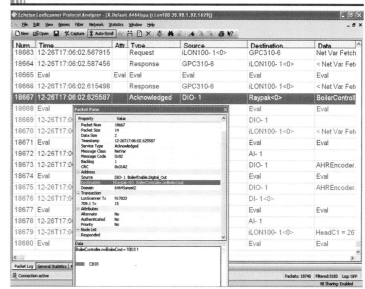

Figure 8-15. Within a protocol analyzer program, individual message packets can be selected for detailed information.

Control Response Measurement

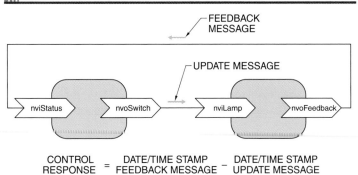

Figure 8-16. By comparing the date/time stamps of network variable update and feedback messages, the control response of a particular binding can be determined.

Channel Capacities*

Channel Type	Average Traffic	Peak Traffic
TP/FT-10	148 to 168	180 to 210
TP/XF-1250	747 to 835	933 to 1043
TP/XF-78	148 to 168	180 to 210
PL-20	14	18 to 20
IP-852	>15,000	
TP-RS485-39	75 to 80	90 to 100
FO-20	750 to 850	950 to 1050

* in approximate number of LonTalk packets per sec

Figure 8-17. Capacity for LonTalk packet traffic varies between channel types and amount of traffic.

Communication between nodes is event driven. Typical network traffic is primarily heartbeats and network variable updates from sensor and controller nodes. As control values change, network variable updates are transmitted onto the network media. However, a small change in a sensor value may be relatively unimportant to a control system, meaning the frequent updates cause unnecessary bandwidth consumption. Optimizing network performance requires managing network variable updates in order to reliably share control information without flooding the network with unnecessary messages.

Human Machine Interface (HMI) Traffic

Human machine interface (HMI) applications can contribute to channel overload if the HMI is excessively polling nodes for control information. Each network variable poll involves at least two messages (request and response), regardless of any change within the node being polled. For example, a user interface that is monitoring 100 data points at a polling rate of once per second causes 200 packets per second of traffic, saturating a TP/FT-10 channel. Increasing poll rates (time between each poll) and reducing the number of polled network variables on active user interfaces substantially decreases HMI traffic. However, care must be exercised when defining poll rates, as critical events may be undetected if they take place between poll intervals.

Alternatively, creating update bindings to HMI tools provides the most efficient bandwidth utilization and monitoring performance. Bound network variable updates are reported directly to the HMI. Therefore, the request message of a polling action is not required. For the same number of monitored network variables, binding updates to HMIs creates half of the additional network traffic as polling.

Network Variable Update Traffic

Manufacturers can use optional configuration properties to regulate how often nodes send network variable updates. Network variable updates represent the majority of the message traffic on the network. These are set during the programming and commissioning process, but may require additional tuning later if the network design is changed or if network performance is lower than expected. Adjusting these configuration properties involves balancing the need to share accurate and up-to-date information with the need to minimize network traffic for redundant messages.

Send Updates on Changes. Configuration properties of an SCPTminDelta type define the minimum change in a network variable value to trigger a network variable update. If the change is less than this value, then an update message is not necessary (unless overridden by another configuration property).

For example, if an SCPTminDeltaTemp configuration property value is set at 1°, a network variable update is only sent if the value of the temperature changes by 1° or more. If an SCPTminDeltaTemp is set to 0.1°, an excessive number of network variable updates would be sent onto the network for temperature changes that are relatively unimportant to the control system. Therefore, the value of this configuration property is influenced by resolution of the network variable that is required by the receiving equipment.

Send Updates on Time Intervals. Some network variables can change values very quickly, which could contribute to excess network traffic. If it is not necessary to report such rapid fluctuations, the rate of network variable updates can be slowed down. SCPTminSendTime (or SCPTminSndT) is a configuration property that regulates the maximum rate of network variable updates by defining a minimum time interval between each update. **See Figure 8-18.** Regardless of a network variable's change in value, updates are not sent until the SCPTminSendTime timer expires.

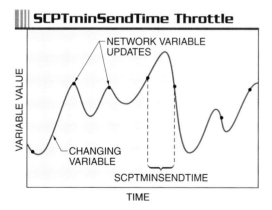

Figure 8-18. The throttle (SCPTminSendTime) defines the minimum time interval between two network variable updates.

Similarly, heartbeat network variable updates are defined by the configuration property SCPTmaxSendTime (or SCPTmaxSndT). SCPTmaxSendTime is a configuration property that regulates the minimum rate of network variable updates by defining a maximum time interval between each update. Lengthening this heartbeat interval also reduces network bandwidth consumption. Changes in the value of SCPTmaxSendTime must also be reflected in the heartbeat receive interval (SCPTmaxRcvTime) for downstream function blocks, which may enter a locked-out condition if an update is not received when expected.

Predicting maximum channel throughput rates allows designers to ensure reliable node communication during peak traffic periods. The effect of these configuration properties on the maximum and minimum network traffic throughput on each network variable connection can be estimated with the following formulas:

$$Z_{max} = \frac{P}{T}$$

$$Z_{min} = \frac{P}{H}$$

where

Z_{max} = maximum throughput (in packet/sec)

Z_{min} = minimum throughput (in packet/sec)

P = number of message packets per transaction

T = SCPTminSendTime (in sec)

H = SCPTmaxSendTime (in sec)

For example, a node measuring the outdoor air temperature sends the temperature value to 20 HVAC control nodes using repeated message service. This service includes 1 update and 3 repeats per transaction. The throttle rate (SCPTminSendTime) is set at 10 seconds and the heartbeat (SCPTmaxSendTime) is set at 30 seconds.

$$Z_{max} = \frac{P}{T}$$

$$Z_{max} = \frac{4}{10}$$

$$Z_{max} = \textbf{0.4 packet/sec}$$

$$Z_{min} = \frac{P}{H}$$

$$Z_{min} = \frac{4}{30}$$

$$Z_{min} = \textbf{0.13 packet/sec}$$

Therefore, the throughput for this connection will likely vary between 0.13 and 0.4 packet/sec. Note that different types of message services, such as acknowledged, change the number of transmitted packets, which then affects the throughput.

Binding Properties

Changing network variable binding properties can also be used to reduce traffic. For example, if a fan-out binding uses acknowledged message service, each receiver in the group sends back an acknowledgement message to the sender, confirming that the message was received. On a large fan-out group, each update causes a burst of acknowledgements from multiple receiving nodes and an increased probability of packet collisions. Also, the sending node must be able to process the received acknowledgments quickly and buffer those that it cannot process fast enough, which may require additional network buffers. Instead, using the repeated message service for large group bindings reduces network traffic, does not require additional network buffers, and provides a high probability of successful message delivery.

High-Speed Backbone Channels

The use of backbone topologies can provide dedicated bandwidth capacity for HMIs and network management tools. High-speed backbones can accommodate monitoring applications where large numbers of network variable values are required to be displayed in a user interface over short polling intervals. A twisted-pair cabling TP/XF-1250 channel has the capacity for up to 700 packet/sec. The FO-20S and FO-20L fiber optic channels have similar bandwidth characteristics. IP-852 channels provide dramatic increases in bandwidth capacity when combined with the LNS operating system. Over 7000 LonTalk packet/sec have been demonstrated on an IP-852 channel consisting of a single LNS Server and multiple IP-852 routers.

Refer to Quick Quiz® on CD-ROM

Summary

- Network management tools are used to verify control functionality and report node error statistics.

- Network designers use node management to simulate control conditions and confirm that the control application operates acceptably within a range of various sensor values and control commands.

- The Node Object includes a mandatory input network variable to request function block status and a mandatory output network variable that reports function block status.

- Testing fail-safe behavior verifies that a node or function block will default to a set of predefined "safe" values for outputs associated with hardware if some part of the control network becomes disabled.

- Node manufacturers can include a software feature to force a node to use its default values on command.

- When network variable monitoring is enabled, a network management tool polls nodes for selected network variable values at a configurable polling interval.

- Each LonWorks node logs node error statistics and status information.

- The statistics of the function block test may include the results of the optional manufacturer-defined self-tests.

- The total conductor resistance of copper media types should be measured with an ohmmeter and compared to the cable specification. A higher resistance indicates breaks in the conductors or bad connections.

- Since they also contain Neuron chips, LonTalk routers and network interfaces include a Neuron error log that can be accessed with the network management tool and examined for network performance problems.

- Protocol analyzers collect and report message packets and traffic statistics on a particular channel segment.
- Protocol analyzer reports include a packet log and statistics on bandwidth consumption percentage, error rate, maximum packets/second, and average packets/second.
- Details of each packet can be individually examined, including sender/receiver address information, message service type, addressing mode, Selector ID, and network variable update values.
- Excessive channel traffic can adversely affect node performance and control response times.
- Optimizing network performance requires managing network variable updates in order to reliably share control information without flooding the network with unnecessary messages.
- Increasing poll rates and reducing the number of polled network variables on active user interfaces substantially decreases HMI traffic.
- Optional configuration properties are used to regulate how often nodes send network variable updates.
- Changing network variable binding properties can also be used to reduce traffic.
- The use of backbone topologies can provide dedicated bandwidth capacity for HMIs and network management tools.

Definitions

- *Node management* (or device management) is the manual forcing of a node's function blocks into specific operating modes in order to send control data onto the network.
- A *protocol analyzer* is software that provides detailed channel traffic statistics, packet logs, individual packet analysis, and other information about network traffic.

Review Questions

1. Why is it impractical to rely on observing the normal operation of the control system to verify control functionality?
2. How can the Node Object be used to determine the status of function blocks?
3. How is the heartbeats feature used to test how function blocks resort to their default values?
4. Why is it important to monitor the outputs and inputs of a network variable binding?
5. How is the integrity of the network media tested and what types of problems can be discovered?
6. What types of network verification does the Wink command provide?
7. Why are protocol analyzers often used on laptop computers?
8. How can packet information be used to measure the control response of a particular binding?
9. What are two ways for optimizing network performance with human machine interfaces (HMIs)?
10. How do configuration properties define the frequency of network variable updates?

Chapter Nine

LonWorks Network Maintenance

Control networks require maintenance to ensure proper operation and to modify the system to address changes in building requirements. The system may need to be expanded to include more control functionality, changed to modify the existing functionality, or repaired to replace failed equipment. These types of network changes involve various additions, moves, and changes of nodes and other network devices. Network changes must be done in both the physical network and the logical network in the network program.

Chapter Objectives

- Describe the types of changes commonly made to LonWorks systems.
- Compare the changes required in the physical network and logical network when making changes to the system.
- Describe the creation, storage, and use of network program backups.
- Differentiate between the requirements and results of network resynchronization and network recovery.
- Describe the recommended components of network documentation to be maintained.

NETWORK MAINTENANCE TASKS

Network maintenance tasks include the replacement of failed nodes, addition of new nodes, relocation of existing nodes, and software updates to control applications. In addition, frequent backup of the network program after each revision ensures against accidental data loss in the event of a computer failure.

Node Replacement

As control systems age, hardware fails and nodes may need to be replaced. With standardized software components and LonMark compliance, nodes can be replaced relatively easily while maintaining network functions.

In addition to physical replacement, node replacement involves logical substitution. The node configurations and network variable binding definitions for the original node are stored in the network database. After a node has been physically replaced, the network management tool downloads the information into the new node. **See Figure 9-1.** If the node Program IDs match, the network management tool applies the same external interface file (XIF), hardware settings, binding definitions, and input/output configurations.

If a replacement node has a newer application program version than the original node, the Program ID is slightly different, so the manufacturer should provide a new XIF. Network variable bindings and configuration property values loaded into the new node are typically left unchanged. However, the original node template may need to be updated to reflect the new XIF. The node template defines the source of a node's XIF.

The replacement of nodes from different manufacturers, even if they include the same functional profile, typically requires some modification of the network program. While the new node may contain the same LonMark functional profiles, it is unlikely that two nodes from different manufacturers have identical hardware and firmware characteristics. Previous network variable bindings, HMI naming references, and configuration properties may need to be redefined.

During the node replacement procedure, the network management tool prompts for the source of configuration property values. If the same network management tool is used to configure nodes, the original node values from the LNS database are selected. Default values are selected in order to use the manufacturer-defined default values within the XIF. However, if the node was preconfigured or was configured by a third-party tool, the values already within the new node are chosen as the source.

Node Additions

A major benefit of the flat network architectures used in LonWorks networks is the ability to add new network nodes without affecting the operation of previously installed nodes. Network management tools can connect to any point on the network media and commission new nodes. New nodes can even be commissioned remotely over an IP network.

When adding new nodes to the network, it is good practice to set the network management tool to OFFNET while defining the new node configuration values and network variable bindings. This minimizes network traffic as each small change adds traffic on the network.

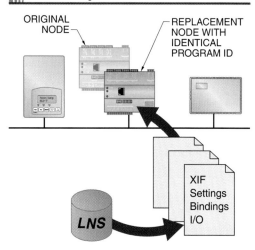

Figure 9-1. Replacement nodes with the same Program ID are loaded with the same XIF, settings, bindings, and input/output information.

When the network programming is complete, the network management tool is placed back to ONNET. The node configurations download to the new node and associated nodes in one batch when the new node is commissioned.

Also, it is good practice to place all nodes that include bindings to the new node in a safe mode while the new node is commissioned. Some nodes may become OFFLINE briefly while new binding information is downloaded, causing unpredictable behavior.

Node Relocation

As the building automation network expands to accommodate additional control subsystems, it may be desirable to move nodes to between existing or new communication channels. While this may not be necessary for operation, it is recommended as a good network design practice to locate nodes that communicate most often with each other on the same channel. **See Figure 9-2.** This helps reduce message traffic between channels.

Segregating nodes for different applications on different channels is particularly important for system expansions. For example, a building may have a network that includes only nodes for HVAC applications. If the network grows to include lighting and access controls, the original channel may become overloaded with traffic, causing response-time performance issues. Instead, creating a backbone channel and separate channels for each control application helps manage message traffic. This reconfiguration, however, may require relocating existing nodes to new channels.

During the relocation process, the node is temporarily placed in the OFFLINE state by the network management tool through device management. Before a node is physically relocated to the new channel, any actuators controlled by it must be placed in a safe position.

> The network database is updated for each change, but the network program file is only saved manually.

Channel Differentiation

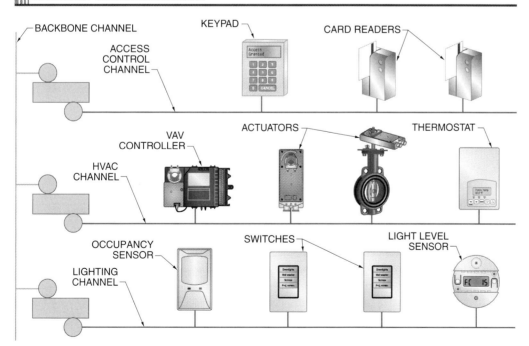

Figure 9-2. Good network design practice includes installing related nodes on application-specific channels.

After the node has been physically installed on the new channel, the network management tool is used to move the node within the network program to a new channel. **See Figure 9-3.** The node is then reconfigured with a new subnet/node address. All binding definitions are preserved. The node typically remains in the OFFLINE state, allowing verification of node communication and other tests with the device management functions. The node is placed ONLINE when testing is complete.

Each LonWorks node application program is associated with a specific Program ID. The external interface file, which includes the same Program ID, is required when loading a new application into a node. The external interface file information may be provided by the manufacturer or retrieved from the node during the application download process. When possible, it is preferable to use the manufacturer's version, which includes more descriptive information about network variables and configuration properties than the self-documentation data uploaded from the node. **See Figure 9-4.**

Figure 9-3. Nodes that are physically relocated must also be moved in the network design to a new channel.

Node Application Updates

Node manufacturers may periodically revise node application programs in order to address programming errors or enhance control algorithms. The loading of a new node application is done by the network management tool, which must be attached to the network at the time. In order to accept a new application over the network, the node must include EEPROM or flash memory that supports application download. The application file is provided in an .apb or .ild format, depending upon the development tool used to compile the application. Legacy .nxe application file types are not recommended.

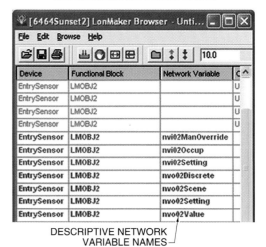

Figure 9-4. The XIF contains more-descriptive names for network variables and configuration properties than those uploaded from a node.

During the application upgrade, the network management tool prompts for the source of configuration property values. In most cases, the original node values stored in the network program database are selected, unless the node was configured with a third-party tool.

Network Management Tool Relocation

Networks with multiple channels may require network management tools to be relocated to different points on the network for maintenance or network modifications.

Network management tools connect to a communication channel through network interface hardware and/or software. The network interface is represented in the network program, and its logical location must match its physical location. **See Figure 9-5.** A mismatch can cause communication issues between the network management tool and nodes. Therefore, like relocating nodes, relocating a network management tool involves both physical and network program design changes. First, the network interface is logically moved in the network program design. Next, the tool prompts for the physical connection to the new channel and verifies that the move was successful.

Careful labeling of communication media with channel identification and documentation of network connection points reduces errors associated with relocating network management tools. In order to minimize the impact of network management traffic, network management tools should be connected to a backbone channel whenever possible.

Network Program Backups

Network management tools typically include a back-up feature, which is used to copy the network database, drawings, and associated files to a single compressed file. **See Figure 9-6.** This network database back-up file can then be restored with the network management tool software. This may be necessary if the original database becomes corrupted or the network program design must be opened by other network management tools.

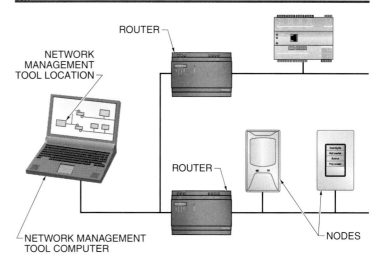

Figure 9-5. The location of the network management tool within the network must match on both the network design and the physical channel wiring.

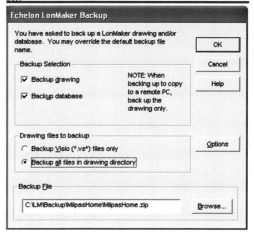

Figure 9-6. Network design backups protect against accidental data loss in the event of computer failures. Backup files should then be stored in a secure location.

Backups may include both on-site and off-site storage. Backups should be copied to a remote or removable storage media, such as a network server, CD, portable hard drive, or USB flash drive that offers secure storage. Subsequent design revisions should be archived as separate files. Then, if a recent change creates a problem in the control operations, an older, more stable backup can be restored.

Remote client connections support integrators connecting to the network temporarily from other locations in order to add, reconfigure, or relocate nodes.

Remote Client Connections

Large LonWorks networks often utilize a client/server architecture where multiple network management and user interface tools access a single LNS Server computer. This architecture allows several technicians to simultaneously perform network maintenance and modifications. **See Figure 9-7.**

The LNS platform can support up to 10 remote clients, each accessing a single LNS database through the LNS Server application. Each client can perform adds, moves, and changes to the network program independently, which updates the network database during each modification.

Client network management tools open the network program, which is distributed to the client computers as a back-up file. This file does not include the network database since that must remain on the LNS Server computer exclusively. However, all remote clients must have access to the network database in order to perform network management tasks such as binding network variables, changing configuration properties, or device management. When the network program file is opened, the client network management tool prompts for the location of the network database. Options include importing the database, attaching as a full client, or attaching as a lightweight client.

Importing should only be selected when the computer running the network management tool must include both the network database and the network program. However, if multiple network management tools working the same network program each include a network database, each design could become inconsistent with the others and the actual network.

Remote Client Connections

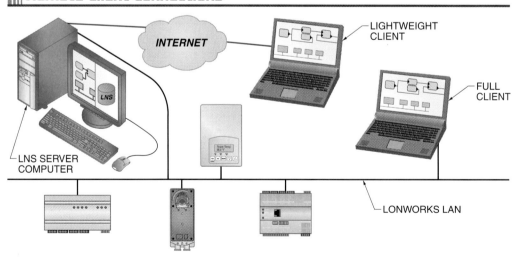

Figure 9-7. Remote clients edit a copy of the network design document and communicate design changes to the single computer that includes the network database.

Clients that connect directly to a network media channel should choose the full client option. Full clients can continue to monitor network variable values even if the LNS Server computer becomes unavailable.

Lightweight clients communicate network management actions through the LNS Server over an IP channel. Therefore, the LNS Server must be present at all times in order for the lightweight client tool to function. Also, multiple lightweight clients may create a bottleneck at the LNS Server, slowing monitoring response times during network management activity.

Network Resynchronization

With multiple clients, situations may occur when the network program file does not match the network database. For example, one integrator may use a client network management tool to make changes to the network program design and the database, but not create a back-up file of the modified design. This file is needed to update the other clients regarding the changes. If another integrator needs to modify the network program on a different client, but does not have the current network program file, the recent changes are not visible.

Or, if the computer running a network management tool lost power or had a major software problem, changes may have been made to the database that were not saved in the network program design. Resynchronization would be necessary to update the network program file.

As long as the current network database is available, resynchronization can be used to update the entire system with the any changes made in any of the components. *Resynchronization* is an LNS network management function used to update the network program, node configurations, and the network database so that they all reflect the same design.

Resynchronization can change information in multiple places. **See Figure 9-8.** Often, it is a recent change in the database that needs to be updated across the network and in the network program. For example, recently added nodes may have objects that reside in the database, but not in the network program, so they are added to the local program file. Similarly, resynchronization can also update the database with changes made only on nodes or in the network program.

Resynchronization updates all of the nodes on the network with any new configurations in the database. If a node has been manually decommissioned, it is recommissioned during the resynchronization process. Nodes may go OFFLINE briefly while the tool updates configuration properties and network variable bindings, so those controlling critical equipment require extra care. The resynchronization process can also be time-consuming, depending upon the size and complexity of the network.

Additional LNS services allow database validation/repair and monitor set synchronization. A monitor set is used by remote LNS clients without access to the LNS Server to monitor and control network variables.

Resynchronization

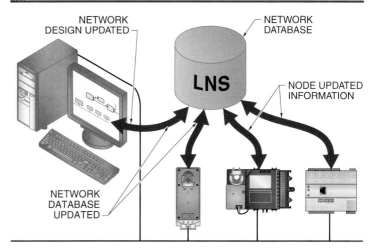

Figure 9-8. Network resynchronization is typically used to match the network design or the nodes to the network database.

Carefully documenting network program changes is vital to avoid many communication and control problems. This is especially important when more than one integrator is working on a program. In this case, any changes must be propagated through the network, and the updated network program file shared with other integrators. Also, a detailed log of changes made by each integrator may be needed if it becomes necessary to troubleshoot a problem later.

Network Recovery

Network recovery is an LNS service used by network management tools to create an LNS network database from information recovered from nodes. Recovery may be required when the network database is lost, corrupted, or unavailable. **See Figure 9-9.**

During the recovery process, the network management tool queries nodes and routers for their Neuron ID, subnet/node address, network variable binding definitions, and external interface file data. The network management tool then creates an LNS database and matching network program.

While recovery is an important solution for a network program that is otherwise unrecoverable, it has disadvantages. Typically, the names of nodes, function blocks, and network variables are in highly abbreviated forms, which may be difficult to identify. Integrators will need to re-enter names in more human-readable forms. The network program file may also require reorganization to arrange graphical elements in a more logical fashion.

Therefore, recovery should not be used as an alternative to regularly backing up a network program design. In addition, successful recovery is contingent on a healthy network. Channels with wiring issues or excess network traffic may prevent the network management tool from accessing node information.

If the network is large and involves many nodes, the network recovery process can take a while and create a lot of network traffic. It may also be incomplete if some parts of the network are having network problems and the process may not distinguish between multiple nodes of the same type. Therefore, network recovery should be used as a last resort.

NETWORK DOCUMENTATION

Accurate network documentation verifies the performance of the control network and enhances the maintenance and management of the control system. Documentation is compiled within an operations and maintenance (O&M) manual.

Network Recovery

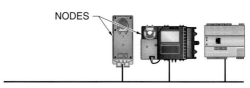
NETWORK WITHOUT DATABASE

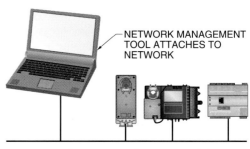
NETWORK MANAGEMENT TOOL QUERIES NODES

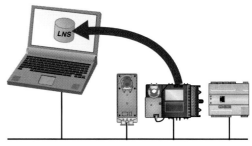
NETWORK MANAGEMENT TOOL CREATES DATABASE

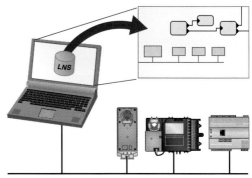
NETWORK MANAGEMENT TOOL CREATES NETWORK DESIGN

Figure 9-9. Network recovery is a process used to create a new network database and network design from information discovered from a network of previously installed nodes.

An operations and maintenance (O&M) manual includes the project description and specification documents, along with complete documentation for the physical and logical design of the network. **See Figure 9-10.** The O&M manual includes a single-line control diagram, physical node and infrastructure information, control sequence descriptions, node wiring diagrams, node and terminator locations, node submittals, input/output datasheets, user guides, and channel/node test reports. A description of the sequence of operations for each subsystem details the control processes and the requirements for implementation.

Operations and Maintenance Documentation

Project Description

Control Specifications

Single-Line Control Diagram
    Network Resource Reports
        Node and Device Inventory
        Infrastructure Inventory
        Network Traffic Reports

Subsystem Drawings
    Sequence of Operation Descriptions
    Subsystem Resource Reports

Node Health Reports

Manufacturer Submittals and Warranties

Vendor/Supplier Information

Computer Hardware and Operating System Documentation

Software Documentation
    Network Management Tool User Guide
    Plug-In User Guides
    HMI User Guides

Back-up Data
    Network Design Files
    Network Database
    HMI Design Files

Figure 9-10. An operations and maintenance (O&M) manual documents all of the information about the physical and logical arrangements and operations of a LonWorks network.

Network Traffic Reports

Verification of channel health involves generating reports of network traffic rates for each channel in the network. A protocol analyzer measures and reports channel bandwidth consumption, error rates, and packet types. Documenting peak traffic, as well as sustained traffic, ensures fast control response times and provides data useful for planning future network expansion. Peak traffic normally occurs when HMIs access the network for monitoring and control tasks. Bandwidth tests should be conducted both with and without HMIs attached to the network.

All operations and maintenance documents should be kept together in an organized binder and stored in a location that is easily accessible to any integrator needing to modify, update, or troubleshoot the system.

Node Health Statistics

Most network management tools provide a device management test feature that reports the contents of each node's Neuron error log. The results of the error log can be used to document node health statistics, including the current state of the node, last cause of reset, and communication errors.

Network management tools based on LNS can utilize plug-ins that automate node health documentation. For example, the Echelon Multi-Test tool produces a comprehensive report of all network node statistics, saved as a plain-text file. **See Figure 9-11.**

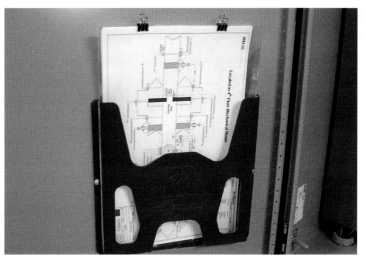

Network documentation should be organized and stored neatly in an area accessible to technicians, such as inside a wiring cabinet.

Node Health Reports

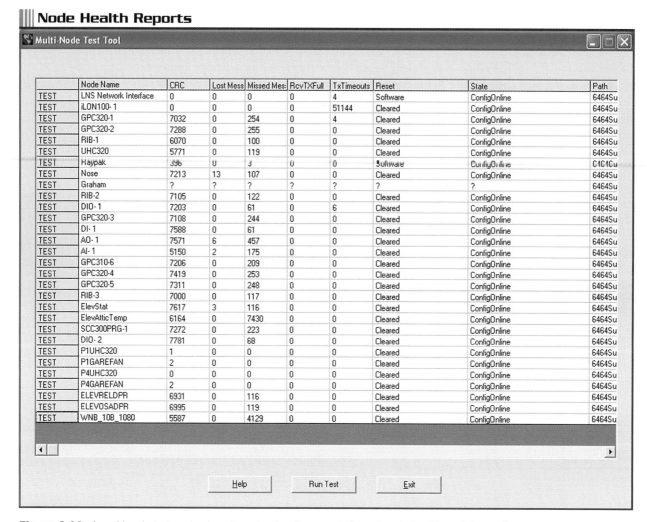

Figure 9-11. A multinode tester plug-in automates the documentation of node health statistics and current status.

Summary

- Network maintenance tasks include the replacement of failed nodes, addition of new nodes, relocation of existing nodes, and software updates to control applications.

- Node replacement or relocation involves both physical and logical network changes.

- If a replacement node has a newer application program version than the original node, the Program ID is slightly different, so the manufacturer should provide a new XIF.

- The replacement of nodes from different manufacturers, even if they include the same functional profile, typically requires some modification of the network program.

- During relocation, a node is temporarily placed in the OFFLINE state by the network management tool through device management.

- A relocated node is recommissioned with a new subnet/node address, but all binding definitions are preserved in the network program.

- The mismatch of a network management tool's physical and logical connections can cause communication issues between the network management tool and nodes.
- The network database, drawings, and associated files can be copied into a single compressed back-up file.
- The LNS platform can support up to 12 remote clients, each accessing a single LNS database through the LNS Server application.
- Client network management tools open the network program, which is distributed to the client computers as a back-up file. The client network management tool then prompts for the location of the network database.
- Resynchronization can be used to update the design file with the changes in the database.
- The network management tool queries nodes and routers for their Neuron ID, subnet/node address, network variable binding definitions, and external interface file data, and can then recover an LNS database and matching network program.
- A maintenance and operations (M&O) manual includes the project description and specification documents, along with complete documentation for the physical and logical design of the network.
- Verification of channel health involves generating reports of network traffic rates for each channel in the network.

Definitions

- **Resynchronization** is an LNS network management function used to update the network program, node configurataions, and the network database so that they all reflect the same design.
- **Network recovery** is an LNS service used by network management tools to create an LNS network database from information recovered from nodes.

Review Questions

1. Under what circumstances can the exact same information be loaded onto a replacement node from the network database?
2. What is the source of configuration property values for replacement nodes?
3. How is a node moved to a different channel in the network program?
4. How is a network interface moved to a new channel?
5. How is the network program backed up?
6. How do remote clients access the network program?
7. When might it be necessary to resynchronize the network?
8. How are a network program and database recovered?
9. What is included in a maintenance and operations (M&O) manual?
10. How are traffic statistics collected for network documentation?

Chapter Ten

BACnet System Overview

BACnet (named for building automation and control network) provides a standardized framework for organizing and sharing control information between networked control devices. This framework supports interoperability through an object-oriented information and function model and flexibility in LAN types and network infrastructure. While BACnet was developed by an organization focused on HVAC systems, it is applicable to many different building systems, even those that do not yet contain control device nodes.

Chapter Objectives

- Describe BACnet's approach to dividing the goals of building automation into separate areas.
- Evaluate the features and limitations of BACnet.
- Compare the implementations of standard and proprietary objects and properties within BACnet's object-oriented information model.
- Identify the roles of services in sharing information between nodes.
- Contrast the standardization and flexibility of BACnet system architectures.
- Describe the roles of BACnet interoperability building blocks (BIBBs) and device profiles in supporting interoperability.

BACNET SYSTEMS

BACnet provides interoperability between cooperating building automation control devices. BACnet is not a specific product or even a type of product. BACnet is a standard set of rules and methods, compiled into a book, that a manufacturer may use to produce nodes that are interoperable with other BACnet nodes. Owners and consulting-specifying engineers may also use BACnet as a tool to add precision to the specification of automation systems, such as particular interoperability and functional requirements.

BACnet was designed to apply to all types of building systems, including perimeter and object security, lighting, HVAC, and elevator systems. **See Figure 10-1.** Products for all of these applications are available today and BACnet allows the possibility of incorporating these and any other systems into an interoperable system if the market demands and makes the applicable products.

BACnet provides interoperability ranging from simple information exchange to complex interactions. While BACnet does not provide "plug and play" interchangeability, it does provide a means for many kinds of interoperations using standardized techniques. This interoperability has proven to be flexible and robust in over a decade of practice in millions of nodes.

Development of BACnet

BACnet was originally developed in 1987 under the auspices of the American Society of Heating, Refrigerating and Air-Conditioning Engineers (ASHRAE) as Standard 135 *BACnet: A Data Communication Protocol for Building Automation and Control Networks*. Subsequent revisions to the standard were published in 1995, 2001, and 2004. BACnet became an American National Standards Institute (ANSI) standard in 1995 and an International Organization for Standards (ISO) standard in 2003.

BACnet Systems

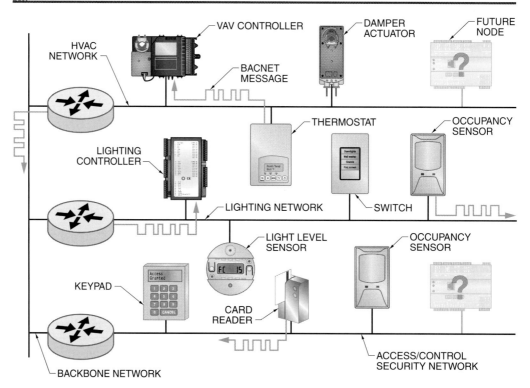

Figure 10-1. BACnet systems can incorporate control device nodes from any building system, including nodes that may be designed in the future.

BACnet was developed by an open consensus process, which continues to maintain the standard. This process is open to all interested parties and there are no fees to participate. ASHRAE (an ANSI-certified standards-making body) oversees the standards activities and ensures that voting members represent a balanced roster of manufacturers, owners, consulting-specifying engineers, academia, government, and general-interest parties.

The Standing Standard Project Committee (SSPC) 135 is the group within ASHRAE that is charged with maintaining and developing the BACnet standard. This committee also answers questions from users regarding the interpretation of the standard, and manages the ongoing revision process, which includes compiling suggested changes, making them available for peer review, and voting on their inclusion in the next revision. SSPC 135 is subdivided into working groups that each address particular aspects of the implementation of BACnet systems. **See Figure 10-2.**

This long history and open process has resulted in an extremely strong standard with wide support and adoption worldwide by a constantly growing number of manufacturers whose products serve the building automation market and related markets.

BACnet Methodology

BACnet defines a generalized framework for all types of automation systems, including how to describe system information and standard methods that nodes can use to ask each other to perform actions. BACnet accomplishes this by dividing automation goals into four specific areas:

- an object-oriented method of describing data and control functions within systems
- a reliable set of services that standardize the method that BACnet nodes used to convey requests and responses
- a selection of local area network (LAN) types that allows the integrator to balance cost and performance
- a scalable internetworking methodology that allows the creation of large and diverse networks

BACnet standardizes a collection of interoperations that incorporate well-defined standards in all of these areas. BACnet separates the goal of interoperability into two aspects, which is the most critical concept of BACnet. One aspect is the standardization of the content of messages exchanged between nodes and the other is the transportation of these messages.

SSPC 135 Working Groups

Working Group	Primary Responsibilities
AP-WG: Applications	Develops applications-oriented profiles of common building automation equipment
IP-WG: Internet Protocol	Extends BACnet/IP for IP-related issues, such as network address translation and IPv6
LA-WG: Lighting Applications	Researches, drafts, and proposes additions to standard for lighting control applications
LSS-WG: Life Safety and Security	Researches, drafts, and proposes additions to standard for life safety and security applications
MS/TP-WG: Master-Slave/Token-Passing	Addresses enhancements and issues related to MS/TP and PTP communications
NS-WG: Network Security	Develops mechanisms for authorization and transfer of control authority and auditing
OS-WG: Objects and Services	Addresses support for new objects, new services, or refinements to existing objects and services
TI-WG: Testing and Interoperability	Maintains Standard 135.1 *Method of Test for Conformance to BACnet*, BACnet interoperability building blocks, and device profiles
UI-WG: Utility Integration	Supports communications between building automation systems and public utility providers
WN-WG: Wireless Networking	Investigates the use of BACnet with wireless communication technologies
XML-WG: XML Applications	Investigates applications of Extensible Markup Language (XML) technology in relation to BACnet systems

Figure 10-2. *The committee responsible for developing and maintaining the BACnet standard is divided into several working groups that concentrate on certain aspects of the technology.*

Standardized Message Content. The content of a message is conveyed using a set of rules about coding, structure, and meaning. These rules represent the language of BACnet. For two nodes to be able to operate together, they must not only agree on the means of conveying messages back and forth, but also on the language and structure of those messages. This is similar to when a person sends a letter written in English to another person; the letter is only understood if the recipient can read English.

BACnet standardization of message content, known as the application language, is divided into objects and services. **See Figure 10-3.** An *object* is an abstract container that organizes related information and makes it accessible to other nodes in a standard way. Objects represent a component of the node, or some collection of information that may be of interest to other nodes. Objects are used to model all information within an interoperable BACnet node.

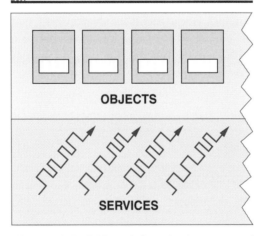

Figure 10-3. BACnet defines the structure and format of control information, which standardizes the messages shared between nodes.

A *service* is a formal procedure for BACnet nodes to make requests of other nodes. For example, a node with an attached temperature sensor may perform the service of reading the temperature and providing this information to another node that needs it.

BACnet nodes exchange information and requests by sending and receiving electronic messages coded in the application language. The numeric codes represent the desired information or functions to be performed. The language of this encoding is common to all BACnet nodes.

Standardized Message Transportation. BACnet provides flexibility in the media and LAN options for transporting messages. However, the content of the messages that are transported is identical, regardless of the transport scheme. To use an analogy, a letter can be sent to another person in a variety of ways. **See Figure 10-4.** The letter can be placed in an envelope with the recipient's address on the outside and sent via mail. However, if both the sender and receiver have fax machines, the same letter could be faxed instead. Or, an image of the letter can be e-mailed. With any method, the received letter is the same size and shape and contains the same symbols, language, and content. In fact, unless one actually watches the receiving process, the receiver may not even know how that content was conveyed. The variety of different transport methods supported by BACnet allows the designer or consulting-specifying engineer to choose the most cost-effective transport method for a given application.

BACnet Features

BACnet has many positive features that are well suited to creating interoperable systems of nearly unlimited size and scope and complexity. Like all technologies, though, BACnet has both strengths and limitations.

Practical Interoperability. BACnet systems and nodes provide some of the broadest choices for building automation system interoperability. This means that nodes with similar features from one manufacturer are largely able to replace or operate with similar components from other manufacturers in performing their day-to-day control tasks. This also means that there is great potential for interoperability and seamless integration across different types of systems, such as lighting and HVAC.

Message Transportation

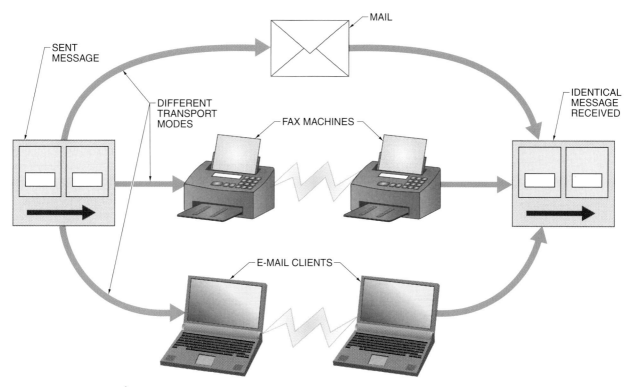

Figure 10-4. The transportation of a BACnet network message is completely independent from the content of the message.

Scalability. Many technologies suffer from degrading performance as the size and scope of the network increases. However, BACnet supports scalability in the key dimensions of cost, performance, and size. This means that BACnet systems can be created whose cost and performance vary according to each task. These systems have the ability to grow or be applied to larger systems. The scaling of BACnet comes mostly without penalty. These are important considerations for growth and change, as the scope and management of facilities tends to be dynamic over time.

Flexibility. Developers of BACnet products have almost no restrictions in their ability to add new products and interoperability features. The object-oriented approach to information and control provided by BACnet means that product developers do not need to rely only on existing information models.

The choice of many LAN options, and the means to connect them easily, adds flexibility. It also makes BACnet adaptable to many different situations common in building automation and related control systems. Since internetworking methods, router behavior, and other details are integrated into the standard, BACnet systems are more robust and less dependent on separate devices that must use tunneling and other less efficient methods.

BACnet's LAN flexibility also allows for integration with existing LAN infrastructure without requiring fundamental changes to existing BACnet nodes. This protects the building owner's investment in not only the infrastructure but also in BACnet nodes and systems.

The flexible internetworking options allow for optimizing the balance between cost and performance. Integrators can choose high-performance networks when required, or lower-cost networks when performance is less critical. Because there is little penalty to this scaling, BACnet systems can grow or be sized at a good value and according to need.

BACnet does not have a concept of hierarchy; so all nodes are peers with all other nodes, even nodes on non-BACnet systems. Gateways into these systems can be located at any point in the BACnet network. From the BACnet side, each non-BACnet system node appears as a single BACnet proprietary object or some collection of standardized BACnet objects.

Standards-Based. All parts of BACnet are based on ANSI and international standards that are controlled by independent, public standards-making bodies. This means that BACnet is not controlled by manufacturers alone and that a objective viewpoint is maintained during the ongoing revisions of the standards. It also means that BACnet is governed by a unified standards suite, not a combination of standard and proprietary parts.

Open Development. The BACnet standard is maintained through an open, transparent, and no-cost consensus process where every interested party can contribute. Voting membership of the standard's maintenance is required to be balanced and represent not only manufacturers, but also end users, consultants, government, academia, and general-interest groups. Anyone may participate in the standards process and the direction of the standard. There are no fees, other than the fees for printed copies of the standards themselves.

Delta Controls
BACnet nodes are available for applications in many different building systems, including lighting.

Wide Adoption. BACnet is endorsed and adopted by nearly every major building automation and controls vendor in North America and in many other countries. BACnet systems are installed in many locations throughout the world.

BACnet Limitations

Making any conceivable control system from seamlessly interchangeable components that have identical features, programming, testing, installation, and configuration is a nearly impossible task. A conscious decision was made while developing BACnet to focus on issues that are important for day-to-day operation of building control systems. Standardization of control in these critical areas has the greatest impact. Because of this emphasis, it has not yet been possible to reach a consensus on many issues that are important during the installation and commissioning of building control systems, such as a standard mechanism for these activities.

BACnet does not allow indiscriminate interchangeability. Replacement controllers must be carefully matched to ensure that they have the appropriate control functionality and communication capabilities. System expansion involves more than just running more cabling and connecting additional nodes. However, within a core set of functionality, BACnet nodes are very much compatible, as many basic features operate together without issue. However, this is still a long way from being completely interchangeable.

While BACnet provides interoperability by defining a common, abstract view of information, each vendor creates its own way to store and use that information within its products. It is still necessary to have vendor-specific configuration tools and programming languages. This is likely to remain true for quite some time.

It is important to note that all of these limitations are also true of today's proprietary systems. BACnet does not make configuration and programming problems necessarily easier or harder to solve. Buyers still have to make choices to determine the best balance between cost and performance.

Common BACnet Questions

There are several questions in the building automation industry about the capabilities of BACnet systems. Most center around perceived limitations in its interoperability and applicability to the growing market of integrated automation systems in commercial buildings. Like many other building automation protocol systems, BACnet systems are quite versatile and capable in a variety of applications.

Does BACnet Work with non-HVAC Systems?
BACnet was developed by ASHRAE, which is a major organization in the HVAC industry. However, BACnet was designed from the beginning to encompass all types of building systems and their interoperability across common communication media and applications.

Can BACnet be Implemented in Small or Large Applications?
BACnet is applicable at every level of an automation system, from sensor or fieldbus devices through controllers and management level workstations. A key feature of BACnet in this regard is scalability that allows large BACnet systems to retain very high efficiency.

Does BACnet Support Flat Architectures?
BACnet systems can be designed in a flat architecture where all nodes are peers with other nodes. Although it is possible to create architectures that are tiered, this is actually a rare and uncommon practice in BACnet.

Will Any Two BACnet Nodes Work Together?
BACnet defines particular interoperations in standard ways, but this kind of standardization does not require every node to include every feature and interoperation. Collections of interoperations are commonly used by nodes of certain types. Nodes must still be matched to the expected functions to perform and to the peer nodes they will communicate with. For example, BACnet standardizes a method for a node with an analog input to make that value available to others. However, a given BACnet node is not required to have an analog input or be able to read the value from another node. There is no single interoperation criterion that defines a BACnet node.

Is BACnet Supported by the Automation Industry?
Hundreds of manufacturers make BACnet products, covering every field of building automation. Thousands of product types are available covering a wide variety of classes of BACnet-capable nodes. Millions of BACnet nodes are installed in many countries, particularly in North America, and that number is growing each year.

INFORMATION ARCHITECTURE

The language of BACnet provides interoperability through object-oriented modeling. *Object-oriented modeling* is the concept of organizing many different types of information into defined and structured units. Objects serve as the basic building blocks of information. In addition, the language of BACnet includes services that define actions and requests that one BACnet node may ask another BACnet node to carry out.

Objects

All information in a BACnet system is represented in terms of objects. Objects can represent physical inputs and outputs, as well as nonphysical concepts like programs, logs, schedules, calculations, or control processes. **See Figure 10-5.** Objects may represent single physical control points or logical collections of control points that perform a specific function. A BACnet node can implement any number of objects of different types, but only the standard Device object is always required. Other objects are included as appropriate to the node's functions.

All objects in BACnet consist of a set of properties. A *property* is one item of information about an object. Properties are used to access object information. An object likely has many properties that are commonly listed in a table. **See Figure 10-6.** Each property is composed of a name and a value. For example, Object_Type is the name of a property that defines the type of information handled by that object. The value of this property is a number that corresponds to a particular object type. Depending on the object type, different types of properties may be needed to handle the different types of information.

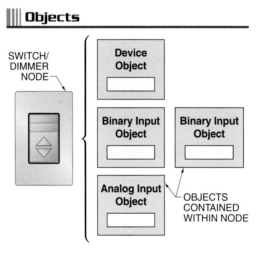

Figure 10-5. Objects represent physical inputs, outputs, and software processes. Any type of information can be modeled as an object.

Properties

Property Name*	Value
Object_Name	"Light Level"
Object_Type	Analog Input
Present_Value	50
Units	percent

* Only selected properties listed. See BACnet standard for complete list.

Figure 10-6. Objects are examined and controlled by accessing a set of properties that belong to each object. Each property has a name and a value.

BACnet provides flexibility for future applications by relying on a rigid standardization of its software components. This methodology can be applied to new objects and services while retaining interoperability with existing ones. By analogy, a glass that is made today can hold drinks that will be invented tomorrow.

An Analog Input is a common type of object often used to model sensors. The Object_Name property identifies its purpose or location within the system. The Present_Value property is equal to the value that the sensor is currently measuring. Other properties contain more information about the sensor function, such as whether it appears to be functioning normally. Other properties also indicate high and low limits used for alarming.

Any kind of information can be modeled with objects. This standardized information architecture provides both backward-compatibility for information sharing and control, as well as forward-compatibility into the future. By conforming to the standard, a node designed today can access and operate with a controller that has not even been invented yet.

Standard Objects and Properties. Most control information can be described by one or more similar types of data points, even between different building systems and device manufacturers. Therefore, BACnet standardized a set of common objects representing these types of information. A *standard object type* is an object type defined in the BACnet standard. The definition includes a description of the object and its expected use, standard properties, and behavior. By defining these objects and properties in detail, there is consistency between implementations by different manufacturers.

There are 25 standard object types in the 2004 edition of the BACnet standard. As of December 2008, another 13 standard object types have been proposed for addition to the standard as addenda. Some have already been approved, while others are in various stages of the proposal and review process. **See Figure 10-7.** The group of standard objects represents much of the functionality in typical building automation systems.

A few properties are required for every object, but most vary depending on the object type. Some are required by the standard, while others are optional. The standard defines a set of required properties and a set of optional properties for each standard object type. For example, an Analog Output object includes properties such as Present_Value, Description, Update_Interval, and Units. **See Figure 10-8.** The required properties for a standard object type must be implemented for each instance. An optional property does not have to be implemented, but if the manufacturer claims that it is implemented, then it must behave as the standard describes.

BACnet Standard Objects

2004 Standard	Approved Additions*
Accumulator	Access Door
Analog Input	Event Log
Analog Output	Global Group
Analog Value	Load Control
Averaging	Structured View
Binary Input	Trend Log Multiple
Binary Output	
Binary Value	
Calendar	
Command	**Proposed Objects***
Device	
Event Enrollment	Access Credential
File	Access Point
Group	Access Rights
Life Safety Point	Access Users
Life Safety Zone	Access Zone
Loop	Credential Data Input
Multi-State Input	Lighting Output
Multi-State Output	
Multi-State Value	
Notification Class	
Program	
Pulse Converter	
Schedule	
Trend Log	

* as of early 2009

Figure 10-7. BACnet defines a collection of standard objects. The current standard contains 25 standard objects. Another 13 standard objects are approved or proposed for addition in the next revision.

Some property values are read only, meaning that other nodes can look at the property value, but not change it. Other properties can be written (changed) by other nodes. However, manufacturers may allow some read-only properties, such as Object_Name or Description, to be written.

The values of properties are standardized into datatypes. A *datatype* is a specification for the type and format of information. For example, a datatype may indicate a TRUE or FALSE, signed integer, floating-point (REAL) number, or a string of characters (such as "ABCDEF"). BACnet defines 13 primitive application datatypes, which cover most types of control information. Many standard properties use only those datatypes, but others are composed of collections of primitive datatypes. For example, the datatype DateTime is a combination of a date datatype and a time datatype. For the most part, these combination datatypes only make sense in a particular context, so they are called context-dependent datatypes.

Properties

Property*	Type		Datatype
Object_Identifier	Required	Read Only	BACnetObjectIdentifier
Object_Name	Required	Read Only	CharacterString
Object_Type	Required	Read Only	BACnetObjectType
Present_Value	Required	Writable	REAL
Description	Optional	Read Only	CharacterString
Status_Flags	Required	Read Only	BACnetStatusFlags
Out_Of_Service	Required	Read Only	BOOLEAN
Update_Interval	Optional	Read Only	Unsigned
Units	Required	Read Only	BACnetEngineeringUnits

* Only selected properties listed. See BACnet standard for complete list.

Figure 10-8. Object properties may be required or optional, read only or writable, and formatted in a certain datatype.

Proprietary Objects and Properties. Standard object types and their standard properties represent a generic and consensus view about common automation information and functions, but this does not encompass every type of functionality or implementation. For example, some nodes implement analog inputs with high and low alarm limits, while others have inner limits and outer limits. Some nodes implement a return-to-normal deadband relative to a limit, while others have a second limit. Some control strategies are simply not represented by any of the standard objects. In these cases, the BACnet standard may not support the feature or its desired implementation.

> It may be more accurate to call proprietary objects and properties "non-standard" since they are not necessarily secret or unique to a single manufacturer. In fact, manufacturer documentation should describe them fully and be freely available.

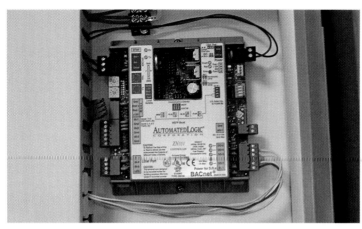

The control points and operation of a specialized equipment controller can be represented in BACnet as either a collection of standard objects or a proprietary object.

Proprietary VAV Controller Object	
Property	Example Values
Object_Identifier	(VAV Controller 3)
Object_Name	"VAV Unit 3"
Object_Type	157 (VAV Controller)
Input_Air_Temp	65.0
Output_Air_Temp	72.0
Temp_Units	degrees-fahrenheit
Damper_Position	55
Position_Units	percent
Heat_Enable	TRUE
Diff_Pressure	1.2
Pressure_Units	inches-of-water
VAV_Status	2
Status_Text	Status_Text[1] = "Air Flow Only" Status_Text[2] = "Heating" Status_Text[3] = "Cooling"

Figure 10-9. Any type of proprietary object can be created by vendors that require specific functionality not available from standard objects.

Therefore, some manufacturers provide a proprietary feature by using BACnet's option for adding proprietary objects and/or properties. Proprietary properties can be added to otherwise standard object types or entirely proprietary object types can be created. BACnet allows manufacturers to implement a large number of proprietary objects or properties with whatever behavior the integrator chooses. BACnet provides a framework for building proprietary objects and properties that allows them to be constructed and operate with other objects in a standardized way.

Building proprietary objects and/or properties is a powerful feature of BACnet that allows the development of new features and technologies, even those not imagined when the standard was written. For example, although an air-handling unit could be represented as a collection of BACnet standard objects whose properties indicate the status and command features of the device, it may be more logical for the integrator to think of the air-handling unit itself as an object. A proprietary air-handling unit object can be created to represent all of the equipment's functions. **See Figure 10-9.**

An object identifier of object type 8 (Device) and instance 4,194,303 (all binary ones) is reserved as a special wildcard value. Nodes reading this value treat this object identifier as if it matched the local node's Device object.

Proprietary objects and properties are not an obstacle to interoperability and need not be avoided altogether. As long as they are documented in the same style as standard objects and use only BACnet primitive datatypes for representing values, manufacturers can be reasonably assured that nearly any other BACnet node is capable of reading and writing to those object's properties if required.

Object Identifiers. In BACnet, at least in terms of the communications between nodes, objects are identified by number. An *object identifier* is a 32-bit number that is unique to each BACnet object within a node. The object identifier contains two components: the object type and the object instance. **See Figure 10-10.**

An *object type* is a 10-bit code number in the object identifier that represents the kind of object. For example, 0 means Analog Input, 1 means Analog Output, 2 means Analog Value, and so forth. The codes from 0 to 127 are reserved for use by BACnet for standard object types. **See Figure 10-11.** The existing and proposed standard object types use only the first 38 code numbers.

Object Identifiers

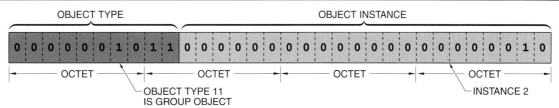

Figure 10-10. Object identifiers are numbers that correspond to a particular instance of a certain object type within a node.

Code numbers from 128 to 1023 are available for proprietary object types. To make use of proprietary objects, one must know the manufacturer of the node containing the object. This is easy to find out, since a vendor identifier code is a required property in the Device object. This unique code is assigned by ASHRAE to every requesting manufacturer. A limitation of this approach is that it is not possible to build a BACnet node that contains proprietary objects from more than one manufacturer.

The *object instance* is a 22-bit code number in the object identifier that assigns a number to each individual object of a certain type within a node. Therefore, each node may contain up to 4,194,303 ($2^{22} - 1$) objects of each type, in any mixture of types. (No object may use instance number 4,194,303, which is all 22 bits equal to 1.) Except for the Device object, object instance codes do not need to be unique across the BACnet network, only within the node.

Often, objects are described by both components. For example, an instance of an Analog Input object might be called "Analog Input, instance 5," or just "Analog Input 5." For well-known standard objects, this is often abbreviated as AI5, or BV7 for "Binary Value 7." Object instances may be designated with any number, including 0 (zero), and in any order.

The Device object is special among all object types. A BACnet device normally contains only one Device object instance. (The only devices that may appear to contain more than one Device object are those that reside on more than one LAN, such as routers.) Unlike other objects, the instance number of the Device object (and its Object_Name) must be unique across the entire BACnet internetwork. The *Device instance* is the unique 22-bit code number of the Device object. Because of this constraint, BACnet is limited to 4,194,303 devices per internetwork.

Object Types*

Object	Object Type Code
Analog Input	0
Analog Output	1
Analog Value	2
Binary Input	3
Binary Output	4
Binary Value	5
Calendar	6
Command	7
Device	8
Event Enrollment	9
File	10
Group	11
Loop	12
Multi-State Input	13
Multi-State Output	14
Notification Class	15
Program	16
Schedule	17
Averaging	18
Multi-State Value	19
Trend Log	20
Life Safety Point	21
Life Safety Zone	22
Accumulator	23
Pulse Converter	24
Event Log	25
Global Group	26
Trend Log Multiple	27
Load Control	28
Structured View	29
Access Door	30
Lighting Output	31
Access Credential	32
Access Point	33
Access Rights	34
Access User	35
Access Zone	36
Credential Data Input	37
Reserved for Future Standard Objects	38–127
Available for Proprietary Objects	128–1023

* Object type code assignments for proposed objects are subject to change before the next edition of standard.

Figure 10-11. Object types are enumerated by an assigned code number.

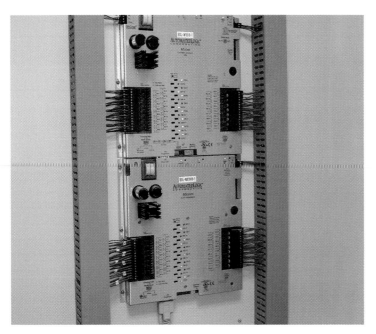

Services are used to share information between nodes on a BACnet network.

The Object_Identifier numbering scheme provides for a theoretical limit of 4,290,772,993 objects in each BACnet node. This corresponds to the use of every instance of every object type (except the Device object), plus the one allowed Device object. However, most nodes contain a few to a few dozen objects.

Property Identifiers. A *property identifier* is a 32-bit code number that identifies a property in an object. The 2004 BACnet standard defines nearly 200 of the 512 property identifier enumeration values reserved for ASHRAE. **See Appendix.** Property identifiers with values of 512 and greater are proprietary properties unique to each vendor through the same vendor identifier code for object types. Over four billion property identifiers, per vendor, are available for nonstandard properties.

Services

Services identify the roles of each node in the interoperation, the parameters needed to carry out the requested interoperation, and the possible replies and their parameters. In effect, the service precisely defines the rules of etiquette for a particular kind of question and reply.

BACnet defines five broad categories of services:
- Object access services allow nodes to read and write properties, create and delete objects, manipulate lists of data, and search for particular objects or properties.
- Device management services let nodes be controlled remotely, for example asking a node to identify itself, to stop communicating for a while, or asking if it requires reloading.
- Alarm and event services allow nodes to communicate alarms, changes of state, and other exception conditions.
- File transfer services allow nodes to send or receive information in bulk form, such as historical trend data or control programs.
- Virtual terminal services allow nodes to provide an interactive human-machine interface, in the form of prompts or menus, for remotely programming or configuring the node.

Most, though not all, service messages include several parameters. For confirmed service requests, in which the receiver is obliged to respond (even if it cannot carry out the request), the first parameter is a message number assigned by the sender to identify the request. Nodes may transmit multiple requests before receiving answers, so the message numbers are used to match requests with responses that are received out of order. Other parameters depend on the particular service.

For example, the simplest BACnet services are the ReadProperty and WriteProperty services, which are object access services. The ReadProperty service is used by one BACnet node to request the value of a certain property in another BACnet node. **See Figure 10-12.** The ReadProperty request requires both an Object Identifier and Property Identifier to specify which property is to be read. For example, if one node needs to know a temperature, it uses a ReadProperty service to request the value of the Present_Value property of the Analog Input object that represents that temperature sensor. There are two possible replies to this service. The service either succeeds and returns the requested value, or it fails. For example, there may be no such object defined in the target node.

ReadProperty Service

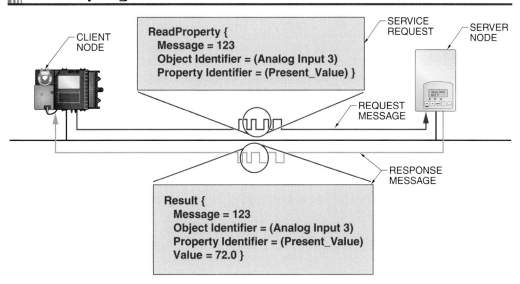

Figure 10-12. The ReadProperty service is a way for one node to request information from another node.

WriteProperty Service

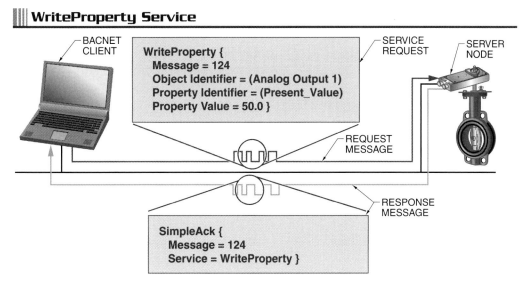

Figure 10-13. The WriteProperty service requests that a property value within a node be changed to the value included in the message.

The WriteProperty service is used by one BACnet node to ask another BACnet node to change the value of a property of a specific object. **See Figure 10-13.** Like ReadProperty, the WriteProperty request requires both an Object Identifier and Property Identifier to specify the property to be written and the Property Value to be written. The result is a reply indicating either success or failure.

Some writable properties are commandable, meaning that a WriteProperty service to this property can include a Priority parameter. If more than one value is written to this property by different client nodes, their relative priorities are used to determine which is more important. The values are stored in a 16-slot array according to their Priority. The most important value is used in the Present_Value.

Service Message Encoding

The service type and its parameters are typically represented in documentation as human-readable words, but the actual messages are transmitted with the numeric codes for the service type, Object Identifier, Property Identifier, and other parameters. This makes for shorter messages that are easier for computer-based devices to understand. Network analyzer tools typically display both versions, which can be helpful for spotting errors or misconfiguration while troubleshooting.

```
ReadProperty {
  Message = 123
  Object Identifier = (Analog Input 3)
  Property Identifier = (Present_Value) }
```

HUMAN-READABLE VERSION

```
01 04 00 00 7B 0C
0C 00 00 00 03
19 55
```

ACTUAL TRANSMITTED CODE

SYSTEM ARCHITECTURE

The architecture of BACnet systems is not very different from many other computer-based networks. The system can incorporate multiple LAN types, often arranged with a central backbone network that branches off into smaller subnetworks. This provides both a physical and logical organization to the system's control device nodes. The nodes have major hardware and software components in common and they can be accessed by network tools located at any point in the network.

Network Infrastructure

BACnet networks can vary in size and scope from very small (only a few nodes) to extremely large (over four million nodes spread across many network segments). As a result, the network architecture can look like almost any other modern computer-based network of workstations, web servers, BACnet nodes, routers, and other network devices. **See Figure 10-14.**

LAN Types. The transport component of BACnet is responsible for physically conveying BACnet messages between nodes. As LAN type options, BACnet allows Ethernet 8802-3, BACnet/IP, ARCNET, Master-Slave/Token-Passing (MS/TP), Point-to-Point (PTP), and LonTalk Foreign Frames.

It may be possible to use any of several different transports effectively. Each option has physical, electrical, and practical considerations. The LAN type is often chosen based on the most appropriate trade-off between cost and performance, which includes speed, allowable distance, and reliability. **See Figure 10-15.** The most common LAN type in BACnet networks is MS/TP. Systems may use more than one LAN type, though different LAN types must be interconnected with routers.

However, sometimes a LAN type provides particular features that are vital to the application, which override the cost and performance considerations. For example, only PTP is appropriate for dial-up applications over telephone. Media options may make a certain LAN more appropriate for special circumstances, such as LonTalk over powerline carrier.

Backbone Architecture. Increasingly, BACnet systems are making use of a building's existing network infrastructure as a backbone to carry control system traffic. With the current trend toward IP-based infrastructure, BACnet/IP is becoming the norm for this application.

From a system viewpoint, even though the IP network often includes multiple IP subnets, it is convenient to think of the IP infrastructure (or at least the BACnet/IP subnets using it) as one logical segment. This is consistent with thinking of multiple 8802-3 segments coupled by switches or hubs also as one logical segment. This segment forms the backbone of a building-wide or campus-wide BACnet system.

BACnet Network Infrastructure

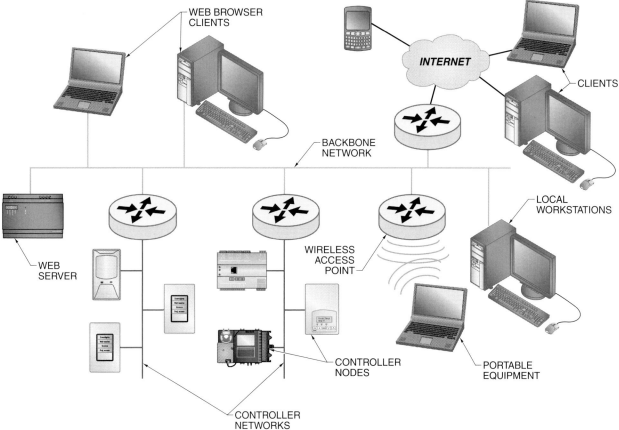

Figure 10-14. The infrastructure of a BACnet network can include many different devices, LAN types, and clients.

Backbones typically run through vertical shafts at the core or corners of a building. **See Figure 10-16.** Earlier backbones were based on 10BASE-5 or 10BASE-2 coaxial cabling, but most today are CAT5 or CAT6 twisted-pair cabling with 8802-3 switches instead of hubs. The switch methodology avoids costly coaxial cabling and greatly improves performance and distance options.

Regardless of the media type, the backbone architecture is common. The backbone branches into smaller subnetworks, typically at each floor or building section. Hubs or switches distribute backbone connections to nodes and routers on that floor. This arrangement makes intrafloor wiring easier and less subject to distance constraints.

BACnet LAN Types

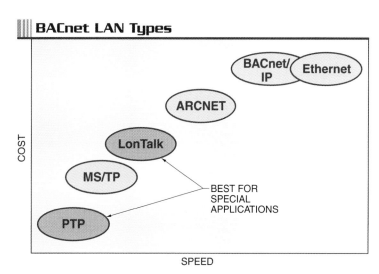

Figure 10-15. Choosing LAN types (transport options) for BACnet requires a trade-off between cost and performance.

Backbone Architecture

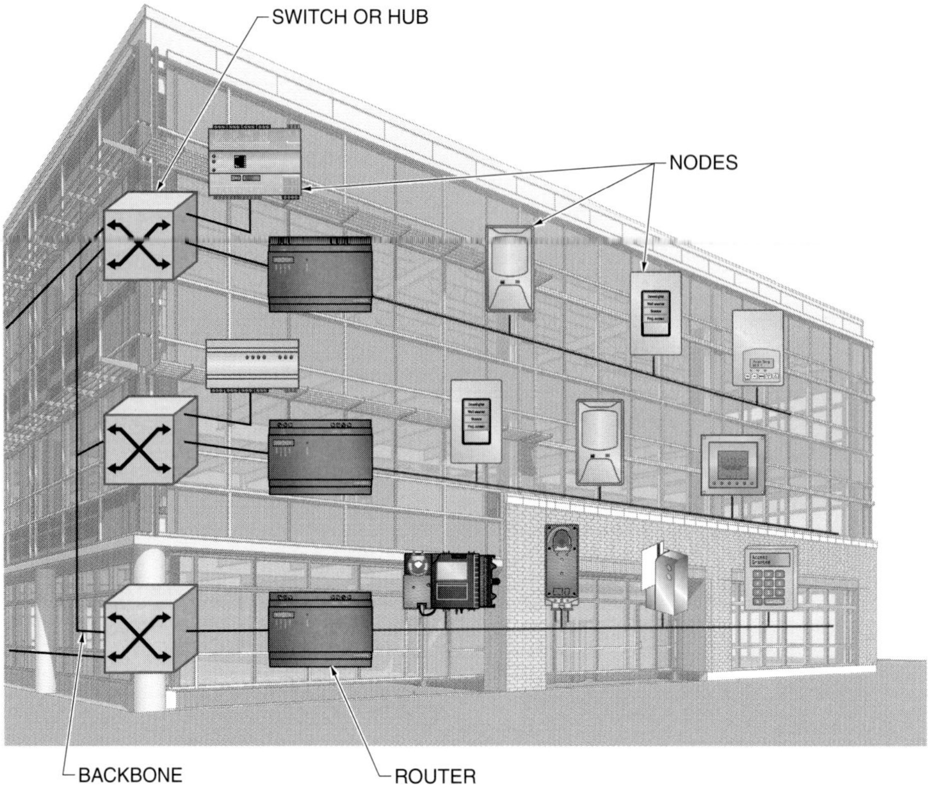

Figure 10-16. BACnet networks are typically arranged in a backbone architecture, which helps manage traffic and provides for future expandability.

Although it is possible to distribute the Ethernet backbone throughout the building, generally the cost of the Ethernet infrastructure makes the simpler wiring of LAN types such as MS/TP more attractive for use at the lowest node level. In those cases, the intrafloor network is further divided using BACnet routers, or controller-routers, into one or more MS/TP segments or ARCNET156 segments.

BACnet Nodes

There are so many possible BACnet nodes that there is no single device architecture. However, BACnet nodes share many of the same basic hardware and software components to provide both the control functionality of the node and its BACnet behavior. **See Figure 10-17.** The software is not visible in any way except to the original programmer, but it is useful to imagine the software in several parts, each of which is responsible for different aspects of the node's behavior. Each of the software components may require set-up information and parameters to customize or configure that part of the software.

Node Hardware. Every BACnet node has at least one embedded microcomputer, which consists of a central processor unit (CPU), memory, a LAN interface, and input/output circuitry. Very simple devices use microcontroller chips that integrate several of these functions into the same chip. Larger controllers typically use multiple chips integrated on a printed circuit board.

Three different types of memory provide information storage for the node. Program memory for storing the software application that operates the node is typically flash memory or EPROM, which is nonvolatile (retains its contents without power). This type of memory is suitable for relatively few write cycles, but this is adequate for storing application programs, which rarely change.

BACnet Node Components

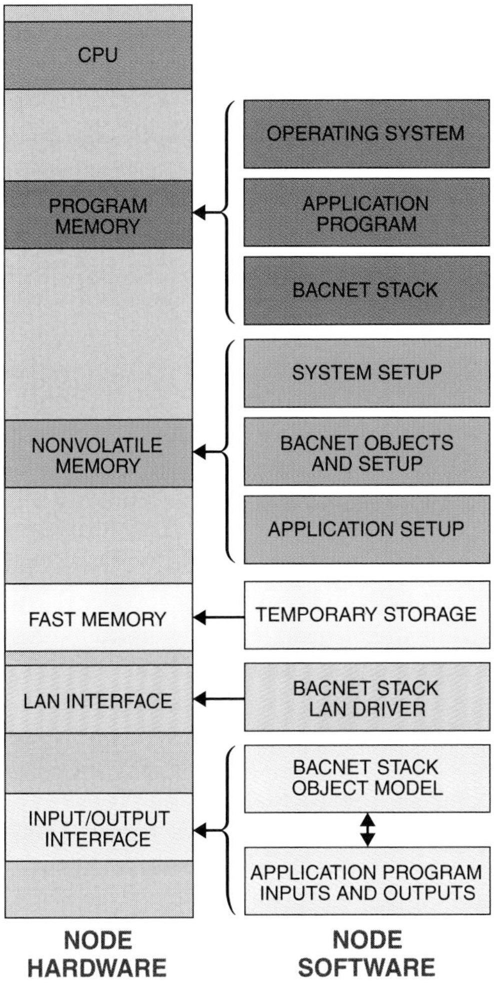

Figure 10-17. The hardware and software architecture of any BACnet node includes common components.

Another type of nonvolatile memory can be written quickly and repeatedly, making it better suited for storing the contents of the rapidly changing object properties. Therefore, it is separate from the program memory. This memory is typically EEPROM or static RAM with battery backup.

Finally, very fast RAM is used by the application program for temporary storage, buffering, and other functions. RAM is volatile, however, losing its contents when power is removed.

The LAN interface is associated with the LAN type supported by the node. The complexity of the interface depends on the LAN type. Ethernet 8802-3 requires a dedicated chipset with MAC and physical layer interface and RJ-45 connector, while MS/TP may be a simple UART with TIA-485 transceiver. Routers, or controllers with router functionality, usually have at least two LAN interfaces.

The input/output interface provides the connection to external hardware. For example, a lighting controller needs circuitry to be able to turn lights ON and OFF.

Operating System. Similar to a desktop computer, the operating system of a BACnet node manages the resources of its microcomputer: memory, CPU time, program execution, and low-level interface to input/output components. In simpler devices, this functionality is integrated into the application program so thoroughly that there is little distinction between the application and the system. In larger controllers, these functions tend to be much more separate as software entities.

Application Program. Some portion of the program memory is dedicated to storing the application program. The *application program* is the software that implements the node's functionality. The application program contains the control logic that makes control decisions based on input information and determines the appropriate output information. If the node is a lighting controller, then its application program implements lighting control. If the node is a VAV terminal box controller, then it performs the VAV control functions. Since it must interact with sensors and actuators, the application program usually manages the interface for the physical input and output ports that are used for monitoring and control.

BACnet Stack. The *BACnet stack* is the portion of the application software that manages BACnet protocol communication over the LAN. **See Figure 10-18.** This involves decoding and interpreting received BACnet messages, and encoding and transmitting BACnet replies and requests to other nodes. The transport component of the stack manages the physical sending and receiving of messages by controlling the LAN interface's initialization and operation, although some of that responsibility may be shared with an operating system.

BACnet Stack

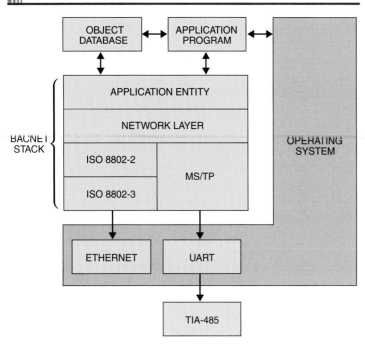

Figure 10-18. The BACnet stack includes the OSI layers used by BACnet to communicate over a network.

The Application Entity (AE) component of the stack is a "go-between" between the node's application software and the BACnet object model. The AE makes the application's private information publicly visible in the form of BACnet objects and properties. To do this, the AE must interpret requests for object property values and locate the appropriate internal information to satisfy each request. For example, if a node has a temperature sensor input, the AE represents this temperature as the Present_Value of an Analog Input object. If other BACnet nodes ask this node about its objects, the AE must report that it has an Analog Input object.

Network Tools

The BACnet standard does not address configuration or troubleshooting procedures, so several vendors have developed their own solutions. These network tools are software applications running on computers that are connected, either temporarily or permanently, into the BACnet network. **See Figure 10-19.** Some network tool software is proprietary, intended to work only with a certain manufacturer's nodes. Most of these network tool software packages include certain common functions and work with most nodes, though their interfaces may vary.

Some BACnet network tools have the ability to discover unknown installed nodes on an existing network and compile a list of their identifying information. Once this is known, the tool can then query nodes for a list of their objects. Basic configuration can often be accomplished using any BACnet client software tool that supports writing to object properties. Individual properties can be individually examined and changed, including the priority array of commandable properties. For day-to-day operations, some general-purpose network tools can monitor object properties in real-time.

For troubleshooting communication and control problems, network tools may include a network analyzer, also known as a sniffer. This software captures network traffic on its segment and saves every message frame for analysis. It can display both the actual hexadecimal code of the message and a human-readable interpretation, so that the various component fields can be analyzed in detail.

TESTING AND CERTIFICATION

After the publication of BACnet as an ANSI standard, the ASHRAE committee responsible for BACnet developed a new companion standard that defines procedures for testing conformance to BACnet in each of its interoperability areas. Experts from member organizations were involved in reviewing the testing standard, evaluating the software testing tools, and developing the test plans and operational procedures. The result is standard ANSI/ASHRAE 135.1, *Method of Test for Conformance to BACnet*, which is now also an ISO standard.

The BACnet Testing Laboratory (BTL) was created as an independent testing agency under the auspices of the BACnet Manufacturer's Association (BMA). BTL began accepting applications for product certification in

January 2001, and actual product testing has been going on since late 2001. Products tested by BTL and found to pass may display the BTL mark. **See Figure 10-20.** BTL is a not-for-profit independent testing agency, essentially managed by a trade organization, and has become the de-facto BACnet certification organization for the United States.

In Europe, the BACnet Interest Group – Europe (BIG-EU) formed a sister organization called WSPLab. The goal is to have reciprocity arrangements that allow certification by one body to be recognized by the other, making it easier for companies to have products tested and certified locally and marketed worldwide.

In 2006, BMA merged its operations with the BACnet Interest Group – North America to become BACnet International (BI). This organization now manages the BTL, which is still an independent agency, as is WSPLab.

Interoperability Criteria

Certification of interoperability involves testing against certain criteria. There are many kinds of interoperability, so the criteria must determine the type of interoperation and the corresponding actions of the node. Specifying interoperability from these criteria requires identifying the BACnet functionality of both the requesting side and the responding side of each interoperation.

When BACnet was initially introduced, the scheme for specifying interoperability requirements was complex and widely misunderstood. After several years, a new scheme was added to BACnet that uses five areas of interoperability. **See Figure 10-21.** This scheme introduces BACnet interoperability building blocks (BIBBs). A *BACnet interoperability building block (BIBB)* is a standardized name that is associated with some BACnet feature or capability. These BIBBs implement the functionality of various specific features that are organized in these five interoperability areas.

BIBBs always come in pairs. One BIBB defines the expectations for the requesting side of an interoperation. The second BIBB defines the expectations for the responding side of the interoperation. BIBB names are constructed using a shorthand notation that includes the interoperability area under which the BIBB is classified, the particular function the BIBB performs, and the letter "A" or "B," which identifies the node as the client (sender) or the server (responder) of the request. For example, within the Data Sharing interoperability area, the ReadProperty service is abbreviated "DS-RP." Therefore, a client must have DS-RP-A capability and a server must have DS-RP-B capability.

Network Tools

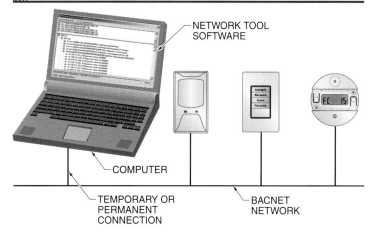

Figure 10-19. Network tools can be connected to the network, either permanently or temporarily, to access information in the nodes.

BTL Mark

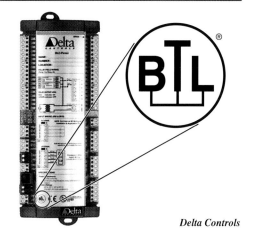

Delta Controls

Figure 10-20. Nodes that pass BACnet Testing Laboratory's (BTL) tests for conformance to the BACnet standard may display the BTL mark.

BACnet Interoperability

Interoperability Area	BACnet Interoperability Building Blocks (BIBBs)			
	Client Side		Server Side	
Data Sharing	ReadProperty-A	DS-RP-A	ReadProperty-B	DS-RP-B
	ReadPropertyMultiple-A	DS-RPM-A	ReadPropertyMultiple-B	DS-RPM-B
	ReadPropertyConditional-A	DS-RPC-A	ReadPropertyConditional-B	DS-RPC-B
	WriteProperty-A	DS-WP-A	WriteProperty-B	DS-WP-B
	WritePropertyMultiple-A	DS-WPM-A	WritePropertyMultiple-B	DS-WPM-B
	COV-A	DS-COV-A	COV-B	DS-COV-B
	COVP-A	DS-COVP-A	COVP-B	DS-COVP-B
	COV-Unsolicited-A	DS-COVU-A	COV-Unsolicited-B	DS-COVU-B
Alarm and Event Management	Notification-A	AE-N-A	Notification-Internal-B	AE-N-I-B
			Notification-External-B	AE-N-E-B
	ACK-A	AE-ACK-A	ACK-B	AE-ACK-B
	Alarm Summary-A	AE-ASUM-A	Alarm Summary-B	AE-ASUM-B
	Enrollment Summary-A	AE-ESUM-A	Enrollment Summary-B	AE-ESUM-B
	Information-A	AE-INFO-A	Information-B	AE-INFO-B
	LifeSafety-A	AE-LS-A	LifeSafety-B	AE-LS-B
Scheduling	Scheduling-A	SCHED-A	Scheduling-Internal-B	SCHED-I-B
			Scheduling-External-B	SCHED-E-B
Trending	Viewing and Modifying Trends-A	T-VMT-A	Viewing and Modifying Trends-Internal-B	T-VMT-I-B
			Viewing and Modifying Trends-External-B	T-VMT-E-B
	Automated Trend Retrieval-A	T-ATR-A	Automated Trend Retrieval-B	T-ATR-B
Device and Network Management	Dynamic Device Binding-A	DM-DDB-A	Dynamic Device Binding-B	DM-DDB-B
	Dynamic Object Binding-A	DM-DOB-A	Dynamic Object Binding-B	DM-DOB-B
	DeviceCommunicationControl-A	DM-DCC-A	DeviceCommunicationControl-B	DM-DCC-B
	PrivateTransfer-A	DM-PT-A	PrivateTransfer-B	DM-PT-B
	Text Message-A	DM-TM-A	Text Message-B	DM-TM-B
	TimeSynchronization-A	DM-TS-A	TimeSynchronization-B	DM-TS-B
	UTCTimeSynchronization-A	DM-UTC-A	UTCTimeSynchronization-B	DM-UTC-B
	ReinitializeDevice-A	DM-RD-A	ReinitializeDevice-B	DM-RD-B
	Backup and Restore-A	DM-BR-A	Backup and Restore-B	DM-BR-B
	Restart-A	DM-R-A	Restart-B	DM-R-B
	List Manipulation-A	DM-LM-A	List Manipulation-B	DM-LM-B
	Object Creation and Deletion-A	DM-OCD-A	Object Creation and Deletion-B	DM-OCD-B
	Virtual Terminal-A	DM-VT-A	Virtual Terminal-B	DM-VT-B
	Connection Establishment-A	NM-CE-A	Connection Establishment-B	NM-CE-B
	Router Configuration-A	NM-RC-A	Router Configuration-B	NM-RC-B

Figure 10-21. BACnet interoperability building blocks (BIBBs) are standardized, individual interoperations that BACnet nodes use to interact with other nodes.

Device profiles do not describe what a device does, only its interoperational capability. These profiles are used to generally describe the features of a node. This may often correspond to very similar node applications, but not always.

For example, a client node may need to know a certain temperature, but the sensor that measures that temperature is attached to a different node. This node may be from a different vendor, but if it has the necessary BIBB, the two nodes can operate together. **See Figure 10-22.** The client has DS-RP-A capability, so it can request the Present_Value of the temperature object with a ReadProperty request. The server has DS-RP-B capability, so it can respond with a ReadProperty reply message that includes the requested value.

Similarly, a client desiring to change a physical output, such as an actuator position, needs to write to a property of an object representing that output. In this case, the client must have the DS-WP-A (Data Sharing, WriteProperty, client) capability and the server must have the DS-WP-B capability. Therefore, each interoperation type has a pair of matching BIBBs that define the capabilities that must be supported by each party in order for that interoperation to succeed.

BACnet Interoperability

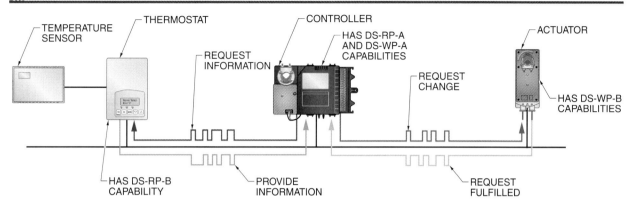

Figure 10-22. BACnet interoperability relies on standardized BIBBs to ensure that nodes from different manufacturers can communicate reliably.

This concept is particularly useful for selecting the interoperability functionalities that must be implemented for each kind of control device. A consulting-specifying engineer can write the appropriate BIBBs into the specification. A product developer can also use BIBBs to describe the capabilities of a BACnet node.

Device Profiles

While it is possible to write specifications that use BIBBs directly to describe the BACnet features of each node, this approach requires a thorough understanding of each BIBB and some expertise in applying them effectively. In practice, many nodes perform similar functions, at least in terms of BACnet interoperability. Therefore, it is practical to define generic device types that represent typical collections of interoperability capabilities.

Annex L of the BACnet standard defines six example device profiles, covering the most common types of nodes. A *device profile* is a collection of BIBBs that are required for the interoperable functionality of a general type of node. **See Figure 10-23.** Any node that satisfies the BIBB requirements for a certain device profile can claim to be a node of that type. These profiles can be used as a starting point for specifying nodes of similar functionality requirements. For each profile, the document defines which BIBBs must be implemented in each of the five interoperability areas.

The standard device profiles in Annex L are only examples of typical groupings of BACnet interoperability features, and are not the only possible or desirable kinds of BACnet nodes. Other device profiles can be defined as needed to describe many nodes of a similar type, such as in specification documents.

Device profiles do not describe a node's function, only its interoperation capability. For example, a lighting controller and an HVAC controller are very different in terms of function, but very similar from an interoperability perspective. They may both be simple BACnet servers with similar BIBB requirements.

Refer to Quick Quiz® on CD-ROM

BACnet node manufacturers can use device profiles as a way to easily describe the general capabilities of their products.

Device Profiles

BACnet Operator Workstation (B-OWS)	BACnet Building Controller (B-BC)	BACnet Advanced Application Controller (B-AAC)	BACnet Application-Specific Controller (B-ASC)	BACnet Smart Actuator (B-SA)	BACnet Smart Sensor (B-SS)
DS-RP-A, B	DS-RP-A, B	DS-RP-B	DS-RP-B	DS-RP-B	DS-RP-B
DS-RPM-A	DS-RPM-A, B	DS-RPM-B			
DS-WP-A	DS-WP-A, B	DS-WP-B	DS-WP-B	DS-WP-B	
DS-WPM-A	DS-WPM-B	DS-WPM-B			
	DS-COVU-A, B				
AE-N-A	AE-N-I-B	AE-N-I-B			
AE-ACK-A	AE-ACK-B	AE-ACK-B			
AE-ESUM-A	AE-ESUM-B				
AE-INFO-A	AE-INFO-B	AE-INFO-B			
SCHED-A	SCHED-E-B	SCHED-I-B			
T-VMT-A	T-VMT-I-B				
T-ATR-A	T-ATR-B				
DM-DDB-A, B	DM-DDB-A, B	DM-DDB-B	DM-DDB-B		
DM-DOB-A, B	DM-DOB-A, B	DM-DOB-B	DM-DOB-B		
DM-DCC-A	DM-DCC-B	DM-DCC-B	DM-DCC-B		
SM-TS-A	DM-TS-B or	DM-TS-B or			
DM-UTC-A	DM-UTC-B	DM-UTC-B			
DM-RD-A	DM-RD-B	DM-RD-B			
DM-BR-A	DM-BR-B				
NM-CE-A	NM-CE-B				

Figure 10-23. Device profiles define collections of BIBBs representing typical node types.

Summary

- BACnet was designed to apply to all types of building systems, including perimeter and object security, lighting, HVAC, and elevator systems.

- BACnet was developed by ASHRAE in an open consensus process.

- BACnet separates the goal of interoperability into two aspects: the standardization of the content of messages exchanged between nodes and the transportation of these messages.

- BACnet standardization of message content, known as the application language, is divided into objects and services. Objects are used to model all information within an interoperable BACnet node. Services are used by nodes to make requests of other nodes.

- BACnet provides flexibility in the media and LAN options for transporting messages, which does not affect the content of the messages.

- While BACnet provides interoperability by defining a common, abstract view of information, each vendor creates their own way to store and use that information within their products.

- All objects in BACnet consist of a set of properties. Depending on the object type, different types of properties may be needed to handle the different types of information.

- Standard object types are defined in the BACnet standard, including a description of the object and its expected use, standard properties, and behavior.

- A BACnet node can implement any number of objects of different types, but only the Device object is always required. Unlike other objects, the instance number of the Device object (and its Object_Name) must be unique across the entire BACnet internetwork.

- Proprietary properties can be added to otherwise standard object types or entirely proprietary object types can be created.

- Objects are identified within a node by a 32-bit number that contains two components: the object type and the object instance.

- A service precisely defines the rules of etiquette for a particular kind of question and reply.

- A service request typically includes several parameters, which vary depending on the particular service.

- BACnet networks can vary in size and scope from very small (only a few nodes) to extremely large (over four million nodes spread across many network segments).

- As LAN type options, BACnet allows Ethernet 8802-3, BACnet/IP, ARCNET, Master-Slave/Token-Passing (MS/TP), Point-to-Point (PTP), and LonTalk Foreign Frames. The most common LAN type in BACnet networks is MS/TP.

- BACnet systems may use more than one LAN type, though different LAN types must be interconnected with routers.

- Increasingly, BACnet systems are making use of a building's existing network infrastructure as a backbone to carry control system traffic, increasing the implementation of BACnet/IP.

- Every BACnet node has at least one embedded microcomputer, which consists of a central processor unit (CPU), memory, a LAN interface, and input/output circuitry.

- The application program contains the control logic that makes control decisions based on input information and determines the appropriate output information.

- The Application Entity (AE) component of the stack is a "go-between" between the node's application software and the BACnet object model.

- Most of BACnet network tool software packages include certain common functions and work with most nodes, though their interfaces may vary.

- The BACnet Testing Laboratory (BTL) is an independent testing agency that tests devices for conformance to BACnet in each of its interoperability areas.

- BACnet interoperability building blocks (BIBBs) are standardized functionalities of various specific features that are organized into five interoperability areas.

- BIBBs always come in pairs. One BIBB defines the expectations for the requesting side of an interoperation. The second BIBB defines the expectations for the responding side of the interoperation.

- Generic device profiles represent typical collections of interoperability capabilities.

Definitions

- An *object* is an abstract container that organizes related information and makes it accessible to other nodes in a standard way.
- A *service* is a formal procedure for BACnet nodes to make requests of other nodes.
- *Object-oriented modeling* is the concept of organizing many different types of information into defined and structured units.
- A *property* is one item of information about an object.
- A *standard object type* is an object type defined in the BACnet standard.
- A *datatype* is a specification for the type and format of information.
- An *object identifier* is a 32-bit number that is unique to each BACnet object within a node.
- An *object type* is a 10-bit code number in the object identifier that represents the kind of object.
- The *object instance* is a 22-bit code number in the object identifier that assigns a number to each individual object of a certain type within a node.
- The *Device instance* is the unique 22-bit code number of the Device object.
- A *property identifier* is a 32-bit code number that identifies a property in an object.
- The *application program* is the software that implements the node's functionality.
- The *BACnet stack* is the portion of the application software that manages BACnet protocol communication over the LAN.
- A *BACnet interoperability building block (BIBB)* is a standardized name that is associated with some BACnet feature or capability.
- A *device profile* is a collection of BIBBs that are required for the interoperable functionality of a general type of node.

Review Questions

1. How does BACnet define a framework that can be applied to all types of automation systems?
2. How is BACnet's application language used to standardize message content?
3. How do properties define the structure of an object?
4. What are the defining characteristics of standard object types?
5. How can manufacturers implement nonstandard functionality into nodes?
6. What information is contained in the Object_Identifier property?
7. What types of information are included in a service message?
8. Why have many BACnet vendors developed their own network tools?
9. How do BACnet interoperability building blocks (BIBBs) standardize node capabilities?
10. What are device profiles?

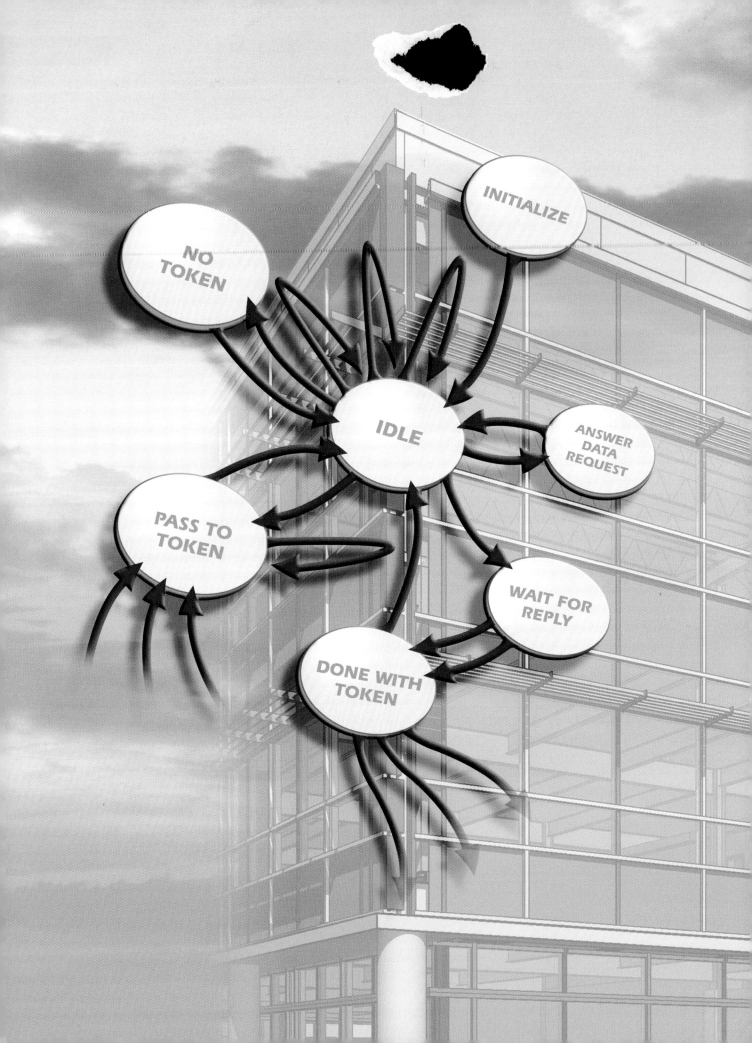

Chapter Eleven

BACnet Transports and Internetworking

BACnet technology provides for a variety of different network infrastructures and LAN types, which are treated as MAC layers by BACnet. Each type has certain strengths and weaknesses that make it appropriate for certain networking applications. When choosing a LAN type for a new building automation system, or when working with an existing network, several matters must be considered that may affect the efficient operation of the network.

Chapter Objectives

- Identify the typical physical and network infrastructures used for BACnet.
- Compare the LAN types that are compatible with BACnet.
- Describe the physical requirements for Master-Slave/Token-Passing (MS/TP) networks.
- Differentiate between the roles and responsibilities of MS/TP master nodes and slave nodes.
- Describe the fields that make up the BACnet network layer header.
- Compare the composition of network layer headers at different points along a routed path.
- Evaluate the solutions for routing and broadcasting BACnet messages on IP infrastructures.

TYPICAL BACNET PHYSICAL ARCHITECTURE

Building wiring for BACnet typically uses a tiered backbone. The backbone wiring runs through vertical risers in the core or corner of the building and connects network devices, such as switches or routers, on each floor. **See Figure 11-1.** From there, straight runs connect to BACnet nodes that are near the physical equipment to be monitored or controlled. In large buildings, wiring based on twisted-pair cabling further distributes communication wiring on each floor. Switches or repeaters are added to extend the segments as needed.

> BACnet can be implemented on a building's existing IP infrastructure, though this may require special devices.

Unlike some other protocols, BACnet requires that the path, or chain of segments connected by routers, never be circular. There must be only one route between any two nodes.

Older BACnet networks used 10BASE-5 or 10BASE-2 cabling for backbones and installed hubs on each floor. Most installations are now based on switches for distribution and use 100BASE-TX or 10BASE-T wiring to daisy chain them together. This architecture increases reliability, lowers wiring costs, and virtually eliminates issues of end-to-end network distance and collisions because all node-to-switch and switch-to-switch connections are point-to-point. The use of managed switches also aids maintenance, though they are significantly more costly.

BACnet Physical Architecture

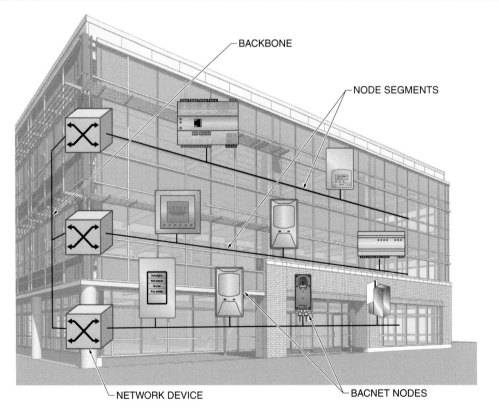

Figure 11-1. The typical BACnet physical architecture arranges network segments by building floor and connects them together vertically with network devices such as switches or routers.

Chapter 11 — BACnet Transports and Internetworking

When there are fewer nodes per floor, some Master-Slave/Token-Passing (MS/TP) or ARCNET segments are extended across multiple floors. This strategy uses fewer routing devices, but may degrade performance. Longer segments with more nodes can decrease network performance because the additional token passing between master nodes increases latency. In general, a design with fewer nodes per segment is the best practice. In addition to better response times, this design isolates network faults and simplifies troubleshooting.

BACNET LAN TYPES

BACnet uses a collapsed architecture that isolates the application messages of BACnet from the transport of those messages. Many LAN types can be used to transport messages between BACnet nodes. Each of these LAN types are treated as a MAC layer by BACnet and provide unique strengths as well as specific issues and limitations.

Six types of LAN technology can be used to transport BACnet messages: Master-Slave/Token-Passing (MS/TP), Ethernet, ARCNET, BACnet/IP, Point-to-Point (PTP), and LonTalk Foreign Frames. **See Figure 11-2.** MS/TP and PTP are unique to BACnet, but the other LAN types make use of other transport protocol standards. As a result, their MAC layers use additional link layer standards or methods to distinguish BACnet frames from other types of frames that might be conveyed using those same MAC layers on the same media.

Master-Slave/Token-Passing (MS/TP)

Master-Slave/Token-Passing (MS/TP) is unique to BACnet and is defined in the BACnet protocol standard. This MAC layer protocol can be used as either a master/slave network or a peer-to-peer token-passing network. When MS/TP is used by multiple cooperating master nodes, it implements token passing.

MS/TP uses TIA-485 signaling at speeds up to 76.8 kbit/s. MS/TP nodes are required to support 9600 bit/s, plus any combination of the data rates 19.2 kbit/s, 38.4 kbit/s, and 76.8 kbit/s. All of the nodes on a network must be configured to the same fixed data rate. MS/TP can theoretically work at higher speeds, and some nodes support 115 kbit/s, but reliable operation requires a dedicated chip or a large portion of CPU resources. Due to this added cost and complexity, other technologies are better choices at these higher speeds.

At 76.8 kbit/s and below, however, MS/TP is the lowest-cost solution for BACnet networks because it is designed for standard single-chip microcontrollers. Accurate MS/TP timing requires a resolution of only 5 ms, which is well within the capabilities of typical single-chip microcontrollers without adding additional hardware.

> Master nodes in a Master-Slave/Token-Passing (MS/TP) network pass a token (signifying permission to use the network) between each other like runners in a relay race pass a baton.

BACnet LAN Types

LAN	Advantages	Disadvantages
Master-Slave/Token-Passing (MS/TP)	Most popular, lowest cost	Lower performance
Ethernet	Popular, fast	Possible routing and broadcasting issues
ARCNET	156K variant better than MS/TP	2.5 Mbit/s variant uncommon, 156K variant costlier than MS/TP
BACnet/IP	Popular, fast, best choice for IP networks	Broadcast issues
Point-to-Point (PTP)	Best for dial-up	Direct line variant uncommon
LonTalk Foreign Frames	Media flexibility	Uncommon

Figure 11-2. The LAN types that can be used with BACnet each have advantages and disadvantages that affect their implementation.

Some LAN types that can be used by BACnet are commonly used for other types of data communications. The Master-Slave/Token-Passing (MS/TP) protocol at the data link layer, however, is unique to BACnet. It was designed specifically for communication between BACnet nodes and was specified in the BACnet standard. While not the fastest network choice, the MS/TP protocol provides efficient communication on a relatively inexpensive infrastructure.

Asynchronous Communication

A universal asynchronous receiver/transmitter (UART) is commonly used by microcontrollers to manage serial communications, such as telephone modem to computer connections over TIA-232. The functionality of a UART may be implemented in its own integrated circuit or as part of the microcontroller itself.

The transmission is asynchronous because the start of the frame is unpredictable and occurs without synchronization to any master clock signal. The line idles in a "1" condition and the transition from 1 to 0 signals the beginning of the frame. Most UARTs sample the signal at some multiple of the data rate and if it remains 0 for more than half of a 1-bit period, the frame is considered validly started. This allows discrimination against short, spurious, noise-induced transients that could otherwise appear to be valid start bits.

MS/TP requires shielded twisted-pair cable with nominal impedance of 100 Ω to 130 Ω, a conductor-to-conductor capacitance of less than 100 pF/m (about 30 pF/ft), a conductor-to-shield capacitance of less than 200 pF/m (about 60 pF/ft), and a maximum length of 1200 m (about 4000′) based on AWG18 wire. Wiring topology is a strict daisy-chain bus with no stubbing. In order to present an optimal balanced load, the extreme ends of the network must have 120 Ω termination resistors across their transceivers.

Message Frames. Most single-chip microcontrollers include a universal asynchronous receiver/transmitter (UART) that can transmit and receive serial data that is individually framed by start and stop bits. MS/TP uses this small-scale framing to enclose octets of data. Multiple octets are transmitted together, surrounded by other MS/TP fields, to form a complete message frame. **See Figure 11-3.**

All MS/TP frames are at least 8 octets in length, which is the size of the header. Two preamble octets (0x55FF) signal the start of a frame. These are followed by a 1-octet frame type code, a 1-octet destination MAC address, and a 1-octet source MAC address. MS/TP MAC addresses range from 0 to 254. The special address 255 (0xFF) is reserved as a broadcast address and is never used as a source MAC address.

Master-Slave/Token-Passing (MS/TP) Frame

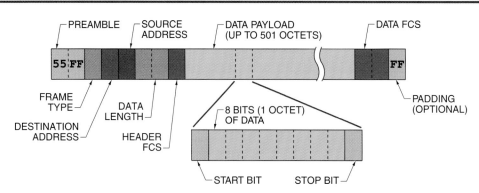

Figure 11-3. The data payload of MS/TP frames consists of individual octets of data surrounded by a start bit and a stop bit.

These octets are followed by a 2-octet data length, transmitted big-endian (most significant octet first). A data payload is optional, so a data length of 0 means that there is no payload. Otherwise, the value of the data-length field is the length of the payload, up to 501 octets. The data payload, if included, is followed by a 2-octet frame check sequence (on the data portion only) and a 1-octet end-of-frame padding of 0xFF. The eighth and last octet in the header is a frame check sequence (FCS) for the preceding seven octets.

The TIA-485 transmitter is enabled prior to or concurrent with the leading edge of the first start bit of the first octet of the message. **See Figure 11-4.** Ideally, the driver is disabled concurrently with the end of the stop bit of the last octet of the message. However, most UARTs cannot determine exactly when the last edge of a stop bit has exited, though even if they could, there may be a latency between the actual stop bit end and when the interrupt software can turn off the transmitter.

BACnet requires that this grace period be less than or equal to a time interval of 15 bits, so nodes may pad the end of a message with a single 0xFF octet, knowing that it may be truncated by a disabled transmitter. The choice of 0xFF is not arbitrary. It is the same value that would be expected if a real 0xFF was prematurely truncated in the middle by a return to the stop bit (idle) signal level.

Biasing. When no TIA-485 signaling drivers are actively asserted, the network floats in an indeterminate state, so receivers on the network may detect signal states (1 or 0) unpredictably and inconsistently. MS/TP networks use biasing to avoid these random signals. *Biasing* is the connection, at a transceiver, of network conductors to the reference voltage and ground through pull-up and pull-down resistors. **See Figure 11-5.**

Biasing

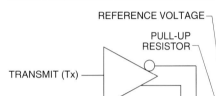

Figure 11-5. Biasing involves actively pulling up and pulling down a transceiver's network connection to provide consistent data levels.

MS/TP Frame Transmission

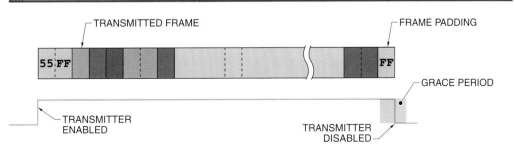

Figure 11-4. The end of a MS/TP frame transmission can occur anytime within a grace period of 15 bit times.

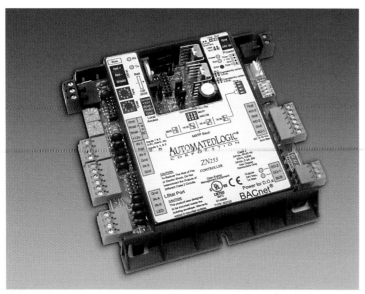
Automated Logic Corporation
Some BACnet nodes can be configured for either MS/TP or ARCNET communication.

Biasing can be done for an entire network segment or for individual nodes. A biased node provides bias for the entire network by using resistors with relatively low values. An unbiased node has no pull-up/pull-down resistors and depends on common network biasing. A locally biased node uses resistors with relatively high values that are sufficient to bias the node itself but are too weak to provide bias for the entire segment. The BACnet standard allows for all three types of node biasing. **See Figure 11-6.**

Some biasing implementations leave most of the nodes in an MS/TP segment unbiased, but instead bias the two nodes at the extreme ends of the segment with 510 Ω pull-up and pull-down resistors. The two biasing nodes have sufficient bias for the entire segment, even if one bias point is accidentally lost. Alternatively, some implementations require every node to use local biasing resistors of 47 kΩ. The BACnet standard indicates a preference for two biased nodes but allows local biasing.

Some node manufacturers build selectable termination and biasing features, usually by DIP switches or jumpers, into every node. **See Figure 11-7.** However, this adds to node costs and is unnecessary when biasing is only implemented in end nodes. Alternatively, external biasing terminators can be used.

Repeaters. The TIA-485 standard specifies that no more than 32 unit loads be present on any segment and models a unit load as a transceiver with an ideal impedance of 12 kΩ. Since newer transceivers are available with ½-load, ¼-load, and even ⅛-load characteristics, more than 32 MS/TP nodes may be allowed on a segment under some circumstances. Nodes must all use fractional unit load transceivers and none can be locally biased. Otherwise, repeaters must be used when more than 32 nodes are required on the segment. MS/TP repeaters are different from simple TIA-485 repeaters because they use a state-machine-driven algorithm for controlling the repetition of signals to ensure that the repeater turns around properly.

MS/TP Biasing Types

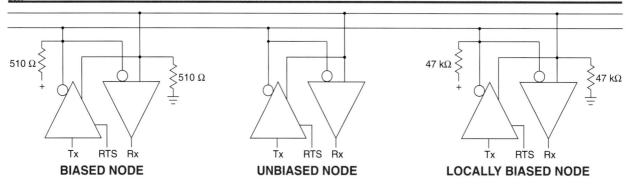

Figure 11-6. MS/TP nodes can be biased, locally biased, or unbiased. No more than two biased nodes can be used in any segment.

Selectable Biasing

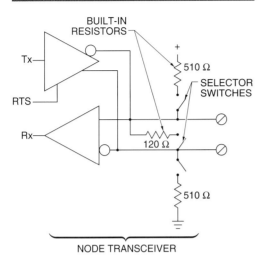

Figure 11-7. Node manufacturers may include selectable biasing and termination within the device that allows for different biasing configurations.

Isolation. When two or more isolated nodes are used together, their power supplies float with respect to each other. When the nodes are in separate buildings, problems occur if the ground potential in the buildings is different by more than a few volts. Therefore, the shield around the twisted-pair cable must be grounded at only one end of an inter-building segment and all of the MS/TP nodes must use optical and magnetic isolation with at least 1500 V isolation. **See Figure 11-8.** Because of lightning hazards, it is even better to use fiber optic media and compatible TIA-485 modems between buildings for optical isolation.

Ethernet 8802-3

An ISO 8802-3 "Ethernet" frame is sometimes still referred to as "802.3" after its IEEE designation. The Ethernet 8802-3 frame includes an 8-octet preamble and start-of-frame delimiter (SFD) field to mark the beginning of the frame. **See Figure 11-9.** This is followed by a 6-octet destination address, a 6-octet source address, a 2-octet payload length, up to 1500 octets of data payload, and a 4-octet frame check sequence (FCS). The FCS is used to detect errors in the frame caused by noise and collisions. The preamble, SFD, and FCS are managed by the Ethernet chip alone, so they are not typically visible to software or network debugging tools.

If the payload length field value is 0x0600 or larger, it is instead interpreted as a code for the protocol subtype of the payload. This indicates that the frame is an Ethernet Version 2 frame, also known as Ethernet II or DIX (for the original implementers of Ethernet: DEC, Intel, and Xerox). This is the oldest and most popular frame type. Other subtype codes are assigned for Ethernet protocols, such as 0x0800 for Internet protocol (IP).

With BACnet, the initial three octets of the Ethernet 8802-3 payload contain ISO 8802-2 data link layer headers, which provide for several types of conveyed link layer protocol data units (LPDUs). The link service access point (LSAP) number for BACnet is 0x82, which provides both the source service access point (SSAP) and destination service access point (DSAP) for BACnet. These are the first two of the three header octets. The effective data payload is up to 1497 octets (1500 octet maximum size for 8802-3, minus the 3-octet 8802-2 header).

> The link service access point (LSAP) for BACnet is the number 0x82, which is assigned by the Internet Assigned Numbers Authority (IANA).

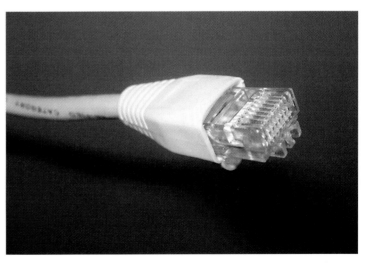

Ethernet over twisted-pair cabling uses 8-conductor modular connectors known as "RJ45," which look similar to common telephone connectors.

Isolation

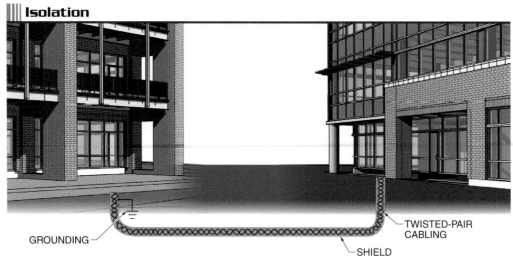
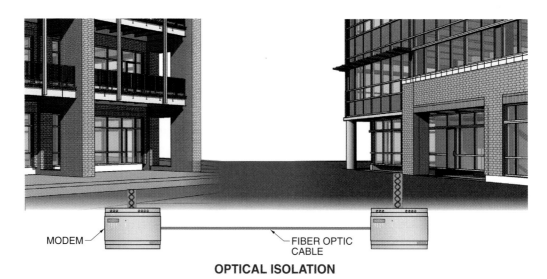

Figure 11-8. When MS/TP segments are connected between buildings, the twisted-pair shield must be connected to ground at only one side, or must use fiber optic cables for optical isolation.

Ethernet 8802-3 Frame

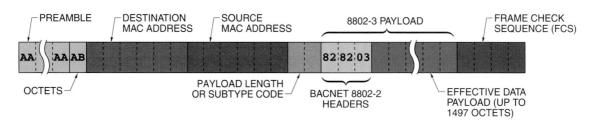

Figure 11-9. The Ethernet frame structure used for BACnet is based on the ISO 8802-3 standard and also includes the ISO 8802-2 header information.

Virtually any of the available Ethernet media types can be used for BACnet, although 100BASE-TX and 10BASE-T are the most common. Ethernet is most commonly used in a star topology based on hubs or switches. These configurations allow up to 100 m of CAT5 wiring from node-to-switch or switch-to-switch, and relatively high speeds of 10 Mbit/s, 100 Mbit/s, or 1000 Mbit/s. **See Figure 11-10.** Ethernet's compatibility with other media types, such as fiber optic cable and radio frequency, make it a popular choice for building-wide and interbuilding networks. Ethernet requires more complex bridging to be used over wide areas.

Ethernet Media Types

Designation	Speed*	Cable Type
10BASE-5	10	RG-8X coaxial (5 conductor)
10BASE-2	10	RG-58 coaxial (2 conductor)
10BASE-T		twisted pair (4 pairs, 2 pairs used)
10BASE-FL		fiber optic
100BASE-TX	100	CAT5 twisted pair (4 pairs, 2 pairs used)
1000BASE-T	1000	CAT6 twisted pair (4 pairs, 4 pairs used)

* in Mbit/s

Figure 11-10. *The media types used with Ethernet affect the speed of the network.*

The cost of the Ethernet chip technology is low, but the requirements for hubs, switches, and repeaters can add significantly to the installed cost of a complete system. Still, Ethernet is widely deployed and provides very high performance for a relatively modest cost. It is common for larger controllers to support Ethernet, although it is still impractical for very small controllers. Smaller controllers tend to have memory constraints, so their implementations of Ethernet are limited in performance by their available buffering (storage capacity for incoming messages). Even controllers that implement Ethernet for its connectivity benefits may not receive the full benefit of higher speed unless sufficient buffering memory is available.

ARCNET

ARCNET is a deterministic token-passing technology. An ARCNET frame includes a 6-bit preamble field and a 1-octet start-of-header (SOH) field. **See Figure 11-11.** This is followed by a 1-octet source address, a 2-octet destination address (a 1-octet address repeated twice), a 2-octet payload length, up to 508 octets of data payload, and a 2-octet frame check sequence (FCS) for error detection and correction. The preamble, SOH, and FCS fields are managed by the ARCNET chip alone. The SOH, source, and destination octets are each preceded by a 3-bit header of "110." Like Ethernet, BACnet over ARCNET includes an 8802-2 frame header, so the effective payload is limited to 505 octets. In practice, ARCNET is limited further to 501 octets in order to be consistent with PTP and MS/TP.

ARCNET Frame

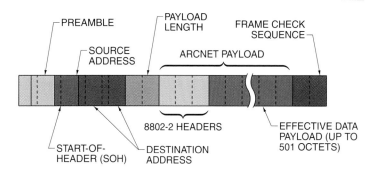

Figure 11-11. *The ARCNET frame structure has an effective data payload of up to 501 octets.*

ARCNET may be used with a variety of types of physical media in either bus or star topologies, and virtually any of these can be used for BACnet. **See Figure 11-12.** Unlike Ethernet, which uses a choice of fixed speeds, ARCNET is scalable to arbitrary speeds as high as 10 Mbit/s. However, in building automation applications, ARCNET is only used in two forms: ARCNET 156K (156 kbit/s) and ANSI/ATA-878.1 (2.5 Mbit/s).

ARCNET 156K. The most popular form of ARCNET scales the speed down to 156 kbit/s and uses relatively inexpensive TIA-485 transceivers.

The resulting media can be used over twisted-pair cabling for up to 1000 m (about 3300′). This version of ARCNET is widely deployed and provides good performance for a moderate cost. However, because all forms of ARCNET require a dedicated chip, there is a cost premium for smaller controllers, nearly on par with Ethernet. In some instances, the benefits outweigh the costs. Compared to MS/TP-based controllers running at 76 kbit/s, one manufacturer claims the throughput of the ARCNET 156K controllers to be nearly 40 times faster, even though the rated bandwidth is only twice as fast.

ARCNET Media Types

Speed	Cable Type	Topology	Maximum Segment Length‡
2.5*	RG-62/U coaxial	bus	305
		star	610
	CAT3 twisted pair	bus	122
		star	100
	fiber optic	star	2000
156†	CAT3 twisted pair	bus	1000

* in Mbit/s
† in kbit/s
‡ in m

Figure 11-12. The media type and topology used to build an ARCNET network affects its speed and maximum segment length.

ANSI/ATA-878.1. The 2.5 Mbit/s form of ARCNET is also known as ANSI/ATA-878.1. This form requires relatively costly transformer-coupled transceivers, and distances of individual segments are limited. ANSI/ATA-878.1 typically uses coaxial cable, although other media can also be used. The falling cost and high performance of Ethernet have reduced the implementation of this technology, and it is increasingly rare.

Point-to-Point (PTP)

Point-to-point (PTP) is a unique BACnet solution for dialed internetwork communications. PTP provides special features that facilitate support for modern modem protocols, notably V.32bis and V.42. These standards allow communications of up to 56 kbit/s with compression over normal voice-grade telephone lines. PTP can also be used over direct-cable connections using TIA-232 signals. In this mode, the speed may be as high as 115 kbit/s.

Arguably, PTP has been eclipsed by dial-up Internet protocols, but it is far more efficient (faster) to use BACnet-over-PTP than BACnet-over-BACnet/IP-over-PPP. Direct-cable PTP has become rare because most BACnet manufacturers prefer to support MS/TP (for lower costs) or Ethernet or BACnet/IP (for better speed and convenience).

BACnet/IP

An Internet protocol (IP) infrastructure is an internetwork composed of multiple segments whose nodes employ one or more IP-based protocols interconnected by IP-based routers. Internet protocols are in widespread use. Nearly every academic and corporate network is based on the TCP/IP suite, which includes many components and applications.

However, IP routers may interfere with BACnet-over-Ethernet messaging, particularly when BACnet 8802-3 and IP-based Ethernet nodes share the same physical segments. BACnet/IP is a form of BACnet that uses one of the standard IP components, the User Datagram Protocol (UDP), in order to transmit messages over IP. BACnet can then treat this option as an alternative MAC layer. This has the benefit of preserving the Ethernet media types, which can be used with IP protocols, and adding some of the special capabilities afforded by IP technology, such as wide area networking (WAN) and virtual private networking (VPN).

BACnet/IP addresses consist of the IP address and the BACnet UDP port number, which is normally 0xBAC0 (47808). **See Figure 11-13.** The UDP payload contains a BACnet Virtual Link Layer (BVLL) that defines one of several functions or message types and their parameters. The BACnet Virtual Link Control (BVLC) header begins with 0x81 followed by a single-octet function code and a 2-octet length, transmitted big-endian. The most common message types are Original-Unicast-NPDU and Original-Broadcast-NPDU, whose payloads are the same NPDU that would normally be the payload of a BACnet MAC layer. The BVLL also has layer management messages identified by BVLC function codes.

LonTalk Foreign Frame

LonTalk is a proprietary technology originally developed by the Echelon Corporation, which has evolved into the standard ANSI/CEA-709.1 (now also ISO/IEC 14908). Like ARCNET and Ethernet, LonTalk has a data link and physical signaling LAN implemented in a special chip. LonTalk provides a lot of flexibility in terms of choosing media, including both wired and wireless LANs. LonTalk also allows scaleable speeds up to 1.25 Mbit/s (depending on media). At lower speeds (below 150 kbit/s), LonTalk is relatively inexpensive, comparable to MS/TP. At higher speeds (above 150 kbit/s), LonTalk becomes a relatively expensive LAN choice.

LonTalk has a multilayer protocol stack architecture with transport and application components. The transport component of LonTalk, which is similar in concept to IP and BACnet's network layer, can transport non-LonTalk frames. This allows LonTalk implementations that use a proprietary application layer (or upper layers) but use LonTalk transports. Such transports are called LonTalk Foreign Frames. BACnet allows the use of LonTalk Foreign Frames as a transport for BACnet messages, essentially treating a LonTalk Foreign Frame as a MAC layer, though this type of implementation is rare.

MS/TP NODES AND TOKEN PASSING

MS/TP is the most commonly implemented LAN type for BACnet networks. It also involves a unique way of managing communication that is important for installers and troubleshooters to understand.

BACnet/IP Frame

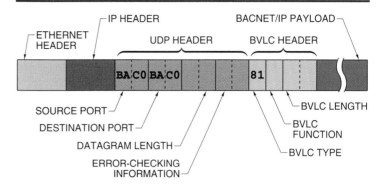

Figure 11-13. The BACnet/IP frame structure encloses the network layer header and application payload inside a BVLL, which is enclosed in a UDP frame.

Master Nodes

The most common usage of MS/TP is between peer nodes, which all have equal status and rights to use the network to communicate. A *master node* is an MS/TP node that shares equal responsibility for administering the network by taking turns to use the network and control the access to the network by other nodes. Mastership is shared by passing a token sequentially around the network.

The usage of the network by a master node is complex and can be divided into several distinct phases of activity: listening, replying, talking to another node, finding other master nodes, and sharing control of the network. The states of the master node are defined by these activities, which are governed by a set of rules for the behavior of the master node. **See Figure 11-14.** An examination of the behavior of master nodes is useful for understanding how MS/TP works.

Token Passing

The fundamental idea behind a token-passing access scheme such as MS/TP is that at any point in time, only one node is in charge, controlling the usage of the network segment. Master nodes share this right among themselves in a cooperative manner. The token holder is the node that has ownership of the segment at any moment.

Master Node States

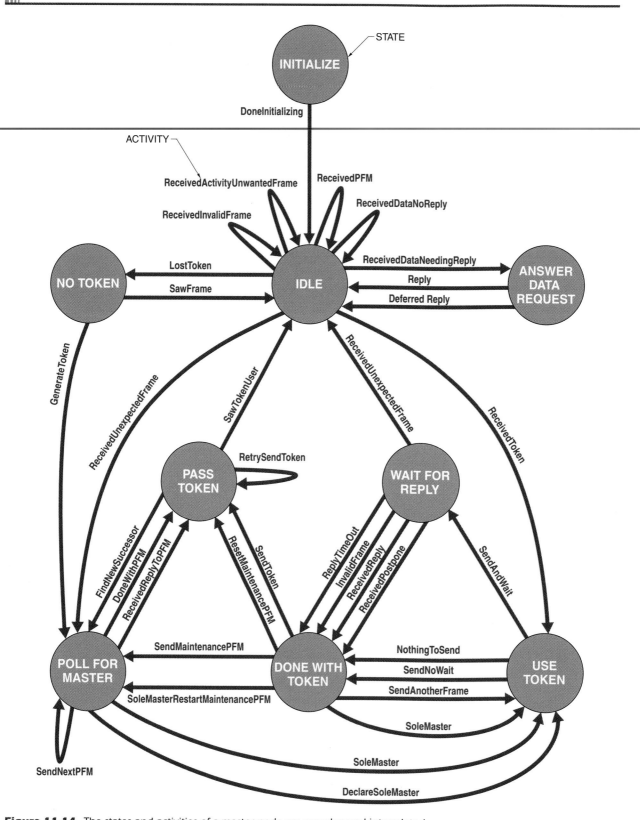

Figure 11-14. The states and activities of a master node are complex and interrelated.

Chapter 11 — BACnet Transports and Internetworking

State Machines

Some processes are hard to explain because they depend on conditions that occur at unknown future times. A familiar example is the protocol that drivers use with traffic lights. In order to avoid accidents, every driver must follow some simple rules:
- If the light is green, the driver may proceed.
- If the light is red, then the driver must stop and wait for green.
- If the light is yellow, the driver must prepare to stop, unless it is unsafe to do so.

These rules dictate a sequence of steps that must be followed in a certain order, and they may require waiting at each step until specific conditions exist in order to proceed.

One way to describe this kind of state-to-state criteria is to use a state diagram. A state diagram shows each possible state and describes the conditions that govern the transition from one state to the next. State diagrams are often used to explain complex state-oriented behavior for protocols of all kinds and are used extensively in BACnet. While it is not critical to understand all of BACnet's rules in this way, it is helpful to visualize some of BACnet's complex mechanisms using state diagrams. A state machine is a program that follows the rules captured in a state diagram in order to implement a particular behavior.

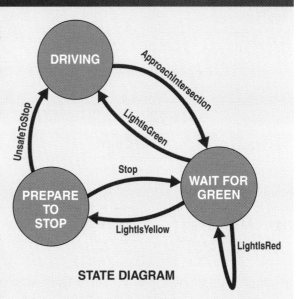

STATE DIAGRAM

Normally, token passing is simple, since most nodes do not have anything to transmit. Nodes continuously pass the token until one needs to transmit a message, such as a Data Expecting Reply (DER) frame, which is responded to after a brief delay by a Data Not Expecting Reply (NER) frame from the receiving node. **See Figure 11-15.** Then the token passing resumes. The current master node sends a Token Pass (TP) to the next node in the sequence, which can then either send a message or forward the token on to the next node. In a busy network, any given node often has one or more messages waiting to be transmitted, so each token pass results in the sending of DER or NER messages.

A node may have more than one message waiting to be transmitted. Master nodes have a MaxInfoFrames parameter that governs how many messages each node is allowed to send for each turn. This setting can be different for different nodes, though most have MaxInfoFrames equal to 1. If MaxInfoFrames is greater than 1, the node is permitted to send up to the value of MaxInfoFrames. Multiple messages may still require individual replies.

Token Recovery

When the token is lost, such as after a power failure, the network is idle and no node is communicating. The master nodes all share responsibility for getting the token passing started again. An integral part of this recovery is knowing how many potential master nodes exist on the segment. MS/TP allows up to 128 master nodes, even though the MAC address allows up to 254 total nodes. Typically, segments are far smaller, so the parameter Max_Master (known to each node) indicates the largest-used master node MAC address in that segment. Some nodes set Max_Master to a default of 127, but in nodes where this setting can be adjusted, it matches the largest MAC address that is expected to be used (though this may be higher than the largest MAC address currently in use).

Each master waits for 500 ms, plus its MAC address multiplied by 10 ms, before attempting token recovery. Therefore, if there is a MAC 00, it is the first to try token recovery after 500 ms of silence. MAC 01 waits for 510 ms, MAC 02 waits for 520 ms, and so on. In this way, the lowest numbered master MAC address always restarts the token passing.

Token Passing

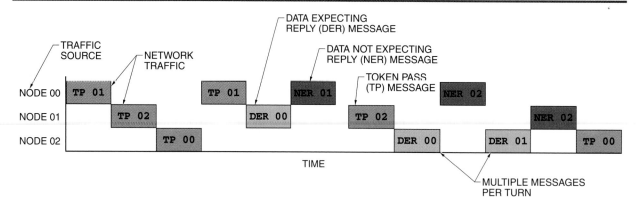

Figure 11-15. Master nodes pass tokens among each other in a highly organized way.

In order to know which node to pass the token to, each node must discover the MAC address of its nearest master neighbor, which is typically its own address plus 1, while taking into account the Max_Master value. The formula for addressing the next master node in the sequence is as follows:

$$MAC_{rcv} = (MAC_{send} + 1) \text{ modulo } (Max\_Master + 1)$$

where

MAC_{rcv} = MAC address of next sequential master node

MAC_{send} = MAC address of sending master node

Max_Master = highest MAC address of all master nodes

The token-recovering master node issues a PollForMaster (PFM) message to the next node addressed based on the formula. As soon as that MAC address answers with a ReplyTo-PollForMaster (RPFM) message, the polling node passes the token on to it using a Token Pass (TP) message.

Modulo

Modulo is a mathematical operation that results in the remainder of division of one number by another. For example, 7 modulo 3 = 1 because 3 divides into 7 twice, with 1 leftover. Likewise, 9 modulo 3 = 0, 8 modulo 3 = 2, and 3 modulo 4 = 0. The modulo operation is commonly abbreviated "mod" or "rem" (remainder), or represented by the symbol %.

For example, a segment of three nodes has its Max_Master parameter set to 3, though the highest MAC address is 02. **See Figure 11-16.** Node 00 starts the token recovery sequence and uses a PFM message addressed to Node 01 ([00 + 1] mod [3 + 1] = 1). When it receives a reply, it passes the token to Node 01. Node 01 uses PFM to find its successor, Node 02, and passes the token to it. Node 02 tries to send a PFM message to Node 03 ([02 + 1] mod [3 + 1] = 3) but gets no response, so after a waiting period, it tries the next higher MAC ([03 + 1] mod [3 + 1] = 0), sends a PFM to Node 00, and finally gets a RPFM reply.

This PollForMaster cycle occurs every 50 token passes to allow nodes to reenter the token-passing ring. Because of this behavior, it is best not to set Max_Master too large or leave big gaps between MAC addresses since this adds to the PFM cycle overhead and slows down the overall network response.

Timing Issues

Timing is an important part of token-passing networks. A master node listens for a token to be used within a certain time interval before giving up and passing it along to the next higher MAC address. **See Figure 11-17.** This parameter is the usage timeout interval ($T_{usage\_timeout}$). The receiving node must answer within 15 ms, plus a 5 ms grace period. Therefore, the usage timeout interval must be at least 20 ms, but, in some implementations, it may take as long as

100 ms. This parameter setting greatly affects network response. If the usage timeout interval is as long as 100 ms, then it significantly delays nodes with a Max_Master that is larger than the actual number of master nodes in use. This is because the maintenance PFM cycle occurs every 50 token passes, or every token pass for situations where the Max_Master is greater than the highest numbered MAC address in use. However, if the usage timeout interval is set to a strict 20 ms, then there could be interoperability problems with any tardy nodes on the same network whose MAC address is larger than the strict node.

One solution is to keep track of the reply times for RPFM messages and individually adjust the usage timeout intervals based on their responsiveness. However, this is complex and requires RAM to remember timings for each node, so it may not be appropriate for smaller nodes. Also, there is no procedure like this defined in the standard.

When a master node sends a message that requires a response, such as a DER or PFM, the responding node must not begin transmitting until at least 40 bit times have elapsed after the end of the last stop bit of the last octet transmitted in the request. This turnaround interval ($T_{turnaround}$) parameter value represents four 10-bit octet frames worth of time to account for a possible padding octet and turnaround time in repeaters that may exist between sender and receiver. This important time delay can cause intermittent message collisions if one or more nodes fail to properly enforce the timing requirement.

MS/TP Timing

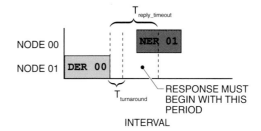

Figure 11-17. Minimum and maximum timing parameters affect the responsiveness of the MS/TP network.

When a master node sends a DER, it must wait for a time interval at least equal to the reply timeout interval ($T_{reply\_timeout}$) parameter for the responding node to begin answering before giving up. When a node receives a DER, it has up to 255 ms to create the reply and begin transmitting before its permission to transmit expires. However, it is best for a node to reply as soon as possible, preferably within 30 ms, because long delays unnecessarily monopolize the network.

Token Recovery

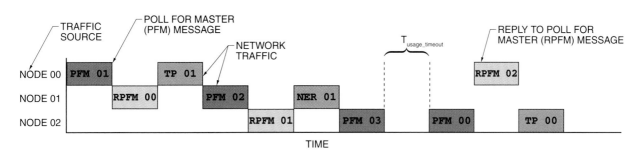

Figure 11-16. Token recovery involves nodes sending each other PollForMaster (PFM) messages.

A master node has the option of postponing the reply until a later time if it is unable to provide the reply immediately and sends a ReplyPostponed (RP) message. **See Figure 11-18.** After a number of token-passing cycles, when a response becomes available, it can be sent anytime that the responding node has the token.

Slave Nodes

Although it is common to have MS/TP segments using only master nodes, it is also possible to have slave nodes. A *slave node* is an MS/TP node that cannot participate in token management or recovery. A slave node can only reply to DER frames sent directly to it or act on NER frames that are sent. Consequently, the states for slaves are much simpler. **See Figure 11-19.** Because they cannot initiate NER frames, slave nodes cannot reply to dynamic device binding requests or be BACnet clients.

Slave nodes can be assigned any MAC address from 0 to 254, and slave and master nodes can be intermixed with respect to their position on the segment and MAC address. For timing reasons, though, it is more efficient to make all master MAC addresses sequential and to start at 0 without gaps.

It is possible to implement a single master node with many slaves, as long as the slave limitations are acceptable to the application. Such a network can be fast because there is no overhead from token passing.

Slave Proxy

A *slave proxy* is an MS/TP master node that acts on behalf of slave nodes on its segment for dynamic device binding. **See Figure 11-20.** Often a slave proxy is also a BACnet router to MS/TP. A slave proxy scans the MS/TP nodes individually using the ReadProperty service. Since initially the Device instance is unknown, the slave proxy uses the wildcard Device instance and reads the Object_Identifier property in order to find out the Device instance of each slave. Once the Device instance and MAC address correspondence is known, the slave proxy can begin to act as a proxy on the slave's behalf.

To support slaves that cannot respond to a wildcard Device instance (pre-2004 nodes), there is a capability for manually configuring which nodes are slaves in the proxy node. Either way, the proxy can read the Protocol_Services_Supported property to find out if Who-Is is supported. Only master nodes can support Who-Is, so the absence of this support indicates a slave node.

After the slaves are identified and their device instances are established, the slave proxy can then act for those slaves when Who-Is requests arrive. In that case, the proxy tries to determine if the range of the Who-Is corresponds to any of the device instances of any of the slave nodes. If so, then the proxy answers with I-Am and the slave's Device instance but uses the MAC address of the slave as the source MAC address.

Postponed Replies

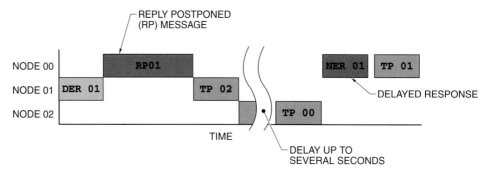

Figure 11-18. Nodes can choose to postpone replies to messages until a later time when they have the token again.

Slave Node States

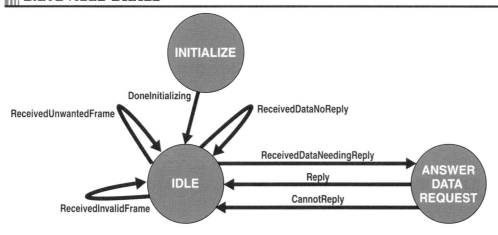

Figure 11-19. A slave node has much simpler states and responsibilities than a master node.

Slave Proxies

Figure 11-20. A MS/TP slave proxy is a master node that automatically finds MS/TP slaves and answers dynamic device binding requests on their behalf.

BACNET NETWORK LAYER

Regardless of the MAC layer technology being used, its payload is always used in BACnet to contain a network layer protocol data unit (NPDU). The NPDU is the envelope that directs an application message payload to its intended recipient.

Network Protocol Header

All messages transmitted between nodes, even those on the same physical segment, include network layer information. However, the header in the BACnet network layer varies in length depending on the circumstances. **See Figure 11-21.** The network layer header information collapses to a very small length for local messages between nodes on the same segment, but grows as needed for messages that require routing information between segments. All but the first two octets are optional.

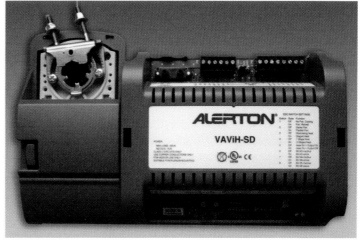

Alerton

All BACnet nodes have the ability to communicate their control information onto the common network.

BACnet Network Layer

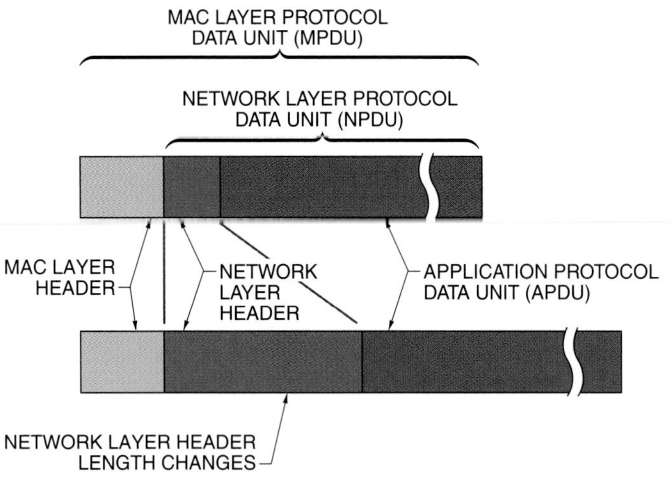

Figure 11-21. The network layer header for BACnet MPDUs can expand or contract as needed for different types of messages.

Local Addressing

The BACnet network layer header is expanded or contracted by routers as needed to accommodate just enough addressing information so that the message reaches its destination. For example, if the destination resides on the same segment, the network numbers are not necessary.

The envelope analogy still applies. For messages between two people in different cities, the envelope must include city, state, and postal code information. However, for a message between two people within the same office, much less routing information is needed. Since the correspondence is local, much of the addressing information is left out. Perhaps all that is needed are the names of the sender and receiver.

The network layer header consists of two mandatory octets and many conditional octets, which are present or absent depending on the type of message. **See Figure 11-22.** The first octet indicates the version of the BACnet protocol, which equals 0x01, and the second octet includes information on whether subsequent fields are included in the header. Conditional fields include destination fields, source fields, message type, and vendor ID.

Network Protocol Control Information (NPCI). The second octet is the network protocol control information (NPCI). The eight bits of this control octet indicate the presence or absence of other portions of the network layer header. **See Figure 11-23.** When a message is sent between two nodes on the same network segment, most of this routing information is not needed, so the network layer collapses to only the version and NPCI octets.

Network Layer Header

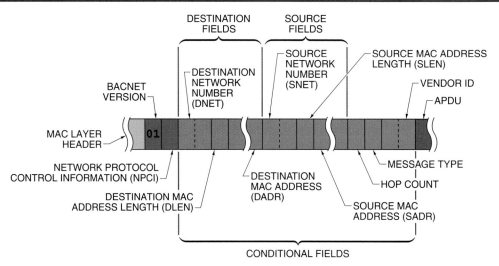

Figure 11-22. The network layer header must include the first two fields, but all others are optional, depending on the type of message.

Network Protocol Control Information (NPCI)

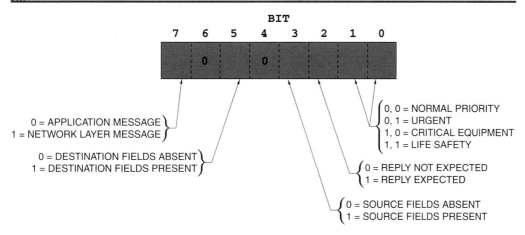

Figure 11-23. The NPCI octet has individual bits that indicate the presence or absence of the other fields in the network layer header.

- Bit 7 (the most significant bit) indicates the type of message. If bit 7 is set to 0, then the message is an application message, which is the most common. These messages include an APDU payload. If bit 7 is set to 1, the message is a network layer message. These kinds of messages are transmitted between routers and in special circumstances by BACnet client devices.
- Bit 6 is reserved for future use and is set to 0.
- Bit 5 indicates whether the destination fields and Hop Count are included in the network layer header. If bit 5 is set to 0, the destination fields are absent. If bit 5 is set to 1, the destination fields are present.
- Bit 4 is reserved for future use and is set to 0.
- Bit 3 indicates whether source fields are included in the network layer header. If bit 3 is set to 0, the source fields are absent. If bit 3 is set to 1, the source fields are present.
- Bit 2 is set to 1 if the message expects a reply. This is only used by MS/TP MAC layers, which need to know whether to wait for a response frame before passing the token.
- Bits 1 and 0 indicate a code for message priority. The network layer protocol orders the processing of incoming messages so that if traffic is high, higher-priority messages are handled first. This allows critical messages such as fire or life-safety messages to pass through routers even if the router has backed up traffic awaiting delivery.

Destination Fields. When the destination bit (bit 5) is set to 1, then the NPCI octet is followed by the following destination fields: DNET (destination network number), DLEN (length of destination MAC address), DADR (destination MAC address), and Hop Count.

DNET is always two octets in length and specifies the destination BACnet network number, in big-endian form. Every BACnet network segment is assigned a unique network number. This allows messages on one network segment to be routed to their correct destination. **See Figure 11-24.** The special DNET address 0xFFFF is a global broadcast, which is sent to all BACnet network segments.

DLEN is a single octet and contains the length of the DADR field, in octets. A DLEN of 0 means a broadcast destination MAC address, in which case there are no DADR octets. For example, if the destination MAC address is on Ethernet 8802-3, then the DLEN would be 6 because 8802-3 MAC addresses are 6 octets long.

DADR is the ultimate MAC layer address of the destination. The length of this field is variable and is specified by the preceding DLEN field.

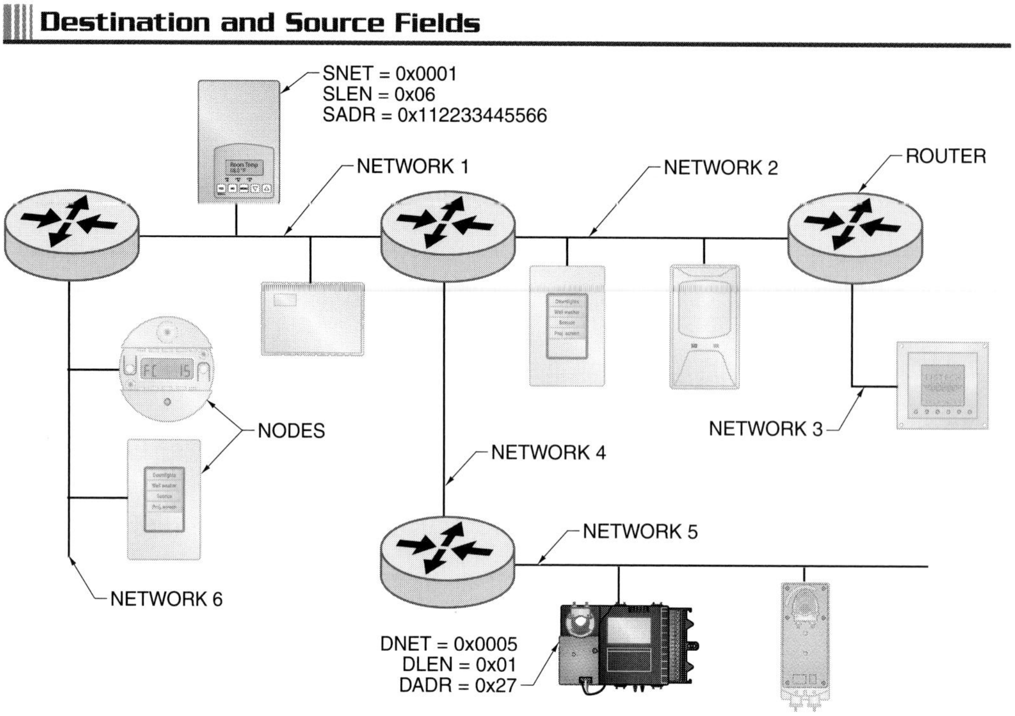

Destination and Source Fields

Figure 11-24. Each BACnet network segment is assigned a unique network number so that messages on one network segment can be routed to messages on other network segments.

Source Fields. When the source bit (bit 3) is set to 1, then the NPCI octet is followed by the following source fields: SNET (source network number), SLEN (length of source MAC address), and SADR (source MAC address). These fields are analogous to the destination fields. SNET is the 2-octet network number of the source node. SLEN is the 1-octet length of the source node's MAC address, which is represented in SADR. Unlike DLEN, a SLEN equal to 0 is invalid.

Hop Count. The Hop Count field is included if the destination fields are present. If source fields are also present, Hop Count follows the SADR field in the message frame. *Hop Count* is the number of segments that the message is allowed to traverse before reaching its destination. The count decreases by 1 each time the message is repeated in a router onto another segment. Hop Count is set to 255 when the message is first sent. Therefore, a message can traverse up to 255 segments before being dropped. This feature prevents messages from circulating forever if a misconfiguration causes a circular network path. Although not part of the standard, some routers allow the configuration of the Hop Count steps, which further limits the depth of the network and the lifetime of messages.

Network Headers in Transit

The network layer header is changed as the message is passed from segment to segment by routers. First, to send a message, a node needs to know the BACnet network number of the destination node and its MAC address on its network segment. **See Figure 11-25.** When the message is sent, its network layer header NPCI control octet includes the DNET, DLEN, DADR, and Hop Count fields. DNET contains the destination's network number and DADR contains the destination's MAC address.

If the message is repeated by a router onto an intermediate network, the router adds source routing information so that the final recipient has the network and MAC address of the message's origin. This new message contains an NPCI with the control octet bit 3 set to 1 to indicate the presence of source information: SNET, SLEN, and SADR.

Network Headers in Transit

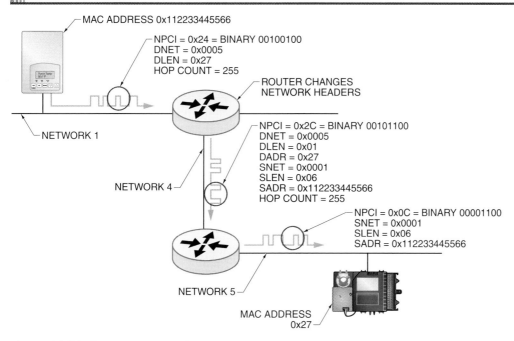

Figure 11-25. The composition of a network layer header is changed by routers as the message is passed from segment to segment.

When the message is repeated by a router onto the destination network, the router removes destination routing information. This new message contains an NPCI with the control octet bit 5 cleared to 0 because there is no longer any destination route information.

Network Layer Messages

The most common messages include APDUs, but there are also instances when routers need to communicate with each other in order to control the operation of the network. For these situations, BACnet provides special network layer messages. These messages all have their NPCI control octet bit 7 set to 1, and all others to 0. After the NPCI control octets, network layer messages include a message type octet, followed by some parameters that vary in length according to the message type. **See Figure 11-26.** For example, the message type 0x05 is the message Router-Available-to-Network and includes as parameters the list of network numbers for each network the router is connected to.

Network Layer Messages

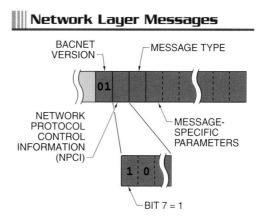

Figure 11-26. Network layer messages include the message type field, along with any message-specific parameters.

All of the BACnet network layer messages have this form, except those used to establish or disconnect a half router PTP connection. Those messages may need to pass through a router to a remote network and therefore may contain additional addressing information in the NPCI and thus additional octets before the message type octet.

BACNET OVER IP INFRASTRUCTURES

Although other IP-based LANs are possible, an Ethernet MAC layer is overwhelmingly the most common kind of IP network. Unlike BACnet, which uses ISO 8802-3 Ethernet, most IP networks use Ethernet II. Although most IP implementations can be configured to support 8802-3, this option is rarely used. When 8802-2 is used, it can carry IP messages using a link service access point (LSAP) of 0x06, instead of BACnet's LSAP of 0x82. As a result, IP traffic and BACnet 8802-3 traffic are easily distinguished. Otherwise, from a signaling and framing perspective, both of these Ethernet types are the same.

Routing Considerations

Ethernet switches are neutral with respect to these different types of messages and forward them equally. However, routers that are designed for IP networks have different functionality with regard to handling non-IP traffic.

IP-Only Routers. An *IP-only router* is a router that only handles IP packets and discards or ignores any others. In an Ethernet network that includes both IP nodes and BACnet 8802-3 nodes, IP-only routers can potentially interfere with BACnet communications. Because their messages are not bridged across the routers, the BACnet 8802-3 nodes on some segments cannot communicate with BACnet 8802-3 nodes on the other segments. **See Figure 11-27.** An additional bridge or switch installed between the segments solves this problem, but also repeats the IP traffic between the segments, which may interfere with proper operation of the IP infrastructure.

Bridging Routers. With a bridging router, incoming IP messages are processed by the router, but all others are simply repeated on the other ports. **See Figure 11-28.** A *bridging router* is a router that bridges non-routed messages at the data link layer. This is similar to how they would be bridged in an Ethernet switch. When BACnet 8802-3 nodes are used in an infrastructure with bridging routers, the router is invisible to BACnet. However, bridging functionality adds complexity and cost to a router because it increases processor requirements.

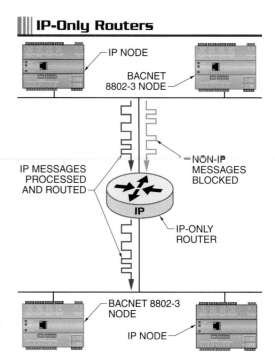

Figure 11-27. IP-only routers interfere with the operation of BACnet 8802-3 Ethernet because they cannot bridge Ethernet segments together.

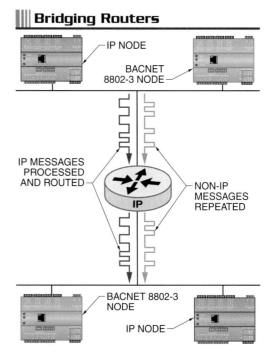

Figure 11-28. Bridging routers process and route IP messages, but non-IP packets are repeated automatically without further processing. The router is effectively invisible to non-IP Ethernet nodes and traffic.

Tunnel Routers. When the BACnet standard was first developed, the bridging problem with BACnet on an IP infrastructure was anticipated. Annex H of the standard includes a description of a special kind of router just for this scenario. A *tunnel router* is a type of BACnet router that bridges segments across IP boundaries by encapsulating the BACnet messages into UDP/IP packets.

When a node transmits a BACnet 8802-3 packet, the tunnel router hears that packet, recognizes it as a BACnet 8802-3 packet, and wraps it as a payload into an UDP/IP packet. The tunnel router then retransmits the packet in a UDP datagram to each peer tunnel router that it knows. **See Figure 11-29.** Each tunnel router must be manually configured to know the IP address of every peer tunnel router in the BACnet network. Each peer tunnel router that receives a datagram removes the wrapper and retransmits the original BACnet message on its local network. A bridging tunnel router takes the further step of pretending to be the Ethernet MAC address of the original sender, a process known as spoofing.

While tunnel routers solve the bridging problem, there are several drawbacks. Each segment that has a BACnet 8802-3 node requires a tunnel router, which adds cost. Also, every BACnet message that is transmitted on any of the segments must be retransmitted twice for each additional segment: once from the source segment's tunnel router to its peer, and again from the peer onto its own local segment. For example, on an IP network with six segments, each BACnet message would be transmitted 11 times. This makes tunnel routers a poor solution when there are more than a few segments.

BACnet/IP. When multiple segments are isolated by IP routers, instead of BACnet 8802-3 nodes, it is much easier to use nodes that support BACnet/IP. BACnet/IP uses UDP datagrams, which can pass through IP routers, so many IP-related issues are avoided. BACnet/IP differs from IP message tunneling in that each BACnet/IP node understands how to use IP directly. Each BACnet/IP node can build its own UDP messages and send them via IP directly to the desired destination node. **See Figure 11-30.**

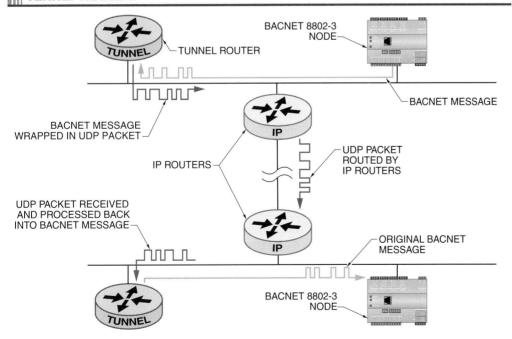

Figure 11-29. Tunnel routers listen for BACnet messages on their respective segments and forward those messages in a UDP wrapper to peer tunnel routers for retransmission.

BACnet/IP

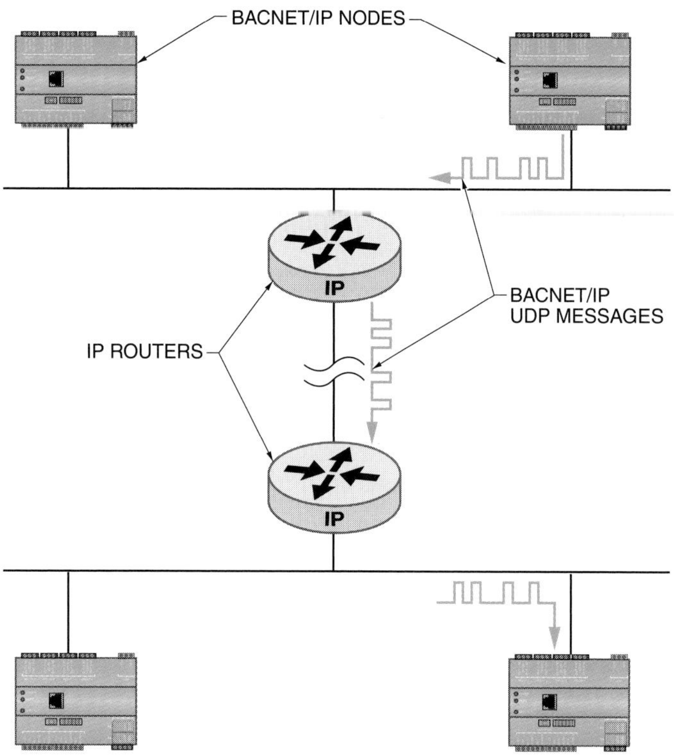

Figure 11-30. BACnet/IP (BIP) nodes use UDP/IP to communicate with each other directly, even through IP-only routers.

Typically, all BACnet/IP nodes share a common BACnet network number even if they reside on different, even widely separated, IP subnets. Therefore, there is usually no need for BACnet routers, in addition to the IP routers, between the IP subnets. To the BACnet nodes, the IP network looks like a large LAN. The collection of related IP subnets is called a BACnet/IP network.

If BACnet 8802-3 nodes must reside on multiple segments of an IP infrastructure, another option is to use a BACnet/IP-to-8802-3 router. **See Figure 11-31.** The 8802-3 nodes on each segment are treated as separate network numbers. For example, the 8802-3 nodes on one segment may be designated as network number 1, while other 8802-3 nodes on another segment are designated as network 3. The BACnet/IP network might then be network number 2.

Broadcasting Considerations

There is, however, an inherent problem with IP as far as BACnet is concerned: broadcast messages (such as those BACnet uses for unconfirmed services) are generally not repeated by standard IP routers. IP routers usually restrict broadcasts to the same IP subnet as the sending node. Therefore, a BACnet/IP broadcast can only reach BACnet/IP nodes in the same subnet.

One way to solve this problem is to use IP multicasting, a technique where each IP node has a second address that corresponds to a group of nodes. All of the IP routers between the source and destination networks must be configured to understand which IP subnets participate in the multicast domain. This can be a problem with currently installed routers, some of which may not support IP multicasting. Even with multicast support, configuration of the IP multicast addresses and their ongoing maintenance are complex and add to the support costs.

BACnet Broadcast Management Devices (BBMDs). The second solution is to use a BACnet broadcast management device (BBMD). A *BACnet broadcast management device (BBMD)* is a BACnet device that embeds broadcast messages in unicast messages (which can pass through IP routers) and sends them to a peer device for further distribution. **See Figure 11-32.** This is similar in concept to a tunnel router.

A BBMD uses a broadcast distribution table (BDT) to identify peer BBMDs and sends special BVLL messages to indicate that the enclosed message is to be broadcast to all BACnet nodes on the remote IP subnet. When a BACnet/IP node transmits a broadcast message to its IP subnet, it is heard by the other BACnet/IP nodes on that subnet and a BBMD. This BBMD wraps the NPDU portion of the Original-Unicast-NPDU in a new Forwarded-NPDU message and resends this to each of the BBMDs in its BDT. Each of these BBMDs receives this Forwarded-NPDU and retransmits it on their own local subnets, where it is heard by all of the local BACnet/IP nodes.

This is an effective solution to the broadcast problem for BACnet/IP, but it requires a BBMD for each BACnet/IP subnet. It has a similar drawback to a tunnel router solution in that many messages are generated, which increases network traffic. However, this is restricted to broadcasts only.

Foreign Device Registration. Sometimes, the requirement for having a BBMD on each subnet is impractical, such as when there is only one BACnet node on a subnet. *Foreign device registration* is a technique that allows a BACnet/IP node to register with a central BBMD for the purpose of sending and receiving broadcast messages.

When a BACnet/IP node wants to use this feature, it only needs the IP address of one BBMD. The "foreign device" (a BACnet/IP node on a subnet without a BBMD) sends a special BVLL message called Register-Foreign-Device. **See Figure 11-33.** If the BBMD accepts the request, the foreign device is added to a Foreign Device Table (FDT) in the BBMD.

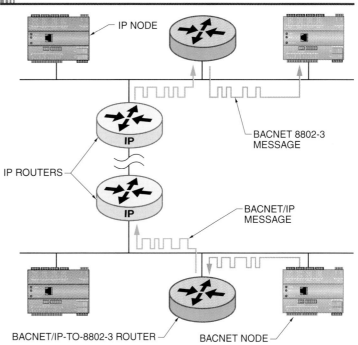

Figure 11-31. BACnet/IP-to-8802-3 routers are used to integrate 8802-3 BACnet nodes into an IP infrastructure.

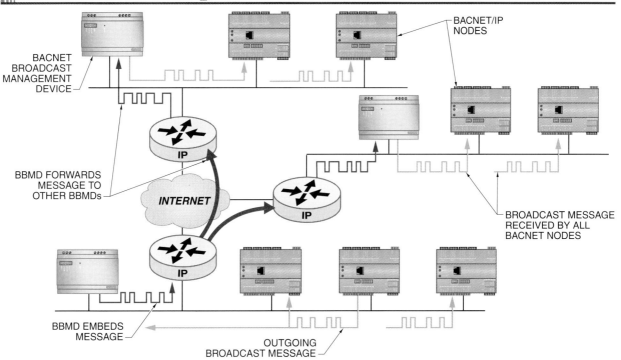

Figure 11-32. BACnet broadcast management devices (BBMDs) are used to forward broadcast messages between IP subnets that contain BACnet/IP nodes.

Foreign Device Registration

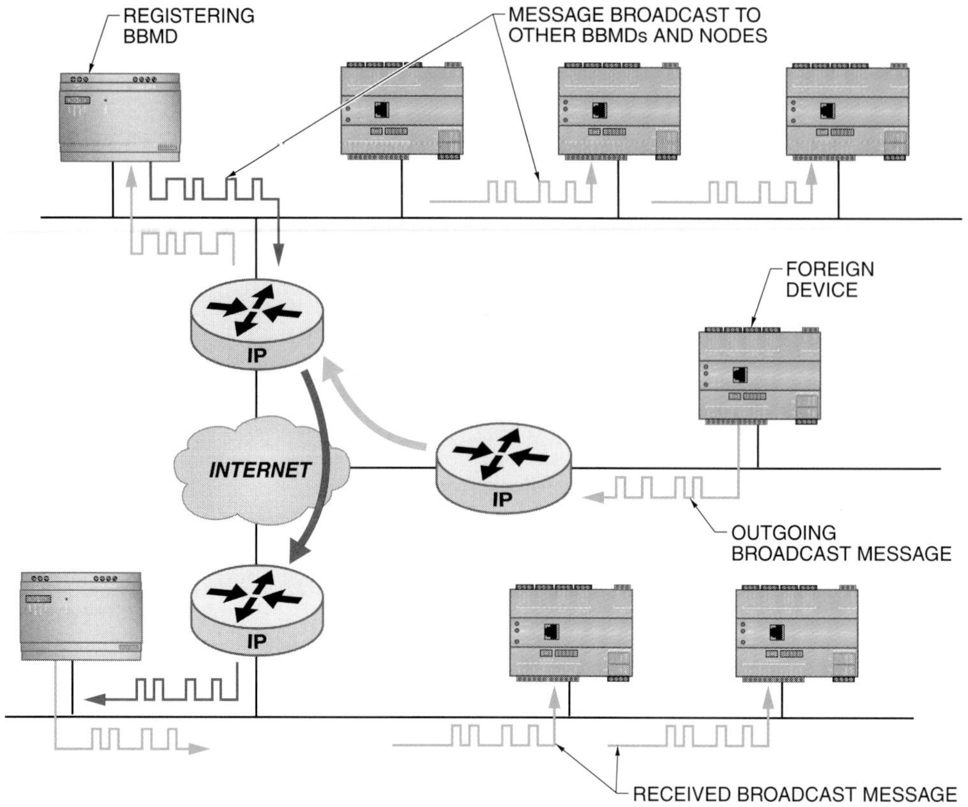

Figure 11-33. A foreign device registers with a BBMD to receive broadcasts and to use the BBMD to send broadcasts on its behalf.

Subsequently, whenever the BBMD receives a Forwarded-NPDU or an Original-Broadcast-NPDU, it will also send that NPDU to every registrant in the FDT as well as peer BBMDs in the BDT. Similarly, if the foreign device needs to send a broadcast, instead of broadcasting to its subnet, it will send the message as a Distribute-Broadcast-To-Network to the BBMD for redistribution.

Network Address Translation (NAT)

In an IP network that connects to the Internet, it may be desirable to restrict the number of "public" IP addresses used. For example, many LANs use multiple IP nodes on the private network, but allow access to the public Internet through only one public IP address. This arrangement uses network address translation (NAT). *Network address translation (NAT)* is a technique that changes source and destination IP addresses, as well as TCP and UDP port numbers, of incoming and outgoing IP packets as they pass through a NAT-capable router. **See Figure 11-34.**

Each node on the private network side of a NAT router has its own IP address, but only one public IP address is available from outside the router. When a message arrives, such as a UDP datagram to a BACnet/IP node, the NAT router uses a translation table to map an outside address to an inside address. For example, 206.211.45.83 UDP port 0xBAC0 (47808) should map to 192.168.2.4:47808. (The IP addresses 192.168.X.X are reserved for private network use.) That allows outside nodes to talk to an inside BACnet/IP node. When that inside BACnet/IP node answers, it uses its internal IP address as the source IP in its answer, but the NAT router automatically changes this to the public IP address.

However, because the NAT router alters the addressing content of an IP message, it does not work if there are two or more BACnet/IP nodes. There is no way to distinguish between them on incoming packets, and their private IP addresses are changed to the one public IP address on outgoing packets.

BACnet/IP-to-BACnet/IP (BIP/BIP) Routers. The easiest solution is to use a BACnet/IP-to-BACnet/IP (BIP/BIP) router. A *BACnet/IP-to-BACnet/IP (BIP/BIP) router* is a BACnet/IP router that routes between two or more BACnet/IP UDP ports. **See Figure 11-35.** For example, an NAT router is configured to map 206.211.45.83:0xBAC1 to 192.168.2.1:0xBAC1, which is one of the BACnet/IP router ports in the BIP/BIP router. This router does normal BACnet routing to other nodes on the 192.168.2.X subnet, but uses port 0xBAC0 for all of those nodes.

In effect, all of the nodes on 192.168.2.X using the 0xBAC0 port are on a different BACnet network segment with its own network number. For example, if 192.168.2.3 sends a global broadcast I-Am unconfirmed message, its UDPsrcport and UDPdstport are 0xBAC0 and its network layer SADR is 192.168.2.3:0xBAC0. The 0xBAC0 port in the BIP/BIP router receives this message and hands it off to the 0xBAC1 port (in effect, changing UDPdstport to 0xBAC1), which transmits it to some BBMD with a UDPsrcport now 0xBAC1. As this UDP datagram passes through the NAT router, the SADR is changed to 206.211.45.83.

Nodes receiving this message cannot get to 192.168.2.3 directly because this address is hidden behind the router. However, if they send a message to the public IP address with a DNET of 28 and DADR of 192.168.2.3:0xBAC0, then the BIP/BIP router can forward it to the correct local address.

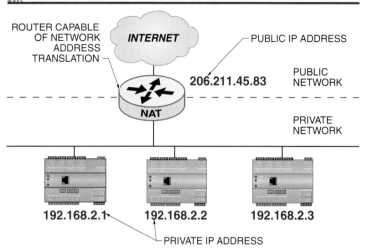

Figure 11-34. Network address translation (NAT) routers allow multiple private IP nodes to share a small number of public IP addresses.

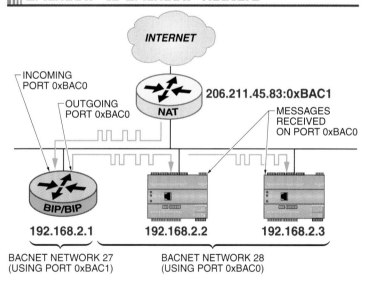

Figure 11-35. A BIP/BIP router routes BACnet/IP packets between different UDP ports, but can share a common public IP address.

Summary

- BACnet physical architectures typically use a tiered backbone, where wiring runs through vertical risers in the core or corner of the building and connects network devices, such as switches or routers, on each floor.
- Six types of LAN technology can be used to transport BACnet messages: Master-Slave/Token-Passing (MS/TP), Ethernet, ARCNET, BACnet/IP, Point-to-Point (PTP), and LonTalk Foreign Frames.
- Master-Slave/Token-Passing (MS/TP) is the most commonly implemented LAN type for BACnet networks.

- The BACnet standard allows for all three types of node biasing.
- MS/TP uses small-scale framing to enclose octets of data, which are transmitted together to form a complete message frame.
- If the Ethernet 8802-3 payload length field value is 0x0600 or larger, it is instead interpreted as a code for the protocol subtype of the payload.
- Ethernet is widely deployed and provides very high performance for a relatively modest cost.
- In building automation applications, ARCNET is only used in two forms: ARCNET 156K (156 kbit/s) and ANSI/ATA-878.1 (2.5 Mbit/s).
- Point-to-Point (PTP) is a unique BACnet solution for dialed internetwork communications.
- BACnet/IP messages use one of the standard IP components, the User Datagram Protocol (UDP), in order to transmit messages over IP.
- BACnet allows the use of LonTalk Foreign Frames as a transport for BACnet messages, essentially treating a LonTalk Foreign Frame as a MAC layer.
- MS/TP nodes take turns becoming master nodes in order to use the network and organize the access to the network by other nodes.
- MS/TP nodes continuously pass the token to the next node in the sequence until one needs to transmit a message.
- The master nodes all share responsibility for getting the token passing started again.
- MS/TP master nodes use timing parameters to manage token passing and message replies.
- A slave node can only reply to DER frames sent directly to it or act on NER frames that are sent.
- A slave proxy acts as a proxy on behalf of slave nodes on its segment.
- The header in the BACnet network layer varies in length depending on circumstances.
- The 8 bits of the network protocol control information (NPCI) octet indicate the presence or absence of other portions of the network layer header.
- Every BACnet network segment is assigned a unique network number.
- The Hop Count decreases by one each time the message is repeated in a router onto another segment.
- The network layer header is changed as the message is passed from segment to segment by routers.
- Routers use network layer messages when they need to communicate with each other in order to control the operation of the network.
- In an Ethernet network that includes both IP nodes and BACnet 8802-3 nodes, IP-only routers can potentially interfere with BACnet communications.
- A tunnel router wraps a BACnet 8802-3 packet as a payload into an IP packet and then retransmits the packet in a UDP datagram to each peer tunnel router.
- Broadcast messages (such as those BACnet uses for unconfirmed services) are generally not repeated by standard IP routers, often necessitating the use of BACnet broadcast management devices (BBMDs).
- Network address translation allows multiple IP nodes on the private network to access the public Internet through only one public IP address.

Definitions

- *Biasing* is the connection, at a transceiver, of network conductors to the reference voltage and ground through pull-up and pull-down resistors.
- A *master node* is an MS/TP node that shares equal responsibility for administering the network by taking turns to use the network and control the access to the network by other nodes.
- A *slave node* is an MS/TP node that cannot participate in token management or recovery.
- A *slave proxy* is an MS/TP master node that acts on behalf of slave nodes on its segment for dynamic device binding.
- *Hop Count* is the number of segments that the message is allowed to traverse before reaching its destination.
- An *IP-only router* is a router that only handles IP packets and discards or ignores any others.
- A *bridging router* is a router that bridges non-routed messages at the data link layer.
- A *tunnel router* is a type of BACnet router that bridges segments across IP boundaries by encapsulating the BACnet messages into UDP/IP packets.
- A *BACnet broadcast management device (BBMD)* is a BACnet device that embeds broadcast messages in unicast messages (which can pass through IP routers) and send them to a peer device for further distribution.
- *Foreign device registration* is a technique that allows a BACnet/IP node to register with a central BBMD for the purpose of sending and receiving broadcast messages.
- *Network address translation (NAT)* is a technique that changes source and destination IP addresses, as well as TCP and UDP port numbers, of incoming and outgoing IP packets as they pass through a NAT-capable router.
- A *BACnet/IP-to-BACnet/IP (BIP/BIP) router* is a BACnet/IP router that routes between two or more BACnet/IP UDP ports.

Review Questions

1. Describe the typical network physical architecture used for BACnet networks.
2. How are MS/TP message frames structured by the UART?
3. Describe the difference between 8802-3 Ethernet and Ethernet II frames.
4. Describe a typical example of token passing.
5. What are the differences between MS/TP master nodes and slave nodes?
6. How does the BACnet network layer header change length to accommodate routing to local or remote nodes?
7. How does the Hop Count field prevent misdirected messages from circling a network forever?
8. Explain the routing issues and possible solutions for BACnet on IP networks.
9. Explain the broadcasting issues and possible solutions for BACnet on IP networks.
10. What is network address translation?

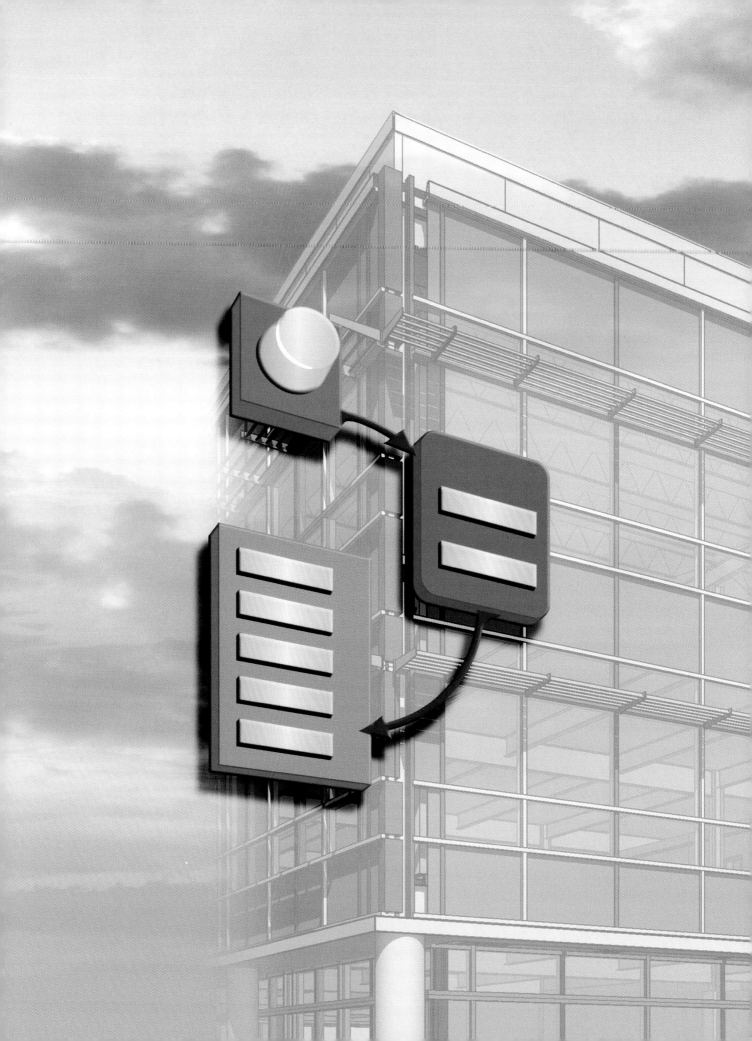

Chapter Twelve

BACnet Basic Objects and Core Services

Much of the information organization and operation of a BACnet control network can be described with a set of basic objects and core services. These components are the simplest to understand, yet are enormously flexible in the ways in which they can be used. A thorough knowledge of these objects and services is essential to working with BACnet networks.

Chapter Objectives

- Identify the common characteristics and requirements for every object.
- Compare the formats and applications of the input, output, and value versions of analog, binary, and multi-state object types.
- Describe the purposes and common properties of the special function objects.
- Identify the object access and remote device management services and their included parameters.
- Compare the variations of the ReadProperty service and how they are used.
- Describe the commandable property feature.
- Explain how services are used to establish device bindings for communication.

BACNET OBJECTS

Objects provide information to other BACnet nodes through the values of their properties. The 2004 BACnet standard defines 25 standard object types and their functionalities. Another 13 standard objects are included in approved or proposed addenda. **See Figure 12-1.** Many object types are generic and can be applied in many different ways, while some have specific uses in particular applications.

BACnet Standard Object Types

Included in Standard 135-2004		Added in Addenda*
Accumulator	Group	Access Door
Analog Input	Life Safety Point	Event Log
Analog Output	Life Safety Zone	Global Group
Analog Value	Loop	Load Control
Averaging	Multi-State Input	Structured View
Binary Input	Multi-State Output	Trend Log Multiple
Binary Output	Multi-State Value	**Proposed***
Binary Value	Notification Class	Access Credential
Calendar	Program	Access Point
Command	Pulse Converter	Access Rights
Device	Schedule	Access User
Event Enrollment	Trend Log	Access Zone
File		Credential Data Input
		Lighting Output

* as of December 2008

Figure 12-1. BACnet defines and describes 38 standard objects for organizing different types of information.

The standard specifies certain properties for each object type that are required for conformance. Many also include optional properties that can be included as needed, depending on the application. Some property values can be written, which can change the behavior of the object. Some of these changes are short-term, like control information to be shared between nodes, and some are configuration changes that affect the way the node operates. All objects in BACnet, even proprietary ones, are required to support at least three properties, the Object_Identifier, the Object_Type, and the Object_Name.

The Object_Identifier is a 32-bit number that uniquely differentiates each object within a node. The upper 10 bits of the Object_Identifier represent an enumeration code for the object type and the lower 22 bits represent the object instance. The object instance identifies one particular object of a given type from others of the same type within a node. For example, there can be only one Analog Input 4 within a given node, but that node may also contain an Analog Input 3 and a Binary Output 4.

The Object_Type property repeats the object type portion of the object identifier.

The Object_Name property contains a character string name for the object, which provides a human-readable way to identify the object. The Object_Name must be unique within the node containing the object.

Every object property represents one particular kind of data or piece of information, such as a number or a name. BACnet allows for many different kinds of data to be represented. The datatype of a property is the kind of data the property represents. In most cases, BACnet object properties are categorized by primitive datatypes, which are also called application datatypes. There are 13 primitive datatypes. **See Figure 12-2.** Many other datatypes are defined in the BACnet standard, each based on two or more of these primitive datatypes.

Most properties contain a single value. However, some properties represent an array of values that all have the same datatype. These array properties can be thought of as a collection of slots that are individually addressed by an array index. The special index number 0 represents the count of the number of slots in the array, and then indices 1 through n (the index number for the last slot) are the array slots themselves. Alternatively, a property can be a list, which means that there are multiple elements, but they cannot be individually addressed.

BASIC OBJECTS

No BACnet node is required to support any of the standard object types except the Device object. However, some of the standard objects are so useful that many nodes implement them as a basic means of sharing information. This group of nine standard objects is so common that they are referred to as the basic object types: Analog Input, Analog Output, Analog Value, Binary Input, Binary Output, Binary Value, Multi-State Input, Multi-State Output, and Multi-State Value.

BACnet Naming

Some aspects of BACnet can be confusing in that very similar names are used for identifying different types of information. Their meanings may be related but used in very different contexts. BACnet addresses this by standardizing a naming style for elements of the protocol based on the capitalization and spacing of words.

The names for object types are written as capitalized words separated by spaces, such as Pulse Converter and Lighting Output.

When two or more words are used in the name of an object property, they are separated by underscore characters (_). For example, Object_Identifier is an object property. However, this name can be confused with Object Identifier, which is a service parameter. Both data containers include the same type of information, but they are used in very different contexts.

Names with capitalized words with no internal spaces are typically used for services, datatypes, and smaller data field elements within properties. For example, ReadProperty and CreateObject are service names, and ObjectIdentifier (or BACnetObjectIdentifier) is a datatype. However, a group of seven services uses hyphens to separate the words, which would otherwise be awkward to read without spaces. These exceptions to the rule are the Who-Is, I-Am, Who-Has, I-Have, VT-Open, VT-Data, and VT-Close services.

Enumerations, such as status, are typically presented in words with all letters capitalized, such as NORMAL or FAILURE. This style is common among many computer-based systems, though not standardized.

When a hierarchy of information extends to several levels, lowercase words separated by hyphens may also be used. For instance, the Trend Log object contains the property Log_Buffer, which contains the data field LogDatum, which may have a value of log-status.

It may be necessary to pay careful attention to the term's context to distinguish between similar names. Names and their meanings are also explained in detail within the BACnet standard.

Device Objects

Every BACnet node is required to support a special type of object, the Device object. There can be only one instance of the Device object per node. Unlike other objects, the Device object's instance and Object_Name must be unique across the whole BACnet internetwork, not just within the node. In practice, there is no way of knowing in advance what other nodes will ever be present in a given network, so each node must allow its Device instance and Object_Name to be configured in the field after the node is installed.

The Object_List is an array property of the Device object that contains the Object_Identifiers of all of the objects within the node. **See Figure 12-3.** For example, a node may contain four objects: Device 39, Analog Input 1, Analog Input 2, and Analog Output 1.

Analog Objects

Analog quantities represent a value within a range of possible values, such as temperature, pressure, humidity, and lighting luminance. BACnet provides for this type of information with the Analog Input, Analog Output, and Analog Value objects. **See Figure 12-4.** They are commonly used with BACnet nodes to hold a sensor reading input or a value that changes an output. The principal property of these objects is the Present_Value property, which always has a datatype of real (floating point) numbers.

BACnet Primitive Datatypes

Datatype	Description
NULL	Empty value
Boolean	TRUE or FALSE
Unsigned Integer	Positive whole numbers
Signed Integer	Positive and negative whole numbers
Real	Floating-point numbers*
Double	Double-precision floating-point numbers*
Octet String	Sequence of octets
Character String	Sequence of characters
Bit String	Sequence of bit values
Enumerated	Unsigned integer code
Date	Month, day, year, day-of-week
Time	Hours, minutes, seconds, hundredths of seconds
BACnetObjectIdentifier	Unique 32-bit number

* defined in ANSI/IEEE-754 Standard

Figure 12-2. All data stored in BACnet objects is held in a form derived from one or more primitive datatypes.

Device Object

Property*	Type	Example Values
Object_Identifier	Required	(Device 39)
Object_Type	Required	8 (Device)
Object_Name	Required	"VAV2"
Object_List	Required	Object_List[0] = 4 Object_List[1] = (Device 39) Object_List[2] = (Analog Input 1) Object_List[3] = (Analog Input 2) Object_List[4] = (Analog Output 1)
Local_Time	Optional	12:34:56.78
Local_Date	Optional	29-SEP-2009, TUESDAY
UTC_Offset	Optional	60
Max_Master	Optional	22
Device_Address_Binding	Required	(Device 1), 1, 0x01; (Device 12), 1, 0x17; (Device 40), 2, 0x02608C41A606

*Only selected properties listed. See BACnet standard for complete list.

Figure 12-3. Every BACnet node must include precisely one Device object, which is unique across its network and contains information for the operation of the entire node.

Analog Objects

Analog Input Object

Property*	Type	Example Values
Object_Identifier	Required	(Analog Input 1)
Object_Type	Required	0 (Analog Input)
Object_Name	Required	"AHU1 Mixed Air Temp"
Present_Value	Required	69.8
Units	Required	degrees-Fahrenheit
Resolution	Optional	0.1

Analog Output Object

Property*	Type	Example Values
Object_Identifier	Required	(Analog Output 2)
Object_Type	Required	1 (Analog Output)
Object_Name	Required	"AHU1 Damper 2"
Present_Value	Required	35.0
Units	Required	percent
Resolution	Optional	0.1

Analog Value Object

Property*	Type	Example Values
Object_Identifier	Required	(Analog Value 5)
Object_Type	Required	2 (Analog Value)
Object_Name	Required	"AHU1 Setpoint"
Present_Value	Required	72.0
Units	Required	degrees-Fahrenheit

*Only selected properties listed. See BACnet standard for complete list.

Figure 12-4. The Analog Input, Analog Output, and Analog Value objects all include similar types of information, but they are used differently for control tasks.

The Present_Value of an Analog Input represents the measured value (or in some cases, the calculated value) of an input or monitored process. By repeatedly reading the Present_Value property of an Analog Input, a BACnet client node receives up-to-date input information.

Similarly, the Present_Value of an Analog Output object represents the desired position or setting of an output. Writing a different value to the Present_Value property of an Analog Output causes the associated physical output (or process) to change.

Analog Value objects work similarly to either Analog Input or Analog Output objects, but typically represent only internal node functions. The Present_Value of an Analog Value object either holds the result of internal calculations or logic, or it is written to in order to affect internal program logic or processes. For example, a control system might use an Analog Value as a setpoint or parameter to define a desired control temperature, or some other variable target value.

All three of these object types include other properties with additional information, such as engineering units, status, and alarm condition.

Binary Objects

Similar to Analog objects, BACnet provides for Binary Input, Binary Output, and Binary Value object types. **See Figure 12-5.** However, since binary quantities have two possible states, the Present_Value for binary objects is an enumerated datatype that uses 0 for INACTIVE and 1 for ACTIVE.

Every binary object includes a Polarity property that defines whether the Present_Value should be interpreted as the actual state of the input or output, or as the reverse state. For example, a Binary Output that controls a light and has a Polarity value of REVERSE would turn the light ON when Present_Value is INACTIVE.

Binary objects have many properties that are different from analog objects. For example, they do not have a Units property, but they do include Inactive_Text and Active_Text properties that provide string datatype names for the two states.

Multi-State Objects

Multi-state objects also come in Input, Output, and Value types. They are similar to binary objects, except that they can represent more than two states. **See Figure 12-6.** In fact, a multi-state object can theoretically support an unlimited number of states. Each instance of a multi-state object can have a different number of states, as defined in the Number_Of_States property. The possible state values in Present_Value are nonzero unsigned integers that represent states such as OFF, LOW, HIGH, and AUTO. The State_Text property is an optional array containing descriptions for each of the states.

SPECIAL FUNCTION OBJECTS

Besides the basic objects, there are other informal groups of objects. One group of objects has more specialized functions than the basic objects, but this group of objects is still commonly used in many different applications.

Accumulator Objects

The Accumulator object is used in applications for counting pulses, and includes scaling and rate information to provide exact conversion between pulses and an equivalent value in engineering units. **See Figure 12-7.** The Present_Value count gradually increases over time, accumulating counts. The Prescale and Scale properties are used to accurately represent the measured quantity, but avoid extremely high Present_Value count numbers. Prescale determines the number of pulses per Present_Value count, and Scale determines the number of counts per engineering unit.

For example, the accumulated count of pulses from a watt-hour meter is used to continuously measure electric power consumption. If the meter pulses once for each kilowatt-hour, the Prescale property can be used to increment the count in Present_Value only once per 100 pulses. To represent the value accurately in kilowatt-hours, though, the Scale property indicates that the Present_Value must be multiplied by 100 (10^2).

Binary Objects

Binary Input Object		
Property*	Type	Example Values
Object_Identifier	Required	(Binary Input 4)
Object_Type	Required	3 (Binary Input)
Object_Name	Required	"Rm 204 Light Switch 2"
Present_Value	Required	INACTIVE
Polarity	Required	REVERSE
Inactive_Text	Optional	"Lights On"
Active_Text	Optional	"Lights Off"
Binary Output Object		
Property*	Type	Example Values
Object_Identifier	Required	(Binary Output 2)
Object_Type	Required	4 (Binary Output)
Object_Name	Required	"Exhaust Fan"
Present_Value	Required	ACTIVE
Polarity	Required	NORMAL
Inactive_Text	Optional	"Fan Off"
Active_Text	Optional	"Fan On"
Binary Value Object		
Property*	Type	Example Values
Object_Identifier	Required	(Binary Value 1)
Object_Type	Required	5 (Binary Value)
Object_Name	Required	"Manual Control Enable"
Present_Value	Required	ACTIVE
Inactive_Text	Optional	"Automatic Control Mode"
Active_Text	Optional	"Manual Control Mode"

* Only selected properties listed. See BACnet standard for complete list.

Figure 12-5. The three types of binary objects hold a Present_Value that is either ACTIVE or INACTIVE.

Accumulator and Pulse Converter objects are commonly used with meters for utility services, such as electrical meters.

Other common Accumulator object applications include water and natural gas metering. The Accumulator object includes additional properties to manage the initialization and synchronization of the object with the physical pulsing device (such as a watt-hour meter), as well as information for managing the accurate trend logging of accumulated values.

Pulse Converter Objects

Like the Accumulator object, the Pulse Converter object is used in situations where the measurement of some analog quantity is represented by pulses or counts that equate to usage. The Count is then multiplied by a Scale_Factor property to derive an actual rate value in engineering units, which is held in Present_Value. **See Figure 12-8.** The pulses may come from an internal measurement in the node or from a referenced Accumulator object's Present_Value property.

Multi-State Objects

Multi-State Input Object

Property*	Type	Example Values
Object_Identifier	Required	(Multi-State Input 1)
Object_Type	Required	13 (Multi-State Input)
Object_Name	Required	"3-Position Control Switch"
Present_Value	Required	1
Number_Of_States	Required	3
State_Text	Optional	State_Text[1] = "Manual" State_Text[2] = "Off" State_Text[3] = "Auto"

Multi-State Output Object

Property*	Type	Example Values
Object_Identifier	Required	(Multi-State Output 2)
Object_Type	Required	14 (Multi-State Output)
Object_Name	Required	"Circulation Pump"
Present_Value	Required	2
Number_Of_States	Required	3
State_Text	Optional	State_Text[1] = "Off" State_Text[2] = "High" State_Text[3] = "Low"

Multi-State Value Object

Property*	Type	Example Values
Object_Identifier	Required	(Multi-State Value 3)
Object_Type	Required	19 (Multi-State Value)
Object_Name	Required	"Rm 39 Status"
Present_Value	Required	4
Number_Of_States	Required	4
State_Text	Optional	State_Text[1] = "Unoccupied" State_Text[2] = "Warmup" State_Text[3] = "Occupied" State_Text[4] = "Setback"

* Only selected properties listed. See BACnet standard for complete list.

Figure 12-6. Multi-state objects can hold the value of one of many different states.

Accumulator Object

Property*	Type	Example Values
Object_Identifier	Required	(Accumulator 2)
Object_Type	Required	23 (Accumulator)
Object_Name	Required	"Bldg 2 Electrical Usage"
Present_Value	Required	163
Scale	Required	2
Units	Required	kilowatt-hours
Prescale	Optional	(1, 100)

* Only selected properties listed. See BACnet standard for complete list.

Figure 12-7. The Accumulator object is used to represent cumulative pulses from a meter as values in engineering units, such as kilowatt-hours.

Pulse Converter Object

Property*	Type	Example Values
Object_Identifier	Required	(Pulse Converter 1)
Object_Type	Required	24 (Pulse Converter)
Object_Name	Required	"Water Flow Rate"
Present_Value	Required	307.5
Units	Required	us-gallons-per-minute
Scale_Factor	Required	2.5
Count	Required	123
Adjust_Value	Required	100
Count_Before_Change	Required	223

* Only selected properties listed. See BACnet standard for complete list.

Figure 12-8. The Pulse Converter object counts pulses similarly to the Accumulator object, but it can be used to determine time-based rates, such as gallons per minute.

The Pulse Converter object is very similar to the Accumulator object, except in two major respects. **See Figure 12-9.** First, the Present_Value is already scaled, so it is already a direct representation of the measured quantity in the indicated engineering units. Second, the Count can be adjusted down periodically by an amount written to the Adjust_Value property. This may be needed in order to synchronize the Present_Value with a real-world value.

Accumulator and Pulse Converter Objects

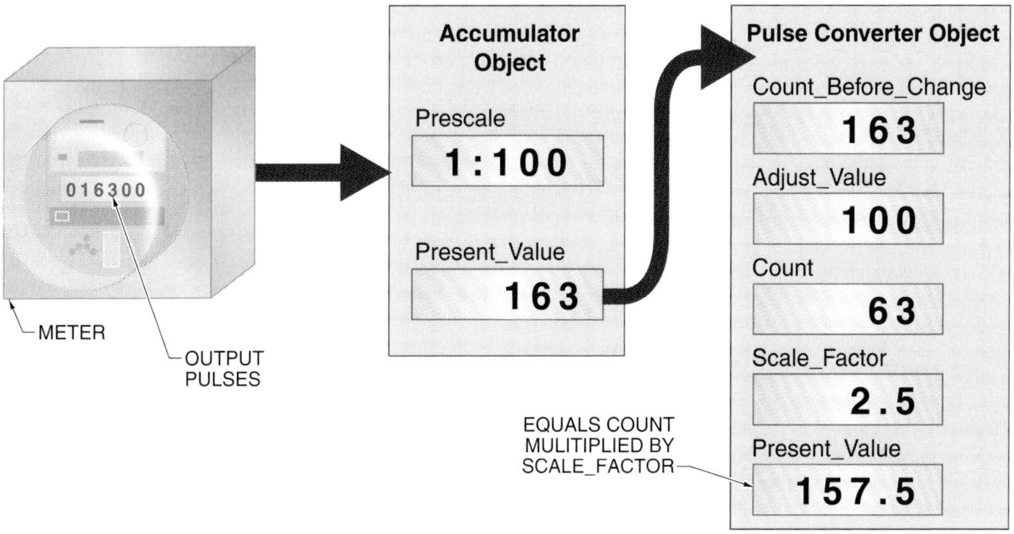

Figure 12-9. The Pulse Converter object may use an Accumulator object as its source of pulse counts, then scale and adjust the count to reflect a value in engineering units.

Averaging Objects

The Averaging object is a sliding window sampling of some property value that calculates the Minimum, Maximum, and Average of the accumulated samples at a fixed interval. **See Figure 12-10.** The Object_Property_Reference property refers to an object property in either the same node as the Averaging object or another node. The Window_Interval property defines the total time period (in sec) of the sample window, while the Window_Samples property defines the number of samples to be taken during the Window_Interval. **See Figure 12-11.**

The method of reading the external samples is not mandated by BACnet, but the ReadProperty service can be used at the fixed interval.

The number of samples used in the calculations remains the same. With each additional sampling, the buffer of samples shifts by one slot, dropping the oldest sample, and the minimum, maximum, average, and variance values are recalculated.

Command Objects

The Command object is used to convert a single command or mode into a sequence of actions that write specific values to specific properties of specific objects. The Command object is similar to the Multi-State Value in that its Present_Value indicates one of a certain number of states, each with a name. For each state, the node is programmed to execute a list of tasks that change the values of properties in other objects. For example, a room may have the states OCCUPIED, UNOCCUPIED, and WARMUP. Depending on the room's state, the temperature setpoints, lighting, security, and other settings may change. The Command object is used to change all associated properties at once for each state. **See Figure 12-12.**

Averaging Object

Property*	Type	Example Values
Object_Identifier	Required	(Averaging 4)
Object_Type	Required	18 (Averaging)
Object_Name	Required	"Electrical Demand"
Minimum_Value	Required	12.5
Average_Value	Required	13.7
Maximum_Value	Required	14.3
Object_Property_Reference	Required	(Analog Input 5, Present_Value)
Window_Interval	Required	60
Window_Samples	Required	15

* Only selected properties listed. See BACnet standard for complete list.

Figure 12-10. The Averaging object samples a property value of another object and continuously calculates its maximum, minimum, and average values.

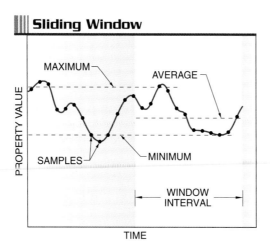

Figure 12-11. The Averaging object uses a sliding window of samples in a first-in, first-out scheme to determine the group of samples used for calculations.

The Action property is an array corresponding to the possible states of Present_Value. Each slot in the Action array is a list of actions for that state. Up to nine components may be included in each action to complete the task of writing a property value: Device_Identifier, Object_Identifier, Property_Identifier, Property_Array_Index, Property_Value, Priority, Post_Delay, Quit_On_Failure, and Write_Successful. The first six components identify the precise property to be written, along with its value and priority. The Post_Delay component specifies how long to wait before going to the next action. The Quit_On_Failure component specifies whether the entire action should stop after a failure in any of the actions. The Write_Success component indicates whether each action was successful.

It is important to note that the actions are not executed by the value of Present_Value, but by the writing of any value to Present_Value. Therefore, if the Present_Value is 3, and a value of 3 is written again to the property, then the node re-executes the corresponding list of actions.

> The Command object is a powerful feature that can make significant and wide-ranging changes. Its In_Process property protects it from writing to itself. However, Command objects can write to Group objects or other Command objects. This must be done carefully to avoid circular references that can cause major problems.

Command Object

Property*	Type	Example Values
Object_Identifier	Required	(Command 1)
Object_Type	Required	7 (Command)
Object_Name	Required	"ZONE CONTROL"
Present_Value	Required	2
Action	Required	Action[1] = (Analog Value 2, Present_Value), 72.0, ...; (Binary Output 5, Present_Value), ACTIVE, ...; Action[2] = (Analog Value 2, Present_Value), 68.0, ...; (Binary Output 5, Present_Value), INACTIVE, ...; Action[3] = (Analog Value 2, Present_Value), 72.0, ...; (Binary Output 5, Present_Value), INACTIVE, ...;
Action_Text	Optional	Action_Text[1] = "Occupied" Action_Text[2] = "Unoccupied" Action_Text[3] = "Warmup"

*Only selected properties listed. See BACnet standard for complete list.

Action[2] Actions

Device_Identifier	—	—
Object_Identifier	Analog Value 2	Binary Output 5
Property_Identifier	Present_Value	Present_Value
Property_Array_Index	—	—
Property_Value	72.0	INACTIVE
Priority	10	10
Post_Delay	1	1
Quit_On_Failure	TRUE	TRUE
Write_Successful	TRUE	TRUE

Figure 12-12. When selected from an array, a list of actions is executed by the Command object. Each action changes a property value in another object.

Group and Global Group Objects

A Group object is used to represent a collection of object properties. **See Figure 12-13.** This provides a simple way to specify a diverse group of object properties from different objects. The members of the group may be of any object type, but must be within the same node as the Group object.

The List_Of_Group_Members property identifies the object properties that form the group. Each member is specified by its Object_Identifier, Property_Identifier, and Array_Index (if needed). The Present_Value is not a single value, but a list of the values of all of the group's members. The Present_Value list is reconstructed each time this property is read in order to reflect any changes in the group member's values.

A Global Group object is related to the Group object, but has three important differences. First, each member of the group may include a Device_Identifier, so group members can then include objects and properties in other BACnet nodes. Second, Global Group objects support intrinsic reporting, meaning that they can report alarms based on changes in the value or status of any group member, and these alarm notifications can convey the values of all members. Third, each member can be assigned a name, independent of its actual Object_Name or Description properties, that identifies that value's meaning within the context of the global group.

> The Global Group object is a new object type described in Addendum 135-2004b. This addendum was approved by ASHRAE in 2008, making it part of the standard.

Group Objects

Group Object		
Property*	**Type**	**Example Values**
Object_Identifier	Required	(Group 1)
Object_Type	Required	11 (Group)
Object_Name	Required	"Zone 5 Temps"
List_Of_Group_Members	Required	(Analog Input 5, Present_Value), (Analog Input 6, Present_Value), (Analog Input 7, Present_Value)
Present_Value	Required	(Analog Input 5, Present_Value), 69.7; (Analog Input 6, Present_Value), 72.1; (Analog Input 7, Present_Value), 70.5

Global Group Object		
Property*	**Type**	**Example Values**
Object_Identifier	Required	(Global Group 1)
Object_Type	Required	26 (Global Group)
Object_Name	Required	"2nd Floor Temps"
Group_Members	Required	Group_Members[1] = (Analog Input 5, Present_Value) Group_Members[2] = (Analog Input 2, Present_Value, Device 2) Group_Members[3] = (Analog Input 1, Present_Value, Device 3)
Group_Member_Names	Optional	Group_Member_Names[1] = "Rm 201 Temp" Group_Member_Names[2] = "Rm 202 Temp" Group_Member_Names[3] = "Rm 203 Temp"
Present_Value	Required	Present_Value[1] = (Analog Input 5, Present_Value), 69.7; Present_Value[2] = (Analog Input 2, Present_Value, Device 2), 71.9; Present_Value[3] = (Analog Input 1, Present_Value, Device 3), 70.7

Figure 12-13. The Group and Global Group objects are used to create a logical collection of related objects within a node or among multiple nodes, respectively.

Loop Objects

In a closed-loop feedback control system, the most common algorithm is the proportional, integral, derivative (PID) loop. This control algorithm measures a variable, compares it to a setpoint, and calculates a control response based on the difference (error), the average error, and rate of change of error. BACnet abstracts this kind of control into the Loop object.

The input into the Loop object is the property Controlled_Variable_Value. **See Figure 12-14.** This value can be written to or coupled internally to another process invisible to BACnet. Alternatively, the Loop object may acquire the value by reading it from another BACnet node or object by using the property Controlled_Variable_Reference. This property identifies the external property, such as the Present_Value of an Analog Input, that is used to set the value of the Controlled_Variable_Value property.

The controlled variable is compared to a Setpoint property, which is similarly provided directly or through the Setpoint_Reference property. The properties Proportional_Constant, Integral_Constant, and Derivative_Constant are gain factors for a classical PID calculation. The result of the calculated correction becomes the Present_Value of the Loop object. This property can be read directly, fed forward internally by some mechanism invisible to BACnet, or actively pushed to the actual control output using WriteProperty to the object property described by the Manipulated_Variable_Reference property.

Program Objects

Just as there can be nonstandard objects, nodes can be programmed with custom control logic to manipulate objects and their properties. However, such proprietary programs require a common mechanism for control or supervision. BACnet standardizes how control programs model their execution and tasking, such as starting, stopping, and waiting. **See Figure 12-15.** The Program object also provides standardization by serving as a "front end" for a custom control program. The Program object is a standard object that may be extended with nonstandard properties and has standard means for controlling program loading, unloading, starting, and stopping.

> The Loop object uses references to associate its properties with properties in other objects. Referencing a property used as an input into the algorithm ensures that the input always reflects that property's changing value. Referencing an output property automatically copies the calculated value into the external property.

Loop Object

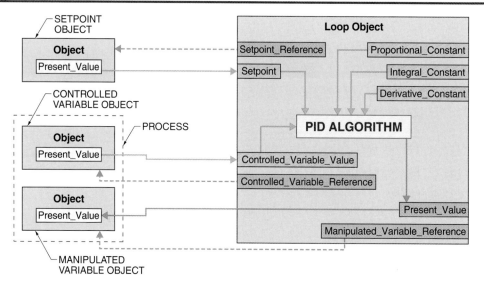

Figure 12-14. The Loop object contains a PID algorithm and the related tuning properties. The object references values in other objects as inputs to and outputs from the algorithm.

Program States and Transitions

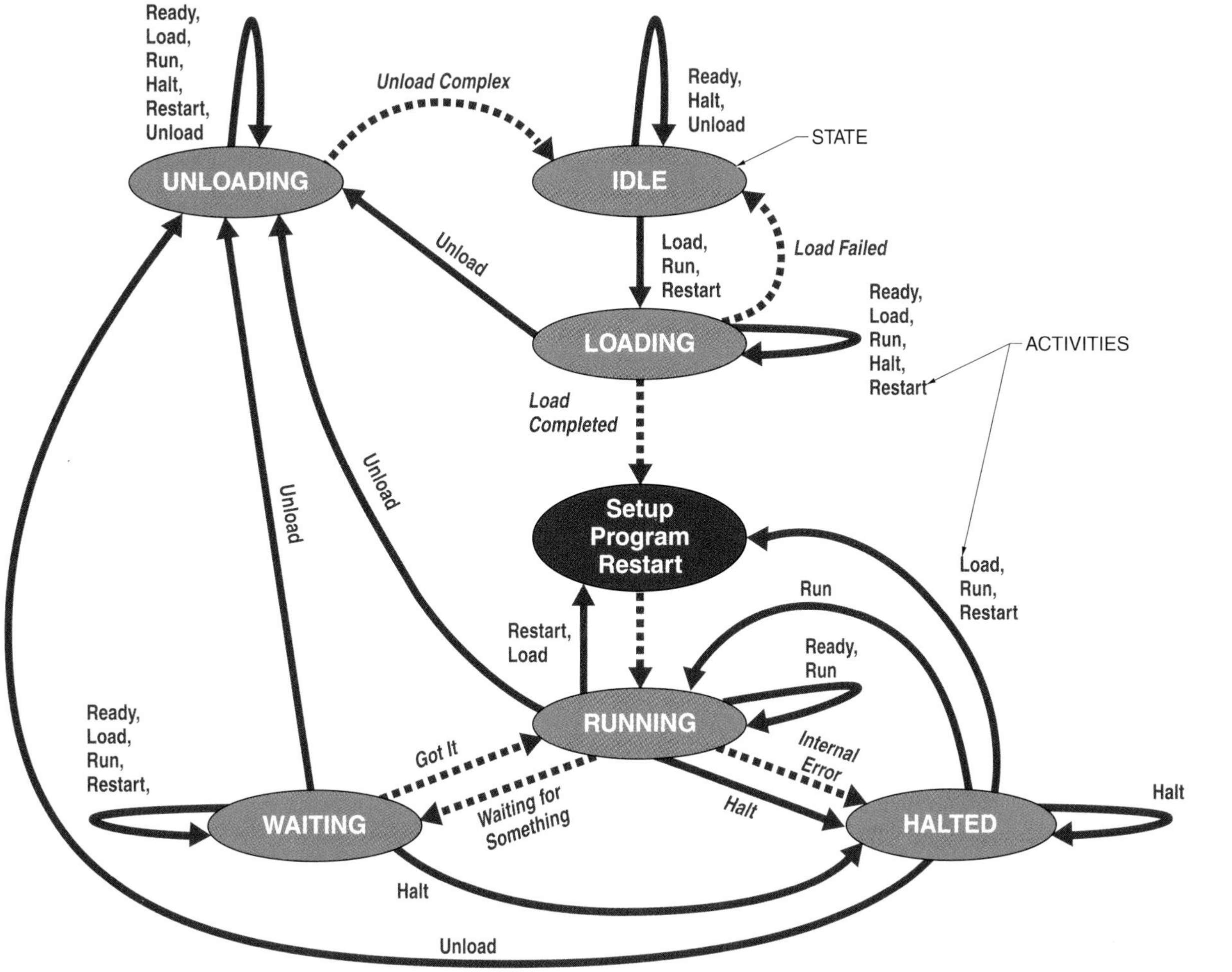

Figure 12-15. BACnet programs operate in and between states in a standardized way.

The read-only property Program_State represents the current logical state of the application program represented by the Program object. **See Figure 12-16.** Its values may be IDLE, LOADING, RUNNING, WAITING, HALTED, or UNLOADING. To make changes to an application process, corresponding READY, LOAD, RUN, HALT, RESTART, or UNLOAD commands are written to the property Program_Change.

Structured View Objects

The Structured View object provides a standard means for organizing a "view" of objects. This is very similar to how a computer's operating system shows an arrangement of data files in folders and subfolders. However, references to the same objects may exist in multiple Structured View hierarchies.

> The general-purpose state machine defined for the Program object can be used as an interoperable model for the control and supervision of multiple processes within a BACnet node. The machine diagram illustrates the valid state transitions that result in a new Program_State.

Program Object		
Property*	Type	Example Values
Object_Identifier	Required	(Program 1)
Object_Type	Required	16 (Program)
Object_Name	Required	"Custom Control"
Program_State	Required	RUNNING
Program_Change	Required	READY

*Only selected properties listed. See BACnet standard for complete list.

Figure 12-16. The Program object provides a way for services to monitor and change the operation of a custom program running on a BACnet node.

The structured view defines a node with a Node_Type, a Node_Subtype, and a list of subordinate objects. **See Figure 12-17.** The Node_Type is a generic classification of the type of information this view represents: Unknown, System, Network, Device, Organizational, Area, Equipment, Point, Collection, Property, Functional, or Other. The Node_Subtype is any kind of descriptive text that summarizes the purpose of the view, such as the type of equipment or logical area.

The Subordinate_List is an array of references for objects that belong to this view. The objects are typically within the same node as the Structured View object, but this is not required, so elements may also include a Device_Identifier instance. The same object can be referenced in more than one Structured View object. Elements may also reference other Structured View objects, resulting in a tiered hierarchy. However, a given Structured View object should not exist in multiple places in the same hierarchy, which would create circular references.

The Subordinate_Annotations is an array of text labels describing each element in the Subordinate_List.

OBJECT ACCESS SERVICES

Since so many ideas in BACnet center on objects and their use, there are a number of services to facilitate and manage access to objects. All of the object access services are confirmed services, meaning that the node is expected to reply, even if the request cannot be fulfilled. If a reply is not received, the request is retried several times.

When a service request is sent, several parameters are conveyed along with the request. The first parameter is a message number assigned by the sender. BACnet does this because the sender is allowed to transmit multiple requests before receiving the answer. The replies may take different lengths of time to carry out, so it is possible that replies will arrive out of order. A reply includes the original message number so that it can be matched with an outstanding request.

Structured View Object		
Property*	Type	Example Values
Object_Identifier	Required	(Structured View 2)
Object_Type	Required	29 (Structured View)
Object_Name	Required	"Air-Handling Unit 2"
Node_Type	Required	equipment
Node_Subtype	Required	"AHU"
Subordinate_List	Required	Subordinate_List[1] = (Schedule 4) Subordinate_List[2] = (Analog Input 3) Subordinate_List[3] = (Device 8, Analog Input 2) Subordinate_List[4] = (Device 2, Structured View 2)
Subordinate_Annotations	Optional	Subordinate_Annotations[1] = "AHU2 Mode Schedule" Subordinate_Annotations[2] = "Mixed Air Temp" Subordinate_Annotations[3] = "Outside Air Temp" Subordinate_Annotations[4] = "Exhaust Fan Subsystem"

*Only selected properties listed. See BACnet standard for complete list.

Figure 12-17. The Structured View object provides a way to organize objects into multilevel hierarchies.

Other parameters conveyed along with the message depend on the service and the specific options involved in the request.

ReadProperty Services

The most basic service is the ReadProperty service. This service is used when one BACnet node needs to look at (read) an object property in another BACnet node. The type of object or property does not typically matter. With only a few special exceptions, every object property can be read using ReadProperty requests.

The ReadProperty request requires both Object Identifier and Property Identifier parameters to specify which object and which property are to be read. **See Figure 12-18.** A Property Array Index parameter is also provided when reading array properties, such as a certain State_Text array slot within a Multi-State Value object.

There are two possible results from a ReadProperty request. Either the service succeeds and the receiving node returns the requested value, or the service fails. A failure returns a message that specifies the reason for the service failure. For example, there may be no such object defined in the target node.

If multiple object properties need to be read from the same node, instead of sending multiple ReadProperty service requests, the ReadPropertyMultiple service can be used. **See Figure 12-19.** This service includes a list of Read Access Specification parameters, which is a list of one or more Object Identifiers, each with one or more Property Identifiers (and Property Array Indexes, when applicable) to be read.

ReadPropertyMultiple also allows the use of three special Property Identifiers ALL, REQUIRED, and OPTIONAL. If the Property Identifier is ALL, then all of the object's properties are returned, including non-standard properties. If REQUIRED, then only the standard properties that are required by this object type are returned. If OPTIONAL, then only the standard optional properties are returned.

ReadProperty Service

```
ReadProperty {
   Message = 123
   Object Identifier = (Analog Input 3)
   Property Identifier = (Present_Value) }
```

EXAMPLE MESSAGES

```
Result {
   Message = 123
   Object Identifier = (Analog Input 3)
   Property Identifier = (Present_Value)
   Property Value = 37.9 }
```

```
Error {
   Message = 123
   Object Identifier = (Analog Input 3)
   Error Class = OBJECT
   Error Code = UNKNOWN_OBJECT }
```

EXAMPLE RESPONSES

Figure 12-18. The ReadProperty service is a basic service used to access the values in object properties.

ReadPropertyMultiple Service

```
ReadPropertyMultiple {
   Message = 123
   List of Read Access Specifications =
          (Analog Input 16, Present_Value),
          (Analog Input 17, Present_Value) }
```

EXAMPLE MESSAGES

```
Result {
   Message = 123
   List of Read Access Results =
          (Analog Input 16, Present_Value), 42.3;
          (Analog Input 17, Present_Value), 437.9 }
```

EXAMPLE RESPONSE

Figure 12-19. ReadPropertyMultiple provides a method for reading the values of two or more properties within a node with a minimum of network traffic.

The ReadPropertyConditional service is particularly useful for reading certain information from a group of nodes that each monitor several environmental variables.

The ReadPropertyConditional service is used to make complex queries based on selection logic criteria. Instead of specifying the exact property to be read, this service returns the values for one or more properties from a group of objects that match certain selection requirements. **See Figure 12-20.** The ReadPropertyConditional service includes two parameters for determining which properties are read: Object Selection Criteria and List of Property References.

The Object Selection Criteria includes multiple elements for determining which objects match certain conditions. First, the Selection Logic element determines how the subsequent List of Selection Criteria will be used to select objects. The possible values are AND, OR, and ALL. For example, if the logic is AND and there are two criteria, then both must be TRUE for an object to be selected. If the logic is OR, then only one of the criteria must be TRUE. If the logic is ALL, all objects are selected and the List of Selection Criteria is ignored.

The List of Selection Criteria includes elements to form one or more selection tests. Put together, the elements of each test resembles a formula that resolves to either TRUE or FALSE. The first part is a Property Identifier (including a Property Array Index if necessary), followed by a Boolean operator (such as greater than or equal to) and a comparison value. For example, selection criteria may be for objects with Units equal to "degrees Fahrenheit" or Present_Value greater than or equal to 70. If both of these tests are criteria, and the Selection Logic is AND, then this service would return only the objects that hold values ≥70°F.

ReadPropertyConditional Service

ReadPropertyConditional {
  Message = 123
  Object Selection Criteria
    Selection Logic = AND
    List of Selection Criteria = (Units = degrees-Fahrenheit),
                                  (Present_Value ≥ 70) }

EXAMPLE MESSAGES

Result {
  Message = 123
  List of Read Access Results = (Analog Input 16, Present_Value), 89.3;
                                 (Analog Input 17, Present_Value), 103.1 }

EXAMPLE RESPONSE

Figure 12-20. ReadPropertyConditional is a powerful service that provides the values of properties selected based on specific criteria. By using multiple criteria, complex conditions can be used to return a very specific group of properties.

For each matching object, the List of Property References optional parameter specifies which properties in that object are to be read. Properties can either be explicitly listed or specified based on their type. The value of REQUIRED for the List of Property References parameter returns only required or writable properties, OPTIONAL returns only optional properties, and ALL returns every property, including proprietary properties.

WriteProperty Services

The WriteProperty service is used to change (write to) object properties. **See Figure 12-21.** Similar to ReadProperty, the WriteProperty service requires an Object_Identifier, Property_Identifier, an Array_Index (when applicable), and the Property_Value to be written. The reply from the receiving node then indicates either a successful write or a failure.

Similar to the ReadPropertyMultiple service, there is a WritePropertyMultiple service. As expected, this service includes a list of Write Access Specification parameters that include Object Identifiers, Property Identifiers, and Property Array Indexes (when applicable). Since this is a type of WriteProperty service, it also includes a list of Property Values.

Commandable Properties. ReadProperty requests can originate from multiple nodes without conflict. However, if two or more nodes each try WriteProperty requests to the same property of the same object in the same node, it is possible to have a conflict. One node could write a property with one value, while another node changes it again shortly thereafter. If the relative importance of each change is not taken into account, this could have damaging or dangerous results.

BACnet solves this problem by using command prioritization. Objects whose Present_Value properties are subject to this type of conflict are required to implement command prioritization. Their Present_Value properties are said to be commandable. A *commandable property* is a property that may be written with a priority on a scale from 1 (most important) to 16 (least important). The WriteProperty service includes the optional parameter Priority to hold this information for each write attempt. (If this parameter is not included, a default priority of 16 is assumed.)

> Priority numbers are also assigned to event notification messages. These are allowed a range of 0 to 255, where the command priority range is 1 to 16. In both cases, however, lower numbers indicate higher priority.

WriteProperty Services

```
WriteProperty {
  Message = 123
  Object Identifier = (Analog Output 14)
  Property Identifier = (Present_Value)
  Property Value = 80.0
  Priority = 10 }
```

```
WritePropertyMultiple {
  Message = 123
  List of Write Access Specifications = (Multi-State Value 4, Present_Value), 4;
                                         (Binary Value 2, Present_Value), ACTIVE;
                                         (Averaging 1, Window_Interval), 90 }
```

EXAMPLE MESSAGES

Figure 12-21. WriteProperty services are used to change the value of one or more properties in another node.

> Command priority 6 is reserved for Binary Output and Binary Value objects with the Minimum_On_Time and Minimum_Off_Time properties. Priority 6 reflects the Present_Value from the moment of its last change until when the applicable time interval expires. This keeps the Present_Value constant for a minimum period, at least at a certain importance. After the intervals expire, this priority level is relinquished.

Each object with a commandable property must also have a matching Relinquish_Default property and Priority_Array property. The Relinquish_Default property holds the value that Present_Value should have if nothing is commanding the output, such as after a power failure. The Priority_Array property is an array of 16 possible Present_Value values, corresponding to the 16 priority levels. **See Figure 12-22.** The Present_Value always reflects the value held in the highest priority array slot. The categorization of most priorities is up to the integrator, but certain applications are assigned a priority, such as life safety actions always having the highest priority.

For example, an Analog Output Present_Value property assumes a Relinquish_Default value of 50%. Then, a WriteProperty to the Present_Value writes a value of 80% at a priority level of 10. Since this is more important than the default level, the Present_Value assumes the new value of 80%. However, the default value is still remembered. Then, another WriteProperty adds the value of 65% at priority level 7. Since this is more important than priority level 10, the Present_Value assumes the new value of 65%.

Values written to each Priority_Array slot are remembered until specifically erased because they may still be used. Using WriteProperty to write the special empty NULL value at any priority effectively relinquishes control at that priority (erasing the value in that slot). The Present_Value then reverts to the next most important output value.

The Present_Value properties of Analog Output, Binary Output, and Multi-State Output objects are required to be commandable, so the related Relinquish_Default and Priority_Array properties must be implemented. Though optional, the Analog Value, Binary Value, and Multi-State Value objects can also implement command prioritization, even when they are writable.

ListElement Services

The services AddListElement and RemoveListElement are used to change the items included in a property list. **See Figure 12-23.** Property lists require these special services because the list items are not individually addressable like array properties. Therefore, WriteProperty services cannot be used to change individual list elements. Likewise, neither AddListElement nor RemoveListElement can be used on array properties.

The AddListElement service is used to add elements to a list property. The addition does not guarantee the order in the list if there are other elements already in the list. Also, the items are added uniquely to the list so that duplicate items are ignored. The only way to ensure list order is to use WriteProperty to rewrite the entire list again.

Command Prioritization

Priority_Array		Example Values		
1	Manual—Life Safety			
2	Automatic—Life Safety			
3	Available			
4	Available			
5	Critical Equipment			
6	Minimum ON/OFF			
7	Available			65%
8	Manual Operator			
9	Available			
10	Available		80%	80%
11	Available			
12	Available			
13	Available			
14	Available			
15	Available			
16	Available			
Relinquish_Default		50%	50%	50%
Present_Value		50%	80%	65%

Figure 12-22. Command prioritization provides a means to assign relative importance to multiple values for the same property.

The RemoveListElement service uses exactly the same parameters, but removes elements from the list instead of adding them.

CreateObject and DeleteObject Services

The CreateObject service is used to dynamically create new objects in a node. This service may use either an Object Identifier or Object Type parameter to specify the type of object to create. **See Figure 12-24.** If only Object Type is provided, the node assigns an instance number to it. The List of Initial Values parameter contains pairs of Property Identifier and Value parameters for initializing properties of the created object with nondefault values.

The DeleteObject service is used to delete objects. While this service is not explicitly limited to deleting certain types of objects, it is expected that most objects are protected from deletion by this method. In practical circumstances, the DeleteObject service is typically used to delete dynamically created objects. Its only parameter is the Object Identifier of the specific object to be deleted.

ReadRange Service

The ReadRange service is used to read a specific set of items from a property that contains a list or array of lists. This is particularly useful for reading only certain data from within lists that may be very long, such as trending logs. Since elements within a list are not individually addressable, ReadRange provides a way to select and read certain items without having to read the entire list, such as with a ReadProperty service.

The Object Identifier, Property Identifier, and Property Array Index (if needed) parameters identify the specific list property to be read. **See Figure 12-25.** The Range parameter then selects certain list items based on position, sequence, or time. If the By Position parameter is used, the message specifies a position index of the first element in the list and the number of elements to read. If the By Sequence parameter is used, the message specifies a sequence number (assigned in the order in which items are added to the list) and the number of elements to read. If the By Time parameter is used, the message specifies a starting timestamp and the number of elements to read. With any of these parameters, the number of elements to be read can be negative, which simply reverses the direction of the reading from the starting point.

ListElement Services

```
AddListElement {
    Object Identifier = (Group 4)
    Property Identifier = (List_Of_Group_Members)
    List of Elements = (Pulse Converter 2, Present_Value);
                       (Binary Input 2, Present_Value, Polarity) }
```

```
RemoveListElement {
    Object Identifier = (Group 4)
    Property Identifier = (List_Of_Group_Members)
    List of Elements = (Pulse Converter 1, Present_Value);
                       (Binary Input 1, Present_Value, Polarity) }
```

EXAMPLE MESSAGES

Figure 12-23. AddListElement and RemoveListElement services can be used to change the composition of lists within properties.

CreateObject and DeleteObject Services

```
CreateObject {
    Object Specifier = (Averaging)
    List of Initial Values = (Object_Name, "Average Temp"),
                             (Object_Property_Reference,
                                (Analog Input 3, Present_Value));
                             (Window_Interval, 60);
                             (Window_Samples, 15) }
```

```
DeleteObject {
    Object Identifier = (Group 7) }
```

EXAMPLE MESSAGES

Figure 12-24. Certain objects can be created in or deleted from a node by sending CreateObject or DeleteObject service messages.

> If the Range parameters are not present in a ReadRange service message, then the responding node attempts to read and return all of the items in the list. If the list is very long, this may result in segmentation of the result into several response messages.

ReadRange Service

```
ReadRange {
  Object Identifier = (Trend Log 2)
  Property Identifier = (Log_Buffer)
  Range
    Reference Time = 28-JUL-2009, 19:52:34.0
    Count = –3 }
```

EXAMPLE MESSAGES

```
Result {
  Object Identifier = (Trend Log 2)
  Property Identifier = (Log_Buffer)
  Result Flags = TRUE, TRUE, FALSE
  Item Count = 3
  Item Data = (28-JUL-2009, 19:52:34.0), 77.0,
              (28-JUL-2009, 19:52:32.0), 75.8,
              (28-JUL-2009, 19:52:30.1), 76.2,
  First Sequence Number = 15 }
```

EXAMPLE RESPONSE

Figure 12-25. The ReadRange service is required to read certain items within a list since they are not individually addressable. Items are selected based on their position, timestamp, or sequence within the list.

REMOTE DEVICE MANAGEMENT SERVICES

There are a number of services used to facilitate and manage access to nodes, rather than their objects. Some of these device management services are confirmed services, meaning that the node is expected to reply immediately, and failing a reply, the request is retried several times. The other device management services are unconfirmed, meaning that the service is unicast to one node or broadcast to all nodes and no immediate reply is expected.

DeviceCommunicationControl Service

The DeviceCommunicationControl service is used to disable (silence) or enable (allow) communication from a node for a period of time, in minutes. **See Figure 12-26.** Once disabled, only the expiration of the time period, or an enabling DeviceCommunicationControl message or a ReinititializeDevice service message can restart communication. Some nodes also require a password to be provided before they honor a DeviceCommunication-Control request.

DeviceCommunicationControl Service

```
DeviceCommunicationControl {
  Time Duration = 5
  Enable/Disable = DISABLE
  Password = "jMw#212aTp" }
```

EXAMPLE MESSAGES

Figure 12-26. Nodes can be temporarily disabled by sending a DeviceCommunicationControl service message.

ReinitializeDevice Service

The ReinitializeDevice service is used to reset or restart a node remotely. **See Figure 12-27.** For example, after new program logic is uploaded into a node, it may require a restart before the new program is active. The Reinitialized State of Device parameter specifies the one of several kinds of resets that can be requested. For example, COLDSTART instructs the node to completely reboot itself, while WARMSTART resets the node to some predefined initial state. Like the DeviceCommunicationControl service, some nodes also require a password to be provided before they honor a reinitialize request.

ReinitializeDevice Service

```
ReinitializeDevice {
  Reinitialized State of Device =
    WARMSTART
  Password = "Wcj?dbs02" }
```

EXAMPLE MESSAGES

Figure 12-27. The ReinitializeDevice service is used to restart a node. Different types of restart methods may be available.

TimeSynchronization Services

Time synchronization is used to ensure that the internal clocks of the nodes are set to the same time and date. There are two time synchronization services. **See Figure 12-28.** Their use depends largely on the size of the control system. For most systems, the TimeSynchronization service is adequate. This service sends out a simple message that from the system's time master node with just one parameter, which includes the current date and time. All receiving nodes then reset their Local_Date and Local_Time properties (within the Device object) to the received values.

TimeSynchronization Services

TimeSynchronization {
Time
Date = 26-Mar-2010
Time = 22:45:30.7 }

UTCTimeSynchronization {
Time
Date = 26-Mar-2010
Time = 22:45:30.7 }

EXAMPLE MESSAGES

Figure 12-28. TimeSynchronization service messages are broadcast to nodes to ensure that each node's internal clock has the same date and time.

Alternatively, the UTCTimeSynchronization service is used when the BACnet network includes nodes in multiple time zones. Therefore, the service broadcasts the coordinated universal time and date. The parameters of the UTCTimeSynchronization message are the same as the TimeSynchronization service. The difference is that each receiving node subtracts a time-zone-dependent UTC_Offset value (in minutes) to determine the correct local time, before storing those values in the Local_Date and Local_Time properties. The UTC_Offset is a property of the Device object that is configured during node commissioning. For example, if the UTC time that is sent in a service message is 22:45:30.7 and a node's UTC_Offset is 60, the node will set its Local_Time property to 21:45:30.7.

TextMessage Services

BACnet provides two services for sending text messages between nodes. These services are used much like an alphanumeric pager for sending text alerts, information, and even alarms. Some routers also use this service for sending administrative notifications.

System interface software may display text message alerts as they are received.

Both the ConfirmedTextMessage and UnconfirmedTextMessage services include the same types of information in the messages: Object Identifier of the source, Message Class, Message Priority, and Message parameters. **See Figure 12-29.** The Message parameter contains the free-form text. The difference is that the ConfirmedTextMessage expects a reply, either a success or failure notification, from the receiving node. Therefore, this service is used for unicast messages. The UnconfirmedTextMessage service does not expect a confirmation of the message, so this service is used for unicast, multicast, or broadcast messages.

TextMessage Services

ConfirmedTextMessage {
Text Message Source Device = (Device 5)
Message Priority = NORMAL
Message = "Maintenance Needed" }

UnconfirmedTextMessage {
Text Message Source Device = (Device 7)
Message Priority = NORMAL
Message = "Check Pump" }

EXAMPLE MESSAGES

Figure 12-29. TextMessage services send human-readable messages to workstation nodes to inform operators of system status and issues.

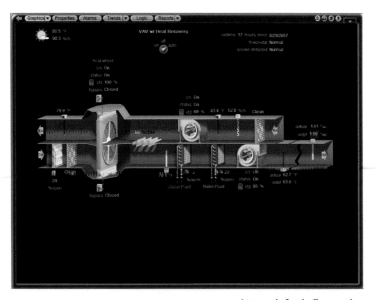

Automatic Logic Corporation

Services are used to share information with other nodes or for displaying on human-machine interfaces (HMIs)

PrivateTransfer Services

BACnet allows manufacturers to extend BACnet functionality with their own proprietary services. These services are not interoperable, unless both nodes in a conversation are aware of the proprietary service and how to use it, but this option provides a way to add special functionality to the control system.

Proprietary services are invoked using the standard private transfer services. There is a ConfirmedPrivateTransfer service and an UnconfirmedPrivateTransfer service, depending on whether a reply is expected from the receiving node. **See Figure 12-30.** Both services require a Vendor ID, a Service Number, and a number of additional parameters that vary depending on the service. It is up to the vendor to define and standardize the format and content of the additional parameters.

The Service Number is unique only in the context of the Vendor ID. Therefore, Vendor ID 4, Service Number 2 is different from Vendor ID 27, Service Number 2.

Who-Is and I-Am Services

The LAN technology used for network communication determines a unique MAC address for each node. A destination MAC address must be known for a BACnet node to send messages to other nodes. If the network includes multiple network segments, the destination BACnet network number must also be known. Since each BACnet node has a unique Device instance number, BACnet client nodes maintain an association, or binding, between the Device instances and the network number/MAC address pairs of the nodes it initiates communication with. The bindings are stored in the Device_Address_Binding property of the Device object. Once these associations are established, a BACnet client node can then refer to other nodes by only their Device instances.

PrivateTransfer Services

```
ConfirmedPrivateTransfer {
  Vendor ID = 7
  Service Number = 28
  Service Parameters = 197.7, ACTIVE, 5 }
```

```
UnconfirmedPrivateTransfer {
  Vendor Identifier = 7
  Service Number = 28
  Service Parameters = 197.7, ACTIVE, 5 }
```

EXAMPLE MESSAGES

Figure 12-30. PrivateTransfer service messages are used to initiate node-manufacturer-specific services.

These bindings can be configured through static device binding. *Static device binding* is the manual association of network numbers and MAC addresses with BACnet Device instances. However, static device binding requires significant effort to configure each node's list of bindings. Also, if a node is replaced with another one that has a different MAC address, which is typical, then the bindings of all the other nodes it communicates with require reconfiguration.

To get around these issues, most BACnet nodes use dynamic device binding. *Dynamic device binding* is the automatic discovery of network numbers and MAC address bindings for corresponding BACnet Device instances. The Who-Is and I-Am services are used to request and receive the network numbers and MAC addresses for one or more Device instances.

The Who-Is message is generally broadcast onto the entire network by a client node. The reply is an I-Am service message that is identified as being from a certain Device instance. The corresponding network number and MAC address are read from the network layer portion of the message frame, so they do not need to be included as a parameter. The I-Am message also indicates the node's maximum APDU length, supported segmentation type, and vendor identifier code.

An unqualified Who-Is broadcast message asks every node to respond with binding information. **See Figure 12-31.** This can cause many nodes to reply within a short time, flooding the network with I-Am messages. Therefore, an optional parameter of the Who-Is service restricts the responding nodes to a certain range of Device instances, or even a single Device instance. Only the nodes whose Device instance falls within the range send an I-Am reply.

The I-Am service produces an unconfirmed broadcast message. An I-Am message must be broadcast in a manner that ensures that it reaches the network that originated the Who-Is. Therefore, the broadcast may be on the local segment only, on a specific remote network segment, or globally on all networks.

Who-Has and I-Have Services

BACnet provides services for nodes to determine which other nodes contain a particular object. *Dynamic object binding* is the identification of nodes containing certain objects with the Who-Has and I-Have services. A Who-Has message is broadcast by the sending node with information on the desired object. **See Figure 12-32.** Either an Object_Identifer number or an Object Name can be used to specify the object. Similar to Who-Is messages, the range of Device instances queried with Who-Has can be restricted.

All nodes within the Device instance range that contain the specified object reply with a I-Have message. The I-Have service produces an unconfirmed broadcast message that indicates the Device instance with the matching object's Object_Identifier and Object Name. The I-Have response must be broadcast to the network that originated the Who-Has.

Who-Is and I-Am Services

Who-Is {
}

Who-Is {
Device Instance Range Low Limit = (Device 2)
Device Instance Range High Limit = (Device 5) }

EXAMPLE MESSAGES

I-Am {
I-Am Device Identifier = (Device 3)
Max APDU Length Accepted = 1024
Segmentation Supported = SEGMENTED_RECEIVE
Vendor Identifier = 7 }

EXAMPLE RESPONSE

Figure 12-31. Dynamic device binding uses broadcast Who-Is and responding I-Am service messages to match Device instances with network numbers and MAC addresses.

Who-Has and I-Have Services

Who-Has {
Object Identifier = (Analog Input 13) }

Who-Has {
Object Name = "Mixed Air Temp" }

Who-Has {
Device Instance Range Low Limit = (Device 2)
Device Instance Range High Limit = (Device 5)
Object Identifier = (Accumulator 2) }

EXAMPLE MESSAGES

I-Have {
Device Identifier = (Device 3)
Object Identifier = (Analog Input 13)
Object Name = "Mixed Air Temp" }

EXAMPLE RESPONSE

Figure 12-32. Who-Has service messages are broadcast onto a network to query the node location of certain objects.

Summary

- The BACnet standard defines 25 standard object types and their functionalities. Another 13 standard objects are included in approved or proposed addenda.

- All objects in BACnet, even proprietary ones, are required to support at least three properties, the Object_Identifier, the Object_Type, and the Object_Name.

- In most cases, BACnet object properties are categorized by one of the 13 different primitive datatypes. Other datatypes are based on two or more of these primitive datatypes.

- Properties may contain a single value, an array of individually addressable values, or a list of values.

- The Device object is the only object that is required for every BACnet node.

- The Analog Input, Analog Output, and Analog Value objects are commonly used with BACnet nodes to hold a sensor reading input or a value that changes an output.

- The Present_Value properties of Binary Input, Binary Output, and Binary Value objects have two possible states: INACTIVE and ACTIVE.

- Multi-State Input, Multi-State Output, and Multi-State Value objects are similar to binary objects, except that they can represent more than two states.

- The Accumulator and Pulse Converter objects are used in applications for counting pulses, and include scaling information to provide exact conversion between pulses and an equivalent value in engineering units.

- The Averaging object is a sliding window sampling of some property value that calculates the Minimum, Maximum, and Average of the accumulated samples at a fixed interval.

- The Command object is used to convert a single command or mode into a sequence of actions that write specific values to specific properties of specific objects.

- The Group and Global Group objects are used to represent a collection of object properties, which may be unrelated in any other way.

- A proportional, integral, derivative (PID) algorithm is provided by the Loop object.

- The Program object provides a standard means for controlling program loading, unloading, starting, and stopping.

- The Structured View object provides a standard means for organizing a hierarchical "view" of objects.

- All of the object access services are confirmed services, meaning that the node is expected to reply, even if the request cannot be fulfilled. If a reply is not received, the request is retried several times.

- Other than a required message number, the parameters conveyed along with a service message vary depending on the service and the specific options involved in the request.

- The ReadProperty and WriteProperty services are the most basic services used to look at (read) or change (write) an object property, respectively.

- The WriteProperty service includes the optional parameter Priority to hold information on the relative importance of each property write attempt.

- The ReadRange service is used to read a specific set of items from a property that contains a list or array of lists.

- Time synchronization services are used to ensure that the internal clocks of the nodes are set to the same time and date.

- Proprietary services are invoked using the private transfer services.
- BACnet client nodes maintain an association, or binding, between the Device instances and the network number/MAC address pairs of the nodes it initiates communication with.
- The Who-Is and I-Am services are used to request and receive the network numbers and MAC addresses for one or more Device instances.

Definitions

- A *commandable property* is a property that may be written with a priority on a scale from 1 (most important) to 16 (least important).
- *Static device binding* is the manual association of network numbers and MAC addresses with BACnet Device instances.
- *Dynamic device binding* is the automatic discovery of network numbers and MAC address bindings for corresponding BACnet Device instances.
- *Dynamic object binding* is the identification of nodes containing certain objects with the Who-Has and I-Have services.

Review Questions

1. What is the difference between an array property and a list property?
2. What are the special circumstances and requirements for Device objects?
3. What are the differences between the Accumulator and Pulse Converter objects?
4. Describe the function of the Command object.
5. What are the differences between Group and Global Group objects?
6. How does the ReadPropertyConditional service specify the objects whose properties are to be read?
7. How does a commandable Present_Value property manage the writing of multiple values of varying importance?
8. In what way is the ReadRange service more capable of reading specific information than a ReadProperty service?
9. Why is a special time synchronization service needed for nodes in multiple time zones?
10. How are the Who-Is and I-Am services used for dynamic device binding?

Chapter Thirteen

BACnet Alarming, Scheduling, and Trending

Alarming, scheduling, and trending are common and indispensible features of building automation systems. In BACnet, these functions are accomplished using combinations of objects and services. These objects and services are used to configure the behavior of these features and respond to certain events by sharing information about the actions and status of control devices. Often there are multiple ways to achieve alarming, scheduling, and trending functions, though the recommended method usually makes the most efficient use of the communication network.

Chapter Objectives

- Compare the various methods of detecting changes in the values of certain object properties.
- Describe the method of subscribing to change-of-value (COV) notifications.
- Differentiate between intrinsic reporting and algorithmic change reporting.
- Identify the objects and services involved in responding to alarm/event conditions.
- Describe the roles of Calendar objects and Schedule objects in BACnet scheduling.
- Describe how Trend Log objects are used to configure and store trend data.

CHANGE-OF-VALUE NOTIFICATION

A BACnet client may need to know when the value of an object property changes by a certain amount. The client can use ReadProperty or ReadPropertyMultiple services to periodically poll properties of objects in other nodes. **See Figure 13-1.** The client repeats these service requests at a certain interval. This method can be effective, but is not an efficient use of the network because it creates network traffic regardless of whether the value has changed. Furthermore, if the interval is long, the client may not react quickly enough to the changes. However, if the interval is short, the excessive polling creates unnecessary network traffic.

Change-of-Value (COV) Detection Methods

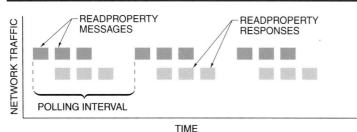

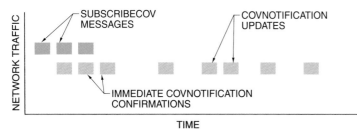

Figure 13-1. Changes in properties values can be determined by using a ReadProperty service, but COV subscriptions are typically a more efficient use of network bandwidth.

Alternatively, change-of-value (COV) notification allows the client to request that the nodes containing the desired properties only send data when the property value changes by a certain amount. This method significantly reduces traffic, but it requires the nodes to support COV notification and subscription services. Some nodes do not support COV subscriptions, or are limited in the number of simultaneous subscriptions they can provide.

SubscribeCOV Services

The SubscribeCOV service is used by a BACnet client to request a notification of changes of value for a specific object. **See Figure 13-2.** The Subscriber Process Identifier parameter is a unique number for the client to identify a subscription among many that may be initiated by the client. The Issue Confirmed Notifications parameter specifies whether COV notifications are confirmed or unconfirmed messages. The Lifetime parameter specifies a period of time (in seconds) during which the subscription is active. If the client does not resubscribe within the Lifetime period, the subscription lapses. With a limited number of simultaneous subscriptions, resources for lapsed subscriptions are made available for reuse.

Each BACnet object that supports COV subscriptions defines the criteria used to detect changes of value. Changes in the Present_Value property by more than a particular amount (as specified in the object's COV_Increment property) trigger change notifications.

If the SubscribeCOV service is accepted, a notification is generated immediately. Thereafter, notifications are only generated when the value changes according to the criteria. Subscriptions are cancelled by sending a SubscribeCOV service message and omitting the Issue Confirmed Notifications and Lifetime parameters.

If a client needs COV notifications for properties that are different from the default properties for a given standard object type, the SubscribeCOVProperty service is used. This service functions the same as SubscribeCOV, except that Monitored Property Identifier and COV Increment parameters are included in the service message.

COVNotification Services

COV notifications are sent with either ConfirmedCOVNotification or UnconfirmedCOVNotification services. Both types of messages contain the same parameters. **See**

Figure 13-3. The Subscriber Process Identifier is repeated so that the client can match the notification to the subscription. The Initiating Device Identifier and Monitored Object Identifier specify the information that has changed. The Time Remaining parameter indicates the subscription's remaining Lifetime.

The List of Values indicates the one or more property values from the monitored object. Responses to SubscribeCOV requests may include several values. This list varies according to the type of object being monitored, but typically includes Status_Flags and Present_Value properties. Responses to SubscribeCOVProperty requests only include the requested property value.

Some BACnet nodes can broadcast UnconfirmedCOVNotification messages without a subscription for announcing changes of value. Thermostats, wall switches, and other simple nodes can efficiently convey their status with these unsubscribed COV notification messages. A value of 0 for the Subscriber Process Identifier indicates that there is no subscription.

ALARMING

A critical concept in BACnet alarming is that an alarm source can only be in one of three fundamental states at any time: NORMAL, OFF-NORMAL, or FAULT. The transition from one state to another can be configured to trigger the notification process. **See Figure 13-4.** However, each state transition can independently require human acknowledgement. Therefore, up to six states are possible: NORMAL, OFFNORMAL, OFFNORMAL ACKNOWLEDGED, FAULT, FAULT ACKNOWLEDGED, and RETURN TO NORMAL. BACnet requires the state to transition through NORMAL.

Clients can monitor these status points using ReadProperty requests or COV subscriptions. Nodes can also report alarms by using a text message service. However, most BACnet nodes use alarm/event notification for alarming. (In BACnet, there is no difference between the concepts of alarms and events, so the terms are often used interchangeably.) The two types of alarming in BACnet are intrinsic reporting and algorithmic change reporting.

SubscribeCOV Services

```
SubscribeCOV {
    Message = 123
    Subscriber Process Identifier = 18
    Monitored Object Identifier = (Binary Input 10)
    Issue Confirmed Notifications = TRUE
    Lifetime = 60 }
```

```
SubscribeCOVProperty {
    Message = 123
    Subscriber Process Identifier = 7
    Monitored Object Identifier = (Averaging 4)
    Issue Confirmed Notifications = FALSE
    Lifetime = 0
    Monitored Property = (Maximum_Value)
    COV Increment = 1.0 }
```

EXAMPLE MESSAGES

Figure 13-2. SubscribeCOV services are used to initiate a subscription for notifications from objects when their properties change by a certain amount.

COVNotification Services

```
ConfirmedCOVNotification {
    Message = 123
    Subscriber Process Identifier = 18
    Initiating Device Identifier = (Device 3)
    Monitored Object Identifier = (Binary Input 10)
    Time Remaining = 15
    List of Values = (Present_Value, ACTIVE),
                    (Status_Flags, (FALSE, FALSE, FALSE, FALSE)) }
```

```
UnconfirmedCOVNotification {
    Message = 123
    Subscriber Process Identifier = 7
    Initiating Device Identifier = (Device 3)
    Monitored Object Identifier = (Averaging 4)
    Time Remaining = 0
    List of Values = (Maximum_Value, 103.0) }
```

EXAMPLE MESSAGES

Figure 13-3. COVNotifications, either confirmed or unconfirmed, are sent by nodes containing objects that have been subscribed to.

> Changes in certain values or states can be designated as "alarms" or "events" according to any criteria of the consulting-specifying engineer. BACnet makes no distinction between the two except in determining the response to the GetAlarmSummary service.

Alarm/Event States

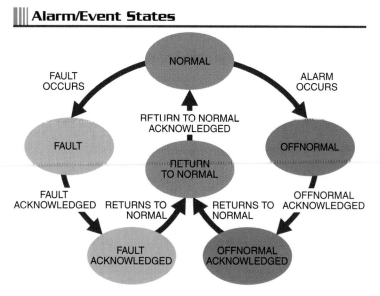

Figure 13-4. BACnet defines up to six distinct states for an object, depending on its status and whether the status has been acknowledged.

Intrinsic Reporting

Intrinsic reporting is an alarming feature of certain BACnet objects where the alarm or event originates from the object. Certain standard object types specify properties for controlling intrinsic reporting behavior. These normally optional properties become required when intrinsic reporting is implemented.

For example, an Analog Input object includes High_Limit, Low_Limit, and Deadband properties that specify a range of acceptable values for Present_Value. Outside these limits, the object is OFFNORMAL and generates alarm notifications. **See Figure 13-5.** The Limit_Enable property determines which limits (high, low, or both) cause Event_State changes. The Event_State property is an enumerated value that indicates the current state, such as NORMAL, FAULT, or OFFNORMAL. There are several special states, and nonstandard states can be added, but any state that is neither NORMAL nor FAULT is considered OFFNORMAL for state tracking purposes.

The Status_Flags property contains four Boolean (TRUE or FALSE) flags that correspond to specific states: IN_ALARM, FAULT, OVERRIDDEN, and OUT_OF_SERVICE (in that order).

The Event_Enable property is a 3-bit string, one bit for each of the three principal states, that determines which alarm/event transitions (TO-NORMAL, TO-OFFNORMAL, or TO-FAULT) are reported. If a bit is 1, then that corresponding transition generates a notification. Otherwise it does not.

The Notify_Type property determines whether the object is an alarm-generating object or event-generating object. (The only distinction here between an alarm and an event is whether the object is returned in a GetAlarmSummary request.)

The Acked_Transitions property is a 3-bit string that indicates whether the transition to each state has been acknowledged. Whenever a transition to a particular state occurs, if no human acknowledgement is required for that kind of transition, then the corresponding bit in Acked_Transitions is set to 1 (implicitly acknowledged). If human acknowledgement for a given transition is required, then when the transition occurs, the corresponding bit is cleared to 0. It remains 0 until the AcknowledgeAlarm service for that transition is received.

For example, an Analog Input object holds the value of a measured space temperature. It is set up with intrinsic reporting to generate alarms if the temperature rises to 85.0°F (the High_Limit property) or falls to 65.0°F (the Low_Limit property). When the temperature is 75.0°F, the Event_State is NORMAL. However, when the temperature rises to 87.0°F, the Event_State changes to OFFNORMAL and the first flag in Status_Flags (for IN_ALARM) changes to TRUE. Since the middle bit in Event_Enable is set to 1, a notification is sent according to the Notification_Class property. The first bit of Acked_Transitions (corresponding to a TO-OFFNORMAL acknowledgement) is cleared to 0. When the acknowledgement is received, it changes back to 1.

As the temperature falls back into the normal range, the Event_State returns to NORMAL, the IN_ALARM part of Status_Flags changes back to FALSE, and the TO-NORMAL bit of Acked_Transitions changes to 0. The first bit in Event_Enable is set to 1, so another notification of the state change is sent. When acknowledged, Acked_Transitions changes back to 1.

Intrinsic Reporting

Property*	Normal Values	Alarm Generated	Alarm Acknowledged	Return to Normal	Return to Normal Acknowledged
Object_Identifier	(Analog Input 11)	(Analog Input 11)	(Analog Input 11)	(Analog Input 11)	(Analog Input 11)
Object_Type	0 (Analog Input)	0 (Analog Input)	0 (Analog Input)	0 (Analog Input)	0 (Analog Input)
Object_Name	"Temp"	"Temp"	"Temp"	"Temp"	"Temp"
Present_Value	75.0	87.0	86.0	80.0	81.0
Status_Flags	FALSE, FALSE, FALSE, FALSE	TRUE, FALSE, FALSE, FALSE	TRUE, FALSE, FALSE, FALSE	FALSE, FALSE, FALSE, FALSE	FALSE, FALSE, FALSE, FALSE
Event_State	NORMAL	OFFNORMAL	OFFNORMAL	NORMAL	NORMAL
Notification_Class	3	3	3	3	3
High_Limit	85.0	85.0	85.0	85.0	85.0
Low_Limit	65.0	65.0	65.0	65.0	65.0
Deadband	2.0	2.0	2.0	2.0	2.0
Limit_Enable	1,1	1,1	1,1	1,1	1,1
Event_Enable	1,1,1	1,1,1	1,1,1	1,1,1	1,1,1
Acked_Transitions	1,1,1	0,1,1	1,1,1	1,1,0	1,1,1
Notify_Type	ALARM	ALARM	ALARM	ALARM	ALARM

* Only selected properties listed. See BACnet standard for complete list.

Figure 13-5. Intrinsic reporting is a feature of certain objects that automatically notifies certain other nodes when changes trigger an alarm.

Algorithmic Change Reporting

Some BACnet nodes, although they contain standard objects, may implement intrinsic reporting differently, or not at all. Even if intrinsic reporting is implemented, some kinds of alarms/events require a different type of algorithm. In these cases, algorithmic change reporting may be implemented. *Algorithmic change reporting* is a BACnet alarming feature that uses Event Enrollment objects to configure the alarm/event behavior of some other object property. **See Figure 13-6.** Algorithmic change reporting can be applied to any property of any object, unlike intrinsic reporting.

Event Enrollment Objects. An Event Enrollment object contains the properties necessary for testing for an alarm/event condition and managing the response. **See Figure 13-7.** The object properties are tested using a standard monitoring algorithm to detect changes in state. The Event Enrollment object also keeps track of the status of an alarm/event, which nodes receive notification of the alarm/event, and the acknowledgements of the notifications.

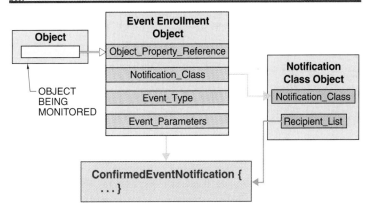

Figure 13-6. Algorithmic change reporting uses Event Enrollment objects to configure alarming behavior. This arrangement enables alarming to be set up for any property of any object.

Each Event Enrollment object refers to only one property of one object. Therefore, multiple Event Enrollment objects may be created, one for each property to be tested. Moreover, multiple Event Enrollment objects can apply different algorithms to the same property.

Event Enrollment Object		
Property*	Type	Example Values
Object_Identifier	Required	(Event Enrollment 1)
Object_Type	Required	9 (Event Enrollment)
Object_Name	Required	"Zone Alarm"
Event_Type	Required	OUT_OF_RANGE
Notify_Type	Required	ALARM
Event_Parameters	Required	10, 65.0, 85.0, 2.0
Object_Property_Reference	Required	(Device 3, Analog Input 11, Present_Value)
Event_State	Required	HIGH_LIMIT
Event_Enable	Required	1, 1, 1
Acked_Transitions	Required	0, 1, 1
Notification_Class	Required	3

* Only selected properties listed. See BACnet standard for complete list.

Figure 13-7. Event Enrollment objects can be used to configure alarming notifications for a particular object property, even one within a different node.

Nine standard algorithms are specified for BACnet alarming. **See Figure 13-8.** It is also possible for nodes to implement Event Enrollment objects that use nonstandard algorithms, but the parameters used by the algorithm may have limited network visibility.

- The CHANGE_OF_STATE algorithm initiates an alarm/event when a binary or multi-state property value equals any one of a list of OFFNORMAL states.
- The COMMAND_FAILURE algorithm initiates an alarm/event when a Command service request fails. Feedback on whether a commanded property has changed is provided by a Feedback_Property_Reference.
- The OUT_OF_RANGE algorithm initiates an alarm/event when an analog property value is outside of a defined range, either high or low.
- The FLOATING_LIMIT algorithm initiates an alarm/event when the difference between a setpoint and an analog property value is outside of a defined range, either high or low. This is similar to the OUT_OF_RANGE algorithm, except in the way the range is defined.
- The CHANGE_OF_BITSTRING algorithm initiates an alarm/event when a bitstring property value equals any one of a list of OFFNORMAL bitstrings. A Bitmask parameter is used to indicate which bits in the string are to be monitored.
- The CHANGE_OF_VALUE algorithm initiates an alarm/event when a property value changes by an amount equal to or greater than a defined Referenced_Property_Increment. If the property is a bitstring, this algorithm operates similarly to the CHANGE_OF_BITSTRING algorithm. However, this algorithm is never OFFNORMAL. This algorithm should not be confused with COV services.
- The CHANGE_OF_STATUS_FLAGS algorithm initiates an alarm/event when any of the selected flags of a property changes to a value of TRUE. Additional TO-OFFNORMAL transitions can occur if other flags change to TRUE.
- The CHANGE_OF_LIFE_SAFETY algorithm initiates an alarm/event when a property value equals any one of a list of alarm values. Life safety applications have both a state and a mode, so this also affects the algorithm. A List_Of_Alarm_Values define conditions for OFFNORMAL states and a List_Of_Life_Safety_Alarm_Values define conditions for LIFE_SAFETY_ALARM states.
- The BUFFER_READY algorithm is used with trending applications to notify other nodes when the trend log buffer contains a certain number of records. An alarm/event initiated by this algorithm generates TO-NORMAL transitions.

Event Enrollment Algorithms

Algorithm	Test*	Event State
CHANGE_OF_STATE	Object_Property_Reference ≠ any List_of_Values	NORMAL
	Object_Property_Reference = any List_of_Values	OFFNORMAL
COMMAND_FAILURE	Object_Property_Reference = Feedback_Property_Reference	NORMAL
	Object_Property_Reference ≠ Feedback_Property_Reference	OFFNORMAL
OUT_OF_RANGE	Low_Limit < Object_Property_Reference < High_Limit	NORMAL†
	Object_Property_Reference < Low_Limit	LOW_LIMIT
	Object_Property_Reference > High_Limit	HIGH_LIMIT
FLOATING_LIMIT	Setpoint_Reference − Low_Diff_Limit < Object_Property_Reference < Setpoint_Reference + High_Diff_Limit	NORMAL†
	Object_Property_Reference < (Setpoint_Reference − Low_Diff_Limit)	LOW_LIMIT
	Object_Property_Reference > (Setpoint_Reference + High_Diff_Limit)	HIGH_LIMIT
CHANGE_OF_BITSTRING	Object_Property_Reference & Bitmask ≠ any List_Of_Bitstring_Values	NORMAL
	Object_Property_Reference & Bitmask = any List_Of_Bitstring_Values	OFFNORMAL
CHANGE_OF_VALUE	Change in Object_Property_Reference ≥ Referenced_Property_Increment	NORMAL
CHANGE_OF_STATUS_FLAGS	All Referenced_Flags = FALSE	NORMAL
	Any Referenced_Flags = TRUE	OFFNORMAL
CHANGE_OF_LIFE_SAFETY	Object_Property_Reference ≠ any List_Of_Alarm_Values OR any List_Of_Life_Safety_Alarm_Values	NORMAL
	Object_Property_Reference = any List_Of_Alarm_Values	OFFNORMAL
	Object_Property_Reference = any List_Of_Life_Safety_Alarm_Values	LIFE_SAFETY_ALARM
BUFFER_READY	(Trend Log, Records_Since_Notification) ≥ Notification_Threshold	NORMAL

* for Time_Delay seconds
† Return to Normal transitions also take Deadband values into account.

Figure 13-8. Nine standard algorithms for detecting alarms/events are defined by the BACnet standard for use with Event Enrollment objects.

EventNotification Services

In either intrinsic or algorithmic change reporting, once an alarm/event has been detected, one or more other nodes are notified of the occurrence using either the ConfirmedEventNotification or UnconfirmedEventNotification services. Both service messages contain the same information about the alarm/event, including the identification of the alarming object, priority, and object status before and after the alarm/event, and whether acknowledgement is required. **See Figure 13-9.**

AcknowledgeAlarm Service

Notification messages are often sent to a workstation or supervisory controller. In these cases, the AcknowledgeAlarm confirmed service is typically used to confirm that a person has not only received the notification, but that the initiating node can stop the alarm. **See Figure 13-10.** The parameter Acknowledgement Source holds a name or other identifier of the person acknowledging the alarm. The receipt of an AcknowledgeAlarm message also generates an EventNotification message, known as an ACK notification. If there are multiple recipients of the original event notification message, each is then aware of the subsequent acknowledgement.

> After receiving an AcknowledgeAlarm message, a node attempts to match the acknowledgement with the object that generated the notification and the timestamp of its most recent event/alarm. If matching, the object's Acked_Transitions bit that corresponds to the notified transition is set to 1.

EventNotification Services

```
ConfirmedEventNotification {
   Message = 123
   Process Identifier = 8
   Initiating Device Identifier = (Device 3)
   Event Object Identifier = (Analog Input 11)
   Time Stamp = (28-Jul-2009, 13:10:30.7)
   Notification Class = 3
   Priority = 100
   Event Type = OUT_OF_RANGE
   Message Text = "Temperature is out of range"
   Notify Type = ALARM
   AckRequired = TRUE
   From State = NORMAL
   To State = HIGH_LIMIT
   Event Values = (Exceeding_Value, 87.0),
                  (Status_Flags, (TRUE, FALSE, FALSE, FALSE)),
                  (Deadband, 2.0),
                  (Exceeded_Limit, 85.0) }
```

```
UnconfirmedEventNotification {
   Message = 123
   Process Identifier = 8
   Initiating Device Identifier = (Device 3)
   Event Object Identifier = (Averaging 4)
   Time Stamp = 16
   Notification Class = 3
   Priority = 50
   Event Type = CHANGE_OF_VALUE
   Message Text = "New maximum flow rate"
   Notify Type = EVENT
   AckRequired = FALSE
   From State = NORMAL
   To State = NORMAL
   Event Values = (New_Value, 103.0),
                  (Status_Flags, (TRUE, FALSE, FALSE, FALSE)) }
```

EXAMPLE MESSAGES

Figure 13-9. When intrinsic reporting or algorithmic change reporting detects an alarm/event, EventNotification service messages are sent to the nodes and objects listed in the appropriate Notification Class object.

AcknowledgeAlarm Services

```
AcknowledgeAlarm {
   Acknowledging Process Identifier = 8
   Event Object Identifier = (Analog Input 11)
   Event State Acknowledged = HIGH_LIMIT
   Time Stamp = 16
   Acknowledgment Source = "J_SMITH"
   Time Of Acknowledgment = (28-Jul-2009, 13:03:41.9) }
```

EXAMPLE MESSAGES

Figure 13-10. EventNotification messages are acknowledged by AcknowledgeAlarm service messages. If a person is required for acknowledgement, their name or identifier is included in the message.

Notification Class Objects

Notification Class objects are used to define which nodes receive EventNotification messages. **See Figure 13-11.** Multiple Notification Class objects can be set up for different lists of notification destinations and circumstances. A Notification Class object is referred to by its instance number, which is also stored in its Notification_Class property.

An alarm/event-initiating object also has a Notification_Class property, which holds the instance number of the Notification Class object with the desired notification list. For intrinsic reporting, the Notification_Class is a property of the objects with intrinsic reporting capabilities. For algorithmic change reporting, the Notification_Class is a property of the Event Enrollment object. A single Notification Class object may be shared by several alarm/event-initiating objects.

The Priority property of the Notification Class object is an array with three slots that correspond to the three possible state transitions: TO-OFFNORMAL, TO-FAULT, and TO-NORMAL. Priority specifies a value between 0 (most important) and 255 (least important). This number is used by recipient nodes to sort incoming event notifications in order of importance.

The Ack_Required property is a 3-bit bitstring, exactly like the Acked_Transitions property (in an intrinsic reporting object), that specifies for each event transition whether an acknowledgement is required for that transition by objects that use this Notification Class. In this context, an acknowledgement means the receipt of an AcknowledgeAlarm service request referencing an event notification generated with this Notification Class.

The Recipient_List property specifies a list of one or more destinations, each of which includes several elements for qualifying and routing event notifications. A destination is only used if the event transition occurs on one of the Valid Days, between the From Time and To Time parameters. At any other time, the destination is ignored and no notification is sent.

Notification Class Objects

Property*	Type	Example Values
Object_Identifier	Required	(Notification Class 3)
Object_Type	Required	15 (Notification Class)
Object_Name	Required	"Critical Alarms"
Notification_Class	Required	3
Priority	Required	15,20,20
Ack_Required	Required	TRUE, TRUE, TRUE
Recipient_List	Required	(Monday, Tuesday, Wednesday, Thursday, Friday, Saturday, Sunday); 0:00, 24:00; (Device 12); 27; TRUE; 1,1,1; (Monday, Tuesday, Wednesday, Thursday, Friday); 6:00, 20:00; (Device 14); 27; TRUE; 1,1,1;

Recipient_List

Valid Days	Monday, Tuesday, Wednesday, Thursday, Friday, Saturday, Sunday	Monday, Tuesday, Wednesday, Thursday, Friday,
From Time, To Time	0:00, 24:00	6:00, 20:00
Recipient	Device 12	Device 14
Process Identifier	27	27
Issue Confirmed Notifications	TRUE	TRUE
Transitions	1,1,1	1,1,1

* Only selected properties listed. See BACnet standard for complete list.

Figure 13-11. Notification Class objects contain information necessary to distribute notification messages for alarms and events.

A Recipient element entry is either a Device Identifier or a BACnet network number and MAC address pair. If the first is used, then the Notification Class is independent of MAC address or network numbering changes, though dynamic device binding must be completed prior to sending the notification. If the second is used, the node does not need dynamic device binding, but this may create issues if the network number or MAC addresses change.

The Process Identifier is a number that is meaningful to the recipient nodes to classify the intended process that receives alarms. This can be useful, for example, in having different processes in a workstation handling life safety alarms, critical equipment alarms, and maintenance alarms.

The Issue Confirmed Notifications element determines whether an UnconfirmedEventNotification or a ConfirmedEventNotification is sent. Since they are more reliable, most alarm reporting devices use confirmed notifications.

The Transitions flags specify which transitions result in a notification. For example, there may be a need to send alarm notifications only for TO-OFFNORMAL and TO-FAULT, but not TO-NORMAL occurrences.

Event Log Objects

If the receiving workstation is not always available, it could miss important alarm/event notifications. Event logging is used to back up this type of system so that no notifications are lost. BACnet nodes not only issue alarm/event notifications normally, but also locally store the notifications in Event Log objects. **See Figure 13-12.** Periodically, some other node (like an alarm workstation) uses the ReadRange service to collect these saved notifications, which can then be deleted. The setup and control of the event gathering process is managed by properties of the Event Log object.

The Log_Buffer property is the storage area in an Event Log object. Each of the log records contains a Timestamp that documents when the record was entered in the log and a LogDatum that contains the details of the alarm/event notification.

The Event Log objects should be read frequently to avoid filling the storage space. As the storage fills, the Event Log object initiates a ConfirmedEventNotification message to alert the collection nodes that the log needs to be retrieved. The Notification_Class property determines which nodes receive this special kind of alarm.

> The Event Log object is a new object added in Addendum 135-2004b. While it logs different information, it functions similarly to Trend Log objects and even includes many of the same properties.

Event Log Object		
Property*	Type	Example Values
Object_Identifier	Required	(Event Log 1)
Object_Type	Required	25 (Event Log)
Object_Name	Required	"Event Log"
Enable	Required	TRUE
Log_Buffer	Required	(28-JUL-2009,13:10:33.6), (8, (Device 3), (Analog Input 11), (28-JUL-2009,13:10:30.7), 3, 100, OUT_OF_RANGE, ALARM, TRUE, NORMAL, HIGH_LIMIT, (87.0, (TRUE, FALSE, FALSE, FALSE), 2.0, 85.0)); (28-JUL-2009,13:12:49.5), LOG_INTERRUPTED; ...
Record_Count	Required	53
Notification_Threshold	Optional	100
Notification_Class	Optional	3

* Only selected properties listed. See BACnet standard for complete list.

Figure 13-12. Event Log objects receive EventNotification messages and save the information in a continuous log.

Some Event Log records are special because they keep a history of certain important events. For example, whenever the Log_Enable is changed, an entry is made in the log. If Record_Count is written with a 0, then a buffer-purged record is made in the log. If the node's time or date is changed, then a time-change record is entered. The Log_Buffer is a running history of everything that has happened to the Event Log object.

Alarm/Event Summary Services

A BACnet client can also discover missed notifications or other notification details by using services. This is particularly useful if the client was unavailable for some time and no longer knows the current state of alarms/events, their priorities, or acknowledgements.

GetAlarmSummary Service. The GetAlarmSummary service is used to retrieve a summary of active alarms, which refers to objects with non-NORMAL states and a Notify_Type of ALARM. Whether an alarm/event is reported in the result of the GetAlarmSummary service is the only distinction BACnet makes between alarms and events. Objects with a Notify_Type of EVENT are not reported in the alarm summary.

The GetAlarmSummary message requires no parameters. **See Figure 13-13.** If there are any alarms to report, the responding message is a List of Alarm Summaries, which includes the Object Identifier, Alarm State, and Acknowledged Transitions parameters.

GetEventInformation Service. The GetEventInformation service is used to retrieve all or part of the list of active event state objects in a node. This includes both alarms and events, as designated by the object's Notify_Type property. An object has an active event state if its Event_State property is not NORMAL or if any of the bits in the Acked_Transitions property are FALSE.

Similar to the GetAlarmSummary service, a GetEventInformation service message may not include any parameters. **See Figure 13-14.** The responding message is a summary of the applicable objects and their current status. More events may fit the criteria than can be included in one response message. In that case, the response includes a More Events parameter set to TRUE. Then, the initiating node sends another GetEventInformation message with the one parameter Last Received Object Identifier to continue the list where it was left off.

GetEnrollmentSummary Service. The GetEnrollmentSummary service is used to retrieve a list of all notification-generating objects that meet certain criteria. **See Figure 13-15.** Summaries can be compiled based on any combination of acknowledgement status, recipient, Process Identifier, Event_State, Event_Type, Priority, and Notification Class number.

GetAlarmSummary Service

GetAlarmSummary {
}

EXAMPLE MESSAGES

Result {
List of Alarm Summaries = (Analog Input 11); HIGH_LIMIT; 0,1,1;
(Binary Value 2); OFFNORMAL; 1,1,1 }

EXAMPLE RESPONSE

Figure 13-13. The GetAlarmSummary service queries a node for a history of its alarm (but not event) occurrences, which are returned as a list in a responding message.

GetEventInformation Service

GetEventInformation {
}

EXAMPLE MESSAGES

Result {
List of Event Summaries = (Analog Input 11); HIGH_LIMIT; 0,1,1;
(10-JAN-2010, 15:32:56.3); ; ;
ALARM; 1,1,1; (100,100,100);
(Averaging 4); NORMAL; 0,1,1;
(10-JAN-2010, 15:35:49.7); ; ;
EVENT; 1,1,1; (50,50,100); }

EXAMPLE RESPONSE

Figure 13-14. The GetEventInformation service queries a node for all of its alarm/event occurrences and related information.

GetEnrollmentSummary Service

GetEnrollmentSummary {
Acknowledgement Filter = NOT-ACKED }

GetEnrollmentSummary {
Enrollment Filter = (Analog Input 11), 8
Event State Filter = ALL
Notification Class Filter = 3 }

EXAMPLE MESSAGES

Result {
List of Enrollment Summaries = (Analog Input 11); OUT_OF_RANGE; HIGH_LIMIT; 100; 3 }

EXAMPLE RESPONSE

Figure 13-15. The GetEnrollmentSummary service message include parameters to query a node for alarms/events matching a specific set of criteria.

> A shortcut to specifying a value applicable to an entire day is to include only one TimeValue, for the time of 00:00. This format is leftover from a previous version of BACnet that supported only whole-day scheduling, though many TimeValues per day are now supported.

SCHEDULING

Scheduling is the process of setting up automated actions to take place at specific times on specific dates. In BACnet, scheduling information is contained by the Calendar object type and Schedule object type.

Calendar Objects

Calendar objects are not required for scheduling, but they are useful for controlling multiple activities that normally have different time schedules but share common dates for special events, holidays, or other dates. They also provide a standardized way of representing calendar information across different vendors.

The Calendar object contains a collection of days of a hypothetical year that are important for a schedule. **See Figure 13-16.** The Calendar object's Present_Value is a Boolean variable that is TRUE or FALSE based on whether the current day is included within a list of dates (the Date_List property). While it is possible, and common, to use the Date_List property for a single date, the real power of the Calendar object is in representing multiple dates.

CalendarEntries. The Date_List property is composed of one or more CalendarEntries, a special datatype that is used in both Calendar and Schedule objects. A CalendarEntry can use one of three types of specification for dates: Date, DateRange, and WeekNDay. In all three forms, several parameters describe the CalendarEntry. A particularly powerful feature is that any or all of the parameters may use an ANY (wildcard) value, under certain circumstances.

The simplest CalendarEntry specification is a Date. A Date has three parameters: month, day of the month, and year. Months have a value from 1 to 12, representing January through December. Days have a value from 1 to 31, as appropriate for the specified month. That is, if the month is January, then the day may be 1 to 31, but if the month is April, then the day may only be 1 to 30. The year parameter is limited to the range of 1900 to 2154. By using the ANY value in one or more parameters, complex dates can be specified. For example, the parameters month = ANY, day = 1, year = ANY represent the first day of any month of any year.

The form DateRange, represents an inclusive range of dates from a startDate parameter to an endDate parameter. These dates are specified in the same way as the Date form. Because the standard is not definitive about the meaning of DateRanges with ANY fields, care must be taken not to use partially wild dates to define a range, as this may not interoperate across all nodes. However, it is generally safe to use all ANY fields in either the startDate or endDate. For example, a startDate of month = ANY, day = ANY, year = ANY and an endDate of month = May, day = 20, year = 2009 means any date up to and including May 20, 2009.

Calendar Objects

Property*	Type	Example Values
Object_Identifier	Required	(Calendar 2)
Object_Type	Required	6 (Calendar)
Object_Name	Required	"Holidays"
Present_Value	Required	TRUE
Date_List	Required	(4-JUL-2009); (NOV, 4, 4); (22-DEC-2009)-(26-DEC-2009)

* Only selected properties listed. See BACnet standard for complete list.

Figure 13-16. Calendar objects determine whether the current date matches any of the CalendarEntries saved in its Date_List. CalendarEntries can be specified in very flexible ways, providing a dynamic set of dates.

The last form is WeekNDay. The parameters are the month, the weekOfMonth, and the dayOfWeek. The month may be 1 to 12, for January through December. The special values of 13 and 14 are designated for odd months and even months, respectively. The weekOfMonth may be 1 to 6 or ANY. Since no month has a sixth week, 6 has a special value, meaning the last week of this month. The dayOfWeek may be 1 to 7, where 1 = Monday, or ANY. This powerful format allows for groups of dates like the second Tuesdays of odd-numbered months.

Schedule Objects

A Schedule object is a network-visible representation of a time-based series of events. The Schedule object provides a collection of TimeValues, each of which is a combination of a particular time and a property value. **See Figure 13-17.** The time specifies when the value should be applied.

The Present_Value of a Schedule object is the most recently applied TimeValue. Programs or other objects within the node can reference the Present_Value, but typically, the Schedule object writes the appropriate TimeValue value to one or more properties in other objects. These properties may be within the same node, or in other nodes.

The Effective_Period property specifies a range of dates when a particular Schedule object is applicable. This is particularly useful for seasonal scheduling by using non-overlapping ranges for several Schedule objects.

Schedule objects are implemented to manage systems that have typical daily ON and OFF times, such as common area lighting.

Schedule Objects

Property*	Type	Example Values
Object_Identifier	Required	(Schedule 1)
Object_Type	Required	17 (Schedule)
Object_Name	Required	"HeatSetpointSched"
Present_Value	Required	72.0
Effective_Period	Required	(1-SEP-2009)-(30-APR-2010)
Weekly_Schedule	Optional	Weekly_Schedule[1] = (6:00, 72.0), (20:00, 68.0) Weekly_Schedule[2] = (6:00, 72.0), (20:00, 68.0) Weekly_Schedule[3] = (6:00, 72.0), (20:00, 68.0) Weekly_Schedule[4] = (6:00, 72.0), (20:00, 68.0) Weekly_Schedule[5] = (6:00, 72.0), (20:00, 68.0) Weekly_Schedule[6] = (0:00, 68.0) Weekly_Schedule[7] = (0:00, 68.0)
Exception_Schedule	Optional	Exception_Schedule[1] = (Holidays, (0:00, 68.0), 7) Exception_Schedule[2] = (30-NOV-2009, (0:00, 68.0), 7)
List_Of_Object_Property_References	Required	(Device 3, Analog Value 1, Present_Value)
Priority_For_Writing	Required	10

* Only selected properties listed. See BACnet standard for complete list.

CALENDARENTRY

Exception_Schedule

Period	Holidays	30-NOV-2009
TimeValue	(0:00, 68.0)	(0:00, 68.0)
EventPriority	7	7

CALENDAR (OBJECT)

Figure 13-17. Schedule objects contain the values of certain object properties that change based on time of day. Calendar object references can be included to further customize time schedules based on special dates.

The Weekly_Schedule and Exception_Schedule properties provide collections of periodic and special dates, with groups of TimeValues for those dates. The Weekly_Schedule property is an array of seven lists of TimeValues for each day of the week. On a normal weekday, the corresponding list from the Weekly_Schedule array specifies times and the values to be written. The Exception_Schedule property is an array of SpecialEvents, which are dates and circumstances under which the Weekly_Schedule is overridden with a different set of TimeValues. A SpecialEvent is a complex datatype that specifies a list of TimeValues, but qualifies them in two important ways. First, it specifies a CalendarEntry or Calendar object for the dates that the SpecialEvent is valid. Second, each SpecialEvent includes an EventPriority from 1 (most important) to 16 (least important). This is used when multiple SpecialEvents are applicable at the same time because the calendar dates overlap. The SpecialEvent with the highest priority (lowest number) takes precedence. A Schedule object is not required to have both a Weekly_Schedule and an Exception_Schedule, but it must have at least one of these properties.

For example, a Schedule object may be used to change the settings of a HVAC system in an office building. During normal workdays, the Weekly_Schedule specifies heating and cooling setpoints for different times of the day, based on when employees are expected to occupy the building. The Weekly_Schedule is also used to implement different setpoints for the weekend, when no one is normally in the building. However, holidays are added to the Schedule object as SpecialEvents, which override the normal weekday schedule and adjust the settings for an unoccupied building. Similarly, other SpecialEvents can be set up with more important priorities that override even the holiday schedule.

The Schedule object's List_Of_Object_Property_References property specifies a list of properties to be written with the value specified by the applicable TimeValue. Each property reference includes an Object Identifier, Property Identifier, and, if applicable, a Property Array Index. A Device Identifier parameter is optional, depending on whether the property to be written resides in another node. However, the ability to write TimeValue information into other nodes may not be available. External scheduling devices are nodes whose Schedule objects can write to other nodes. Internal scheduling devices are nodes that can only write to their own objects.

When the Schedule object writes a property, it is done at the priority specified by the Priority_For_Writing property. This corresponds to the Priority parameter of the WriteProperty service. However, it is different from the EventPriority parameter, which arbitrates among simultaneously applicable SpecialEvents.

Schedules are not required to have any property references in the list. Some nodes allow the list to be written using the WriteProperty or AddListElement services. Other nodes may simply present the list as read-only.

TRENDING

Trending records values of one or more variables over time. A data point is sampled repeatedly at a certain time interval and the results are stored for later use. Each data sample usually includes a timestamp, which helps deal with missing data due to network traffic, power outages, or other issues. Trending involves first gathering the data, then saving it in an organized format.

Data Sampling

Data can be sampled through periodic polling or change-of-value reporting. **See Figure 13-18.** With periodic polling, ReadProperty or ReadPropertyMultiple services are used to read properties at a certain interval. However, like polling for other types of information, this method requires a balance between update frequency and increased network traffic.

Alternatively, COV subscriptions can be used to gather data samples. This significantly reduces network traffic but requires the BACnet nodes involved to support COV services. Many do not, or are limited in the number of simultaneous subscriptions they can provide.

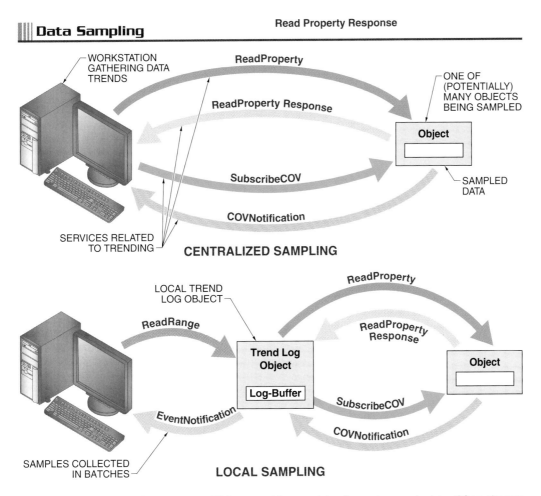

Figure 13-18. For trending purposes, BACnet provides a variety of ways to sample data. Often, the use of Trend Log objects is the most efficient method.

Also, the saved trend data is usually transferred from where it has been recorded to another node, such as a workstation, where it can be examined or archived for later use. Therefore, it is often not practical to use centralized sampling for a significant number of trends (sampled properties). An alternative is for multiple BACnet nodes to sample and collect trend data. Pushing the sampling of trend data out to as close to the data source as possible is typically a more efficient use of network resources. Although there are proprietary ways to do this, the standard method uses Trend Log objects to store the samples. The Trend Log object can be thought of as a container for locally sampled trend data, but it also standardizes how nodes are set up for trending and how their collected data is accessed.

Trend Log Objects

The Trend Log object can be used to sample properties in objects within the same node (internal sampling) or in other nodes (external sampling). **See Figure 13-19.** External sampling is typically used by supervisory controllers to gather information from their constituent nodes. As with centralized sampling, the Trend Log object may implement either polling or COV methods for gathering data samples.

A node samples and saves the trend data in a Trend Log object. The setup and control of the trending process is defined by the properties of the Trend Log object. Another node then uses the ReadRange service to collect a number of the samples, which may then be deleted by the sampling node.

Trend Log Objects

Property*	Type	Example Values
Object_Identifier	Required	(Trend Log 3)
Object_Type	Required	20 (Trend Log)
Object_Name	Required	"TempTrends"
Log_Enable	Required	TRUE
Log_DeviceObjectProperty	Optional	(Device 3, Analog Input 11, Present_Value)
Log_Buffer	Required	(19-MAY-2009, 12:38:19.0); 71.5; (FALSE, FALSE, FALSE, FALSE); (19-MAY-2009, 12:38:49.0); 71.4; (FALSE, FALSE, FALSE, FALSE); (19-MAY-2009, 12:39:19.0); 71.3; (FALSE, FALSE, FALSE, FALSE); ...
Record_Count	Required	276
Notification_Class	Optional	3

* Only selected properties listed. See BACnet standard for complete list.

Figure 13-19. In addition to the Log_Buffer property, which can hold hundreds of data samples, the Trend Log object includes information to manage and configure the data sampling actions.

The Log_Buffer property is the storage area for the trended sample data and is organized similarly to the Log_Buffer property in an Event Log object. Each of the log records contains a Timestamp, StatusFlags (representing the Status_Flags property of the monitored object, if available), and the LogDatum. The LogDatum is typically a value of the sampled property, but may instead be a record of Trend Log events. For example, log entries are made when the Log_Enable is changed, the Record_Count is written with a 0, a sampling error occurs, or the node's time or date is changed. The Log_Buffer is a running history of everything that has happened to the Trend Log object.

Samples should be collected more frequently than they are accumulated in the sampling node so that it does not run out of room to save them. However, the Trend Log object can initiate ConfirmedEventNotification messages to the collection node when its internal storage is full, or close to full. The Notification_Class property of the Trend Log object identifies a Notification Class object that lists which nodes are recipients of this "need to collect" notification.

Sometimes more than one property must be sampled at a time, even from different objects. It may be possible to employ more than one Trend Log object, but BACnet provides the Trend Log Multiple object to make this easier. This object simplifies the data gathering and consolidation of samples from multiple trended sources.

Refer to Quick Quiz® on CD-ROM

Summary

- Change-of-value (COV) notification allows the client to request that the nodes containing the desired properties only send data when the property value changes by a certain amount.

- Each BACnet object that supports COV subscriptions defines the criteria used to detect changes of value.

- The SubscribeCOVProperty service initiates COV notifications for properties that are different from the default properties for a given standard object type.

- The Subscriber Process Identifier is repeated in COV notifications so that the client can match the notification to the subscription.

- Some BACnet nodes can broadcast UnconfirmedCOVNotification messages without a subscription for announcing changes of value.

- The transition from one alarm/event state to another can be configured to trigger the notification process.

- Each of the three states can independently require human acknowledgement, so up to six states are possible: NORMAL, OFFNORMAL, OFFNORMAL ACKNOWLEDGED, FAULT, FAULT ACKNOWLEDGED, and RETURN TO NORMAL.

- Objects that provide intrinsic reporting provide certain properties to control alarming behavior.

- The only distinction between an alarm and an event is whether the object is returned in a GetAlarmSummary request.

- Algorithmic change reporting can be applied to any property of any object, but it requires the use of Event Enrollment objects.

- In algorithmic change reporting, the object properties are tested using a standard monitoring algorithm to detect changes in state.

- Nine standard algorithms are specified for BACnet alarming, though nonstandard algorithms can be implemented instead.

- Once an alarm/event has been detected, one or more other nodes are notified of the occurrence using either the ConfirmedEventNotification or UnconfirmedEventNotification services.

- Notification Class objects are used to define which nodes receive EventNotification messages.

- A Notification Class object is referred to by its instance number by intrinsic reporting objects and Event Enrollment objects.

- When BACnet nodes issue alarm/event notifications, they also locally store the notifications in Event Log objects. This log can be read periodically by some other node (like an alarm workstation) to collect these saved notifications.

- The GetAlarmSummary service is used to retrieve a summary of active alarms, which refers to objects with non-NORMAL states and a Notify_Type of ALARM.

- The GetEventInformation service is used to retrieve all or part of the list of active event state objects in a node, including both alarms and events, as designated by the object's Notify_Type property.

- The Calendar object contains a collection of days of a hypothetical year that are important for a schedule. Its Date_List can use any of three types of specification for dates: Date, DateRange, and WeekNDay.

- The Schedule object provides a collection of TimeValues, each of which is a combination of a particular time and a property value.

- The Weekly_Schedule property is an array of seven lists of TimeValues for each day of the week.

- The Exception_Schedule property is an array of SpecialEvents, which are dates and circumstances under which the Weekly_Schedule is overridden with a different set of TimeValues.

- The Trend Log object stores locally sampled trend data and standardizes how nodes are set up for trending and how their collected data is accessed.

- The Trend Log object can be used to sample properties in objects within the same node (internal sampling) or in other nodes (external sampling).

- The Log_Buffer is a running history of everything that has happened to the Trend Log object.

Definitions

- *Intrinsic reporting* is an alarming feature of certain BACnet objects where the alarm or event originates from the object.
- *Algorithmic change reporting* is a BACnet alarming feature that uses Event Enrollment objects to configure the alarm/event behavior of some other object property.

Review Questions

1. How is change-of-value (COV) notification different from using ReadProperty services to detect changes in a property value?

2. What information is needed to complete a COV subscription?

3. Describe the general function of intrinsic reporting.

4. What is the role of the Event Enrollment object in algorithmic change reporting?

5. How is the instance number of a Notification Class object used to specify the recipients of an alarm/event notification?

6. What are the primary differences between the list items returned by the GetAlarmSummary service and the GetEventInformation service?

7. In what ways can a CalendarEntry be specified?

8. What is the difference between a Schedule object's Weekly_Schedule and Exception_Schedule properties?

9. What three methods can be used for periodic data sampling?

10. What types of information is stored in a Trend Log object's Log_Buffer?

Chapter Fourteen

BACnet Special Applications

For some special applications, the use of basic objects and core services is not an efficient way to model the necessary control behaviors. An alternative is implementing proprietary objects, though this can negatively affect interoperability. To address this issue, BACnet provides a group of objects and services designed for special applications. These components include built-in properties or parameters that are specifically designed for certain control features. Many of these objects and services were added to the standard through addenda.

Chapter Objectives

- Identify the special objects used to organize the features of an access control system.
- Describe the special features available in the Lighting Output object and how they are configured and initiated.
- Describe the properties and function of the Load Control object.
- Differentiate between Life Safety Point and Life Safety Zone objects, and explain the role of the LifeSafetyOperation service.
- Compare the organization of data in File objects and how they are accessed.
- Describe how virtual terminal services are used to establish special communication links between BACnet nodes.

SPECIAL APPLICATION OBJECTS AND SERVICES

The basic objects and core services were specifically designed to be flexible enough to be used in a wide variety of applications, even very specialized ones. However, BACnet is continually subject to suggestions for improvement and expansion. As part of the periodic changes to the standard, new objects and services are added to provide better control solutions for certain applications. **See Appendix.** For example, the Addendum 135-2004*e* added a new Load Control object type as a BACnet standard object. Other addenda added several other new object types, in addition to making other additions and changes that improve BACnet functionality. By standardizing their form and function, these new components preserve BACnet's interoperability.

Access Control Objects

An access control system is abstracted into the components needed for access control, interfacing these special objects with other standard objects such as Binary Outputs for actuating door strikes and Binary Inputs for sensing door contacts. **See Figure 14-1.**

Access Point objects represent an access-controlled area's entry or exit points, such as doors with credential readers and electric door strikes.

An access point is used to control entry into, or exit from, any space designated as an access-controlled area. An access-controlled area may involve several access points. BACnet provides Access Point objects to represent the relatively simple function of a physical access control device, such as a door or gate. Also, the Access Zone object represents a collection of multiple Access Point and/or Access Zone objects that logically form an access-controlled area.

Access User objects define people with Access Credentials, which define the types of devices, such as magnetic stripe cards, they may use. Access Rights objects define the rights a kind of user has, so each Access User is associated with some number of Access Rights. Access Rights are also associated with BACnet schedules so that rights or zones may be restricted to specific times and dates.

Credential Data Input objects define types of numeric strings used with authenticating devices such as magnetic stripe or fingerprint readers. These can be associated with specific Access Users and Access Rights.

Access Door objects are used to abstract a collection of hardware points associated with the door sensing and opening mechanism and related points.

Lighting Output Objects

Basic lighting control is easily accomplished using the nine binary, analog, and multi-state object types. However, lighting applications can get complex, requiring additional functionality and parameters that cannot be modeled with basic object types. Some systems use a combination of standard objects and nonstandard properties to accomplish these special lighting functions, but a better solution is a standard BACnet object that already has properties to represent these concepts. Therefore, BACnet added a special Lighting Output object type that includes properties to configure these special features. **See Figure 14-2.**

Similar to an Analog Output object, the Lighting Output object's Present_Value represents a 0% to 100% lighting intensity output. However, several other properties are used to control how the Present_Value changes, especially over time, to replicate special lighting functions.

Access Control Objects

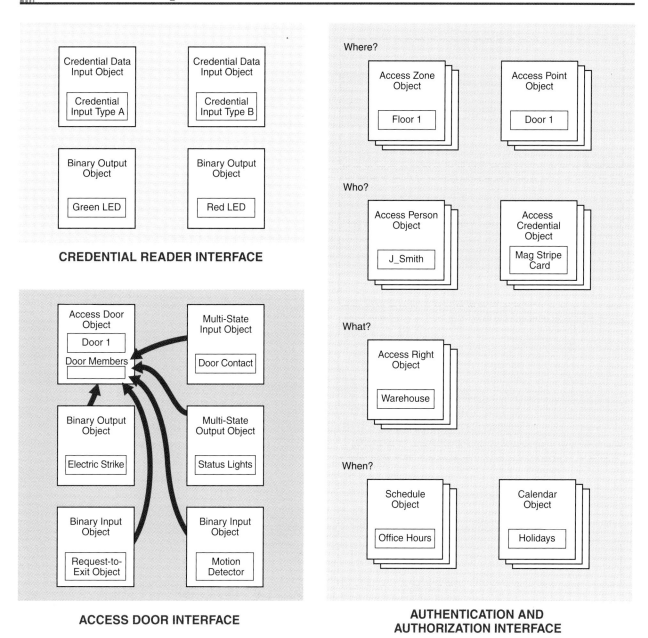

Figure 14-1. A variety of access control objects are available to organize the abstract concepts of authentication, authorization, and access.

Blink Warning. A common feature in ON/OFF lighting control is a blink warning. This feature blinks the lights briefly to alert occupants that the lights are due to turn OFF again soon, unless they take action to override the normal control. Lighting control nodes typically initiate a blink warning at the end of a scheduled occupancy period.

> The blink warning feature may be implemented with one or multiple blinks during the Off_Delay period, depending on the node manufacturer's preference. There is currently no standard object property to directly control the number of blinks, though manufacturers can define proprietary properties to configure this and other behaviors.

Lighting Output Objects

Property*	Type	Example Values
Object_Identifier	Required	(Lighting Output 3)
Object_Type	Required	31 (Lighting Output)
Object_Name	Required	"TrainingRoom"
Present_Value	Required	78%
Lighting_Command	Required	FADE_TO_OVER, 30, 20
Fade_Time	Optional	30
Ramp_Rate	Optional	3.0
Blink_Time	Optional	60
Blink_Priority_Threshold	Optional	9
Off_Delay	Optional	60
Step_Increment	Optional	5.0
Min_Pres_Value	Optional	10.0
Max_Pres_Value	Optional	100.0
Lighting_Command_Priority	Required	10
Binary_Active_Value	Optional	50%
Binary_Inactive_Value	Optional	45%
Minimum_Off_Time	Optional	60
Minimum_On_Time	Optional	0

* Only selected properties listed. See BACnet standard for complete list.

Figure 14-2. The Lighting Output object provides many properties for defining complex lighting behavior.

The blink warning is supported in the Lighting Output object through properties including Blink_Time, Off_Delay, and Blink_Priority_Threshold. **See Figure 14-3.** If the Blink_Time property has a nonzero time value, it indicates that a blink warning is desired and the output turns OFF for Blink_Time. Then the light turns ON again for Off_Delay time, until it turns OFF again. If it is necessary to override the blink warning feature and turn the lights OFF immediately, the Present_Value is written with a priority greater (lower priority number) than the Blink_Priority_Threshold. If the priority number that accompanies the write to Present_Value is higher (less important) than Blink_Priority_Threshold, then the blink warning operates as configured.

ON/OFF Operation. The Lighting Output object provides a unified interface to lighting outputs, regardless of whether the end actuation is analog or binary. To facilitate ON/OFF outputs with analog inputs, the properties Binary_Active_Value and Binary_Inactive_Value define the thresholds at which the Present_Value becomes ACTIVE and INACTIVE.

The output is ACTIVE if the Present_Value exceeds the Binary_Active_Value and INACTIVE is the Present_Value is lower than Binary_Inactive_Value. If these properties are not implemented, then Present_Value is INACTIVE when it equals 0, and ACTIVE otherwise. The Polarity property defines whether the ACTIVE state represents ON (normal) or OFF (reverse).

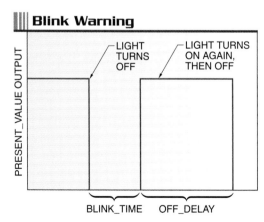

Figure 14-3. A blink warning is a lighting behavior that alerts occupants that the lights will shortly shut off automatically.

The Minimum_On_Time and Minimum_Off_Time properties, though the names can cause confusion, indicate the minimum ACTIVE time and minimum INACTIVE time. When the output switches between ACTIVE and INACTIVE, these minimum times keep changes in Present_Value from affecting the output for a short time.

Clamping. *Clamping* is the restriction of lighting level between minimum and maximum levels, as long as the light is ON. **See Figure 14-4.** Although the total range for dimmable lights is typically 0% to 100%, there may be a need to distinguish between OFF (0%) and a minimum allowable level, such as 15%. Similarly, the maximum level in a particular area under particular conditions may be restricted to less than 100%. The configured minimum and maximum levels vary with application and scene presets.

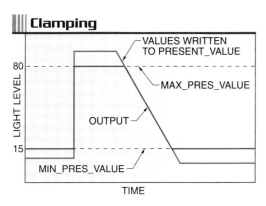

Figure 14-4. *Clamping restricts actual lighting level output between two values, even if the Present_Value is written with a value outside of this range.*

The optional properties Min_Pres_Value and Max_Pres_Value allow the Present_Value to be clamped. For example, if Min_Pres_Value equals 15 and Max_Pres_Value equals 80, when Present_Value is written with 10, the Present_Value and output are clamped to 15% instead. When Present_Value is written with 100, Present_Value and output are clamped to 80%.

Fading, Ramping, and Stepping. It may also be necessary to control the rate at which the level changes from one setting to another. The three common lighting operations for level-based dimmers are fading, ramping, and stepping. **See Figure 14-5.** *Fading* is the changing from one lighting level to another over a fixed period. For example, a light is dimmed from 100% to 50% over a 10-sec period. *Ramping* is the changing from one lighting level to another at a fixed rate of change of level per second. For example, a light is ramped up to 100% at a rate of 5% per second. The output effect of fading and ramping is the same, only the way they are specified is different. *Stepping* is the incremental increasing or decreasing of lighting level in fixed steps.

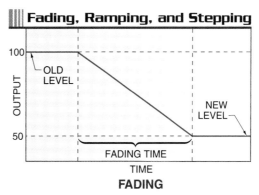
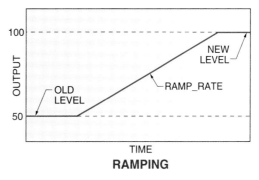
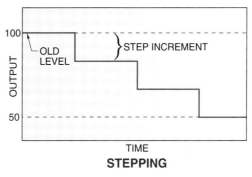

Figure 14-5. *Lighting can be set to transition gradually between two levels using fading, ramping, and stepping features.*

Fading, ramping, and stepping are special control functions for two reasons. First, these functions are initiated by writing to a property, but are carried out over time. Second, each operation requires additional parameters, which are also specific to the desired effect.

Fading, ramping, and stepping are initiated with the property Lighting_Command, which is written with a compound value that specifies the lighting operation and configures its operation. When Lighting_Command is written, any in-progress operation is cancelled first. Lighting_Command is not commandable and no attempt is made to synchronize multiple writers, so the most recent write prevails. However, when lighting operations occur, they affect the Present_Value at the priority level specified in the Lighting_Command_Priority property, and the Present_Value is commandable. The Lighting_Command property allows for several special operations. **See Figure 14-6.**

The simplest operation is STOP, which stops any previously initiated operation regardless of whether it has completed. For example, if lights are fading from 100% to 50% over 10 sec, and a STOP is received 4 sec into the fade, further fading is cancelled, leaving the output at 80%.

The operation GOTO_LEVEL has a single parameter for the desired target level. This has the same effect as writing the level directly to Present_Value.

There are two operations for fading. The FADE_TO operation specifies only a target level to fade to. A default value in the Fade_Time property specifies the period of time over which the fade occurs. Alternatively, the FADE_TO_OVER operation includes as parameters the fade time as well as the target level.

There are six ramping operations. The RAMP_TO operation specifies only a target level. The Ramp_Rate property specifies a default rate of change for ramping. When the desired level is reached, the ramping stops. The RAMP_TO_AT_RATE operation specifies both the target level and the ramping rate.

The RAMP_UP operation increases Present_Value with no specific target level (in effect, the target level is Max_Pres_Value) and the Ramp_Rate specifies a default rate of change. The RAMP_UP_AT_RATE operation also has no target level, but does specify the rate. Similarly, the RAMP_DOWN operation decreases Present_Value with no specific target level (in effect, the target level is Min_Pres_Value) at the Ramp_Rate. The RAMP_DOWN_AT_RATE operation also has no target level, but specifies the rate.

There are four stepping operations. The STEP_UP operation increases the output by an amount specified in the Step_Increment property. This has the same effect as if Present_Value was written with a value of Present_Value plus Step_Increment. The STEP_DOWN operation decreases the output by an amount specified in the Step_Increment property. The STEP_UP_BY and STEP_DOWN_BY operations have similar effects, except that they specify the increment amount as a parameter in the operation.

Lighting_Command Property Operations

Operation	Parameters
STOP	none
GOTO_LEVEL	level
FADE_TO	level
FADE_TO_OVER	level, fade-time
RAMP_TO	level
RAMP_TO_AT_RATE	level, ramp-rate
RAMP_UP	none
RAMP_UP_AT_RATE	ramp-rate
RAMP_DOWN	none
RAMP_DOWN_AT_RATE	ramp-rate
STEP_UP	none
STEP_DOWN	none
STEP_UP_BY	step-increment
STEP_DOWN_BY	step-increment
RELINQUISH	none

Figure 14-6. The Lighting_Command property of the Lighting Output object can be written with any of 15 commands and their associated parameters to initiate different lighting features.

Nodes are not required to support every Lighting_Command operation. If the property is written with an unsupported value, the write operation fails and returns an error message.

An important lighting command is RELINQUISH. This writes a NULL value (relinquish) to the Present_Value at the priority specified by Lighting_Command_Priority. In effect, the control of the Present_Value is released at that priority level.

Load Control Objects

The Load Control object is used to implement electrical load shedding. **See Figure 14-7.** Load control is state driven based on whether the Load Control object is ENABLED and able to achieve the requested shed level. The Load Control logic is provided target information through the Requested_Shed_Level, Start_Time, Shed_Duration, and Duty_Window properties. The Present_Value indicates the current state of the shedding logic.

The Requested_Shed_Level property expresses the level requested in terms of PERCENT, LEVEL, or AMOUNT parameters. PERCENT is a percentage of the current baseline load, such as 80%, which is 20% less than the current baseline. LEVEL is an index into the Shed_Levels array of preset target levels. AMOUNT is an actual amount subtracted from the current baseline, such as 15 kW.

The Start_Time indicates a future time by which the shedding target is achieved. This defines the beginning of the duty window. Shed_Duration is the time, in minutes, that the system must maintain the shed level. Duty_Window is the length of time in minutes used for load-shed accounting. The average power consumed during the duty window must be less than or equal to the requested reduced level of consumption.

Life Safety Objects and Service

Although all BACnet objects and services can be used in fire and life safety applications, BACnet includes specific object types and a special service for these applications that address their special characteristics. Life safety points must maintain both a mode and a state. The operating mode of a life safety node selects the logic to be applied under different circumstances, which subsequently affects the evaluation of the state of the node. Also, it is common for life safety nodes to lock into a particular state until deliberately reset.

For example, a smoke detector behaves differently in test mode than it does in service mode or normal mode. Once smoke is detected and the device is in the ALARM state, the device remains locked in the ALARM state until deliberately reset, even if the smoke clears.

Life Safety Point Objects. The simplest kind of life safety point, such as a smoke detector, pull station, notification appliance, or annunciator, is represented with a Life Safety Point object. **See Figure 14-8.** Its Present_Value is a special enumeration called BACnetLifeSafetyState with values such as QUIET, PRE-ALARM, ALARM, FAULT, NOT-READY, ACTIVE, TAMPER, TEST-ALARM, HOLDUP, DURESS, and ABNORMAL. A Mode property, which is always writable, may include enumerations such as OFF, ON, TEST, MANNED, UNMANNED, ARMED, and DISARMED. The properties Life_Safety_Alarm_Values, Alarm_Values, and Fault_Values define the states that correspond to major alarm types.

Load Control Objects

Property*	Type	Example Values
Object_Identifier	Required	(Load Control 1)
Object_Type	Required	28 (Load Control)
Object_Name	Required	"Chiller Load Control"
Present_Value	Required	SHED_COMPLIANT
Requested_Shed_Level	Required	Percent, 80
Start_Time	Required	11-JUL-2009, 12:30:00.0
Shed_Duration	Required	120.0
Duty_Window	Required	30
Enable	Required	TRUE

* Only selected properties listed. See BACnet standard for complete list.

Figure 14-7. Scheduled load shedding for reducing electrical consumption can be controlled by the Load Control object.

Life Safety Point Objects

Property*	Type	Example Values
Object_Identifier	Required	(Life Safety Point 3)
Object_Type	Required	21 (Life Safety Point)
Object_Name	Required	"SmokeDetector23"
Present_Value	Required	PREALARM
Life_Safety_Alarm_Values	Optional	ALARM
Alarm_Values	Optional	PREALARM
Fault_Values	Optional	FAULT

* Only selected properties listed. See BACnet standard for complete list.

Figure 14-8. A Life Safety Point object represents a single input or output device, such as a smoke detector.

Life Safety Zone Objects. The Life Safety Zone object is used to represent logical zones that group one or more Life Safety Points together. **See Figure 14-9.** Its Present_Value is a reflection of the combined status of its constituent members. For example, if a zone has three members and one of them is in an ALARM state, the zone is also in an ALARM state. The properties of the Life Safety Zone object are essentially the same as Life Safety Point but also include a Zone_Members property that lists the Life Safety Point (and Life Safety Zone) objects that are members of this zone. Life Safety Zone objects can be a nested hierarchy of other zones and points. It is also possible for the same Life Safety Point to exist in multiple Life Safety Zones.

The writable Mode property of Life Safety Point and Life Safety Zone objects influences the evaluation of their states. Mode examples include MANNED, UNMANNED, ARMED, DISABLED, and TEST.

LifeSafetyOperation Service. When Life Safety Zone or Life Safety Point objects need to be reset, or when they represent annunciators or notification appliances that need to be silenced, BACnet provides a special service message to convey the information. The LifeSafetyOperation service conveys additional parameters necessary for life safety functions. **See Figure 14-10.**

LifeSafetyOperation Service

```
LifeSafetyOperation {
    Message = 123
    Requesting Process Identifier = 23
    Requesting Source = "J_Smith"
    Requesting = SILENCE
    Object Identifier = (Life Safety Point 2)
```

EXAMPLE MESSAGE

Figure 14-10. The LifeSafetyOperation service is used to initiate some action in a life safety device, such as silencing an alarming smoke detector.

Life Safety Zone Objects

Property*	Type	Example Values
Object_Identifier	Required	(Life Safety Zone 3)
Object_Type	Required	22 (Life Safety Zone)
Object_Name	Required	"3rdFloorSmokeDetectors"
Present_Value	Required	ALARM
Zone_Members	Required	(Life Safety Point 3), (Life Safety Point 4), (Life Safety Zone 2), …
Member_Of	Optional	(Life Safety Zone 6)

* Only selected properties listed. See BACnet standard for complete list.

Figure 14-9. Life Safety Zone objects are collections of Life Safety Point or other Life Safety Zone objects that are related by type, location, or function.

The Requesting Source parameter is the text-based name of the human operator making the request. Given the importance of life safety building systems, it is vital to document the name of the person making requests of the system. This information may then be displayed on other workstations and recorded in logs. The Request is the type of operation being requested, such as RESET, SILENCE_AUDIBLE, SILENCE_VISUAL, or SILENCE (both audible and visual notification appliances). The Object Identifier is for the Life Safety Point or Life Safety Zone object to which the operation should be applied. The Object Identifier is omitted if the operation applies to all of the objects in the node.

File Objects and Services

In BACnet, a file is a conceptual container for data. The content and meaning are not defined by BACnet. Only the conveyance and organization of the file itself are defined by BACnet. Files do not necessarily represent computer data files in the common sense of the term. Instead, BACnet files are a metaphor for a collection of data that can be accessed in any order and quantity.

File Objects. Every file in a BACnet node is represented by a File object, which corresponds roughly to the file folder or directory entry for a file in a computer file system. **See Figure 14-11.** The File object provides a name and description for the file through the Object_Name property, optional Description property, and File_Type property. The File object also provides properties for the File_Size, Modification_Date, and Read_Only information about the file.

Files are also distinguished by their organization, which affects the way that the file is accessed. The two types of file access are RECORD_ACCESS and STREAM_ACCESS. A given File object can only be one access type.

RECORD_ACCESS files are read or written in discrete chunks called records. For files of this type, the Record_Count property contains the number of records that the file contains. Records are like pages in a book. Each page may contain more or less data than another page, but they can only be accessed as these units. A portion of a file is accessed by specifying the starting record and number of records.

File Objects		
Property*	Type	Example Values
Object_Identifier	Required	(File 10)
Object_Type	Required	10 (File)
Object_Name	Required	"AnalogTrend"
File_Type	Required	"TREND"
File_Size	Required	834
Modification_Date	Required	12-DEC-2009, 18:45:37.2
Read_Only	Required	TRUE
File_Access_Method	Required	RECORD_ACCESS

* Only selected properties listed. See BACnet standard for complete list.

Figure 14-11. Large collections of data grouped into files are defined and accessed via File objects.

In contrast, a STREAM_ACCESS file is modeled as one long continuous stream of octets instead of a sequence of records. A portion of the file is read or written in blocks of a certain number of octets, starting at a particular octet offset from the beginning of the file.

Files are accessed using two special services: AtomicReadFile and AtomicWriteFile. The services use the term "atomic" to mean "uninterrupted." The BACnet client makes a request to read or write a specific section of the data represented by the File object, and during the time that the data is being read or written, no other changes are allowed to take place in the file data, including reads or writes by other clients to the same file.

AtomicReadFile Service. When using the AtomicReadFile service, the client specifies the starting position within the file and a requested count. **See Figure 14-12.** For RECORD_ACCESS files, the starting position is the starting record number (0 is the beginning), and the count is the number of records. For STREAM_ACCESS files, the starting position is the offset in octets relative to the beginning of the file data, and the number of octets.

AtomicReadFile Service

```
AtomicReadFile {
  File Identifier = (File 10)
  Record Access
    File Start Record = 15
    Requesting Record Count = 10 }
```

```
AtomicReadFile {
  File Identifier = (File 11)
  Stream Access
    File Start Position = 120
    Requested Octet Count = 27 }
```

EXAMPLE MESSAGES

```
Result {
  End of File = FALSE
  Record Access
    File Start Record = 15
    Returned Record Count = 10
    File Record Data = "42 75 69 6c 64 69 6e 67 20 41 75 74 6f 6d 61 74 69 6f 6e..." }
```

```
Result {
  End of File = TRUE
  Stream Access
    File Start Position = 120
    File Data = "42 75 69 6c 64 69 6e 67 20" }
```

EXAMPLE RESPONSES

Figure 14-12. Selected data can be read from a file by using the AtomicReadFile service.

The service returns the account of actual records or octets read, since it is possible that the client mistakenly tries to read more data than the file contains. The result also contains an indicator of whether the read includes the end of the file, meaning no more data is available.

AtomicWriteFile Service. The AtomicWriteFile service is similar to the AtomicReadFile service, except that it also provides the file data to be written. **See Figure 14-13.** The service includes parameters for both RECORD_ACCESS and STREAM_ACCESS files that specify a starting position and count for inserting the new data into the file.

As a special feature of the AtomicWriteFile service, the File Start Record or File Start Position may be specified as –1, which means the end of the file. This has the effect of appending the new data to the end of the file without requiring that the client know the offset to the end of the file.

Virtual Terminal Services

Many microcontroller-based devices provide a simple user interface through a standard TIA-232 serial port and a terminal. A *terminal* is a human-machine interface (HMI) that serves as an input and display device for text-based communication with another computer-based device. An operator interface program on the microcontroller transmits ANSI characters through the serial port, which are displayed on a terminal monitor. Any keyboard inputs are received on the serial port by the interface program. **See Figure 14-14.**

AtomicWriteFile Service

```
AtomicWriteFile {
  File Identifier = (File 10)
  Record Access
    File Start Record = 25
    Record Count = 3
    File Record Data = "3a 20 53 79 73 74 65 6d 20 49 6e 74 65 67 72 61 74 69 6f 6e..." }
```

```
AtomicWriteFile {
  File Identifier = (File 11)
  Stream Access
    File Start Position = -1
    File Data = "41 75 74 6f 6d 61 74 69 6f 6e" }
```

EXAMPLE MESSAGES

```
Result {
  Record Access
    File Start Record = 25
```

```
Result {
  Stream Access
    File Start Position = 129
```

EXAMPLE RESPONSES

Figure 14-13. The AtomicWriteFile service is used to add new data to a file. Service parameters define the precise location within the file to insert the new data.

In practice, the interface program can typically generate output much faster than it can be transmitted and displayed, and sometimes a human operator is typing ahead of the program's processing. Therefore, most interface programs use queues for saving outgoing data and incoming data until they can be used.

In BACnet, it is possible to create virtual terminals that extend the input and output queue concept across BACnet nodes and networks. The interface program resides in one BACnet node and the matching terminal resides in another BACnet node, such as a laptop computer acting as a BACnet workstation. Virtual terminals do not need to be actual physical terminals. For example, a BACnet node may send diagnostic output to a virtual logging terminal in another BACnet node.

To establish a virtual terminal connection, BACnet provides the VT-Open, VT-Data, and VT-Close services. **See Figure 14-15.**

Virtual Terminals

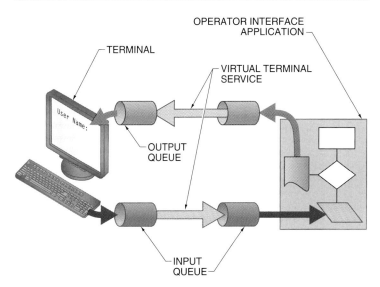

Figure 14-14. BACnet's virtual terminal services allow communication between two computer-based devices in order to provide a simple, text-based human interface into a remote device.

Virtual Terminal Service

```
VT-Open {
    VT-class = ANSI_X3.64
    Local VT Session Identifier = 5 }

        Result {
            Remote VT Session Identifier = 29 }

VT-Data {
    VT-session Identifier = 5
    VT-new Data = "User Name:"
    VT-data Flag = 0 }

        Result {
            All new Data Accepted = TRUE }

        VT-Data {
            VT-session Identifier = 29
            VT-new Data = "J_SMITH"
            VT-data Flag = 0 }

Result {
    All New Data Accepted = TRUE }

        VT-Close {
            List of Remote VT Session Identifiers = 29 }
```

EXAMPLE SESSION MESSAGES

Figure 14-15. The VT-Open, VT-Data, and VT-Close services are used to manage virtual terminal sessions and transfer data between the two participating devices.

VT-Open Service. The VT-Open service is used to establish a connection to a virtual terminal in a BACnet node. The two nodes involved in a virtual terminal session are known as VT-users. If the connection can be established, each VT-user independently unloads content from its input queue and transmits content to the other VT-user using a VT-Data service request.

It is possible for BACnet nodes to provide and manage more than one virtual terminal at a time. Each virtual terminal session is identified by a VT-session Identifier. A successful VT-Open returns a VT-session Identifier that is meaningful to the peer VT-user as a means of distinguishing the sessions and data. Incoming VT-Data identifies both the destination session and the source session so data can be directed to the proper receiving session.

VT-Data Service. VT-Data is a confirmed service to exchange data between VT-users through an established VT-session. The parameter VT-data Flag indicates the expected sequence of VT-Data transmissions, which detects input or output data that is lost or delivered out of sequence.

VT-Close Service. The VT-Close service is used to terminate a virtual terminal connection. The parameter List of Remote VT Session Identifiers includes one or more VT-session Identifier numbers for the sessions that are to be cancelled.

Summary

- As part of the periodic changes to the BACnet standard, new objects and services are added to provide better control solutions for special applications.
- An access control system is abstracted into the many components needed for access control.
- The components needed for access control are represented by both basic objects and special access control objects.
- The Lighting Output object was created to include properties to control special lighting features.
- A blink warning is a feature that blinks the lights briefly to alert occupants that the lights are due to turn OFF again soon, unless they take action to override the normal control.
- Fading, ramping, and stepping are ways to control the rate at which the level changes from one setting to another.
- Fading, ramping, and stepping are initiated with the property Lighting_Command, which is written with a compound value that specifies the lighting operation and configures its operation.

- The Load Control object is used to implement standardized electrical load shedding.
- Life safety points must maintain both a mode and a state. The operating mode of a life safety node selects the logic to be applied under different circumstances, which subsequently affects the evaluation of the state of the node.
- BACnet files represent a collection of data that can be accessed in any order and quantity.
- Files are also distinguished by their organization, RECORD_ACCESS or STREAM_ACCESS, which affects the way that the file is accessed.
- Files are accessed using two special services: AtomicReadFile and AtomicWriteFile. The services use the term "atomic" to mean "uninterrupted."
- An operator interface program can transmit ANSI characters through the serial port that are displayed on a terminal monitor, and any keyboard inputs are received on the serial port by the interface program.
- In BACnet, it is possible to create virtual terminals that extend the input and output queue concept across BACnet nodes and networks.

Definitions

- *Clamping* is the restriction of lighting level between minimum and maximum levels, as long as the light is ON.
- *Fading* is the changing from one lighting level to another over a fixed period.
- *Ramping* is the changing from one lighting level to another at a fixed rate of change of level per second.
- *Stepping* is the incremental increasing or decreasing of lighting level in fixed steps.
- A *terminal* is a human-machine interface (HMI) that serves as an input and display device for text-based communication with another computer-based device.

Review Questions

1. How are access control system functions represented by objects?
2. In what ways can different types of objects be used to control lighting systems?
3. How does a Lighting Output object control ON/OFF outputs when its Present_Value is analog?
4. What is the difference between how fading and ramping effects are specified?
5. Explain how the properties of the Load Control object configure load shedding.
6. What is the difference between a Life Safety Point object and a Life Safety Zone object?
7. Why does the LifeSafetyOperation service include the Requesting Source parameter?
8. What is the difference between RECORD_ACCESS and STREAM_ACCESS File objects?
9. Why are file services referred to as "atomic?"
10. How do BACnet nodes establish and manage virtual terminal sessions?

Chapter Fifteen

BACnet Installation, Configuration, and Troubleshooting

While the BACnet standard carefully details the organization and sharing of control information for interoperability, it does not specify the methods and procedures for configuring a system. ASHRAE, the organization responsible for the protocol, does not provide any software tools for managing or troubleshooting networks. However, many comprehensive software solutions that fill this need are available from a number of BACnet vendors. The general procedures for hardware installation are also similar to those for other types of network systems.

Chapter Objectives

- Compare the types of network tools available for installing, configuring, operating, and troubleshooting BACnet systems.
- Describe the basic tasks involved in installing BACnet infrastructure and devices.
- Identify the major parameters that must be set during node configuration.
- Evaluate the effects of different settings for major parameters.
- Detail some of the common problems with BACnet networks and systems, along with possible causes and remedies.

NETWORK TOOLS

A variety of network tools is used with BACnet nodes and networks. As with all types of building automation systems, there are proprietary tools that are unique to specific manufacturers and products. There are also third-party tools that can be used with most or all BACnet nodes, regardless of manufacturer or type. These network tools typically run on workstation computers that are connected, either temporarily or permanently, into the BACnet network. The three general categories of tools are installation and configuration tools, operation tools, and troubleshooting tools.

Installation and Configuration Tools

Procedures for installing and configuring BACnet nodes are not described in the BACnet standard because it focuses on day-to-day interoperability. Therefore, many node manufacturers have developed proprietary tools to fully configure each feature of their nodes. **See Figure 15-1.** However, many BACnet nodes can also be configured with general-purpose network tools. General-purpose network tools all contain certain basic functions that allow an integrator to configure at least most of a node's network and operational features.

Automated Discovery. One of the first tasks in configuring a new BACnet node is determining an appropriate MAC address and Device instance. Therefore, the integrator must know which MAC address and Device instance pairs are already in use. This information is also needed for assigning network numbers to router ports and configuring BACnet clients to know the locations of other nodes. Address information for each node in the network should be well documented, but this is not always the case.

Automated discovery is a method of automatically locating and identifying the node and object components of a BACnet network. **See Figure 15-2.** Automated discovery uses dynamic device binding to identify Device instances by their network numbers and MAC addresses, building a list of associations. Automated discovery also uses dynamic object binding to identify particular objects associated with BACnet nodes.

BACnet Network Tools

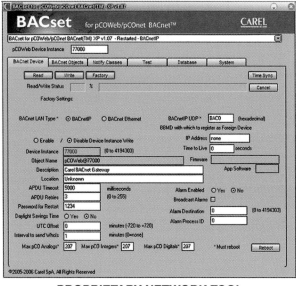

PROPRIETARY NETWORK TOOL

GENERAL-PURPOSE NETWORK TOOL

Figure 15-1. BACnet nodes may require proprietary network tools for configuring every feature, but many features can usually be configured with general-purpose network tools.

Configuration. Basic node configuration can often be accomplished using almost any BACnet client software tool that supports writing to object properties. The interfaces of these tools vary, but all allow a node's object properties to be individually examined and changed, including Priority_Array properties, when applicable. **See Figure 15-3.** Some tools include a statistics display that shows summaries of successful messages, time-out (no response) errors, and other information. Some network configuration tool functions, such as router or device configuration, may also be accessible through web browser-based interfaces.

Operation Tools

There are many day-to-day operations for which general-purpose BACnet tools can be used. The most common example is real-time monitoring of the status of sensor and controller information. This kind of information can be presented to a user in a number of ways.

Tree Structure. A very common format for monitoring BACnet control information is a text detail view, where logical areas or systems are displayed in a hierarchical tree structure. **See Figure 15-4.** Choosing a branch of the tree displays the component status and controller values in a list that is refreshed continuously or at some predefined frequency.

Graphical Displays. Graphical displays show status and controller information in pictorial form, often in color. **See Figure 15-5.** An illustration of the controlled system shows the major equipment and control devices. Most of this remains fixed, but portions representing sensors, actuators, or calculated information change according to actual values received over the network from the controllers. For example, temperature displays reflect actual temperatures, valve and damper graphics change position, and fans spin at different speeds.

Automated Discovery

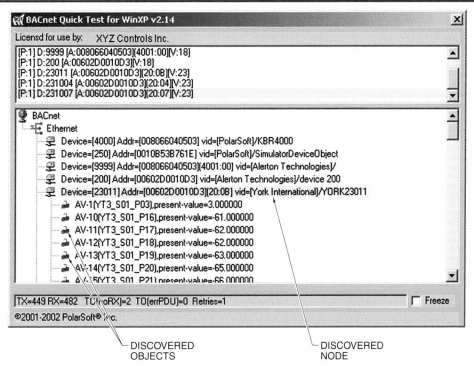

Figure 15-2. The automated discovery feature of a network tool can locate and identify BACnet nodes on a network.

Property Configuration

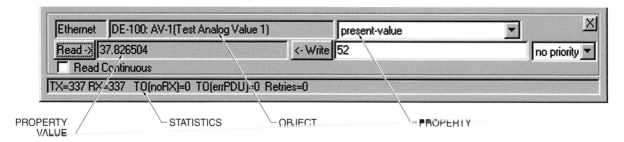

Figure 15-3. Many BACnet node functions can be configured by writing to certain object properties.

Operating Information in Tree Structure

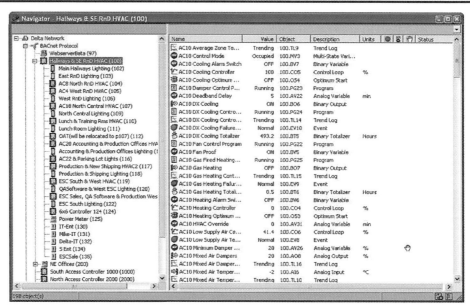

Figure 15-4. Day-to-day operational monitoring can be displayed as a hierarchy of logical nodes and areas that show status in real time.

The use of color is also effective at visually conveying general information, such as trouble areas. Color mapping changes the color of floor plan areas or equipment icons to indicate temperature or status changes.

A BACnet network analyzer, also known as a sniffer or packet analyzer, is comparable to a LonWorks protocol analyzer. In fact, network analyzers are available that can collect, display, and interpret network messages in any of many different protocols, such as Ethernet and 802.11 wireless. Most of these tools have many similar features.

The manipulation of Calendar and Schedule objects is usually managed with graphical tools. **See Figure 15-6.** Typically, the entire schedule is displayed and individual scheduling parameters can be adjusted by adjusting the bars representing time periods on individual days.

Logs. Logs accumulate information over time. Many general-purpose network tools can handle logs for alarms, schedules, and trends. Incoming alarm and event notifications are displayed, usually in table form, and ranked according to time and importance. Under specific conditions, more complex alarm management tools are capable of triggering responses,

Chapter 15—BACnet Installation, Configuration, and Troubleshooting

such as playing sounds, instructional videos, and printing. Trend Log object data can be retrieved from BACnet nodes that accumulate and sample control values. Often this kind of data is displayed as plots, along with historical peak information. **See Figure 15-7.**

> Graphical displays are especially useful for conveying important system information at a glance, such as out-of-range temperatures, security alarms, and lighting status. These displays may be stand-alone units or software running on typical computer workstations.

Operating Information in Graphical Displays

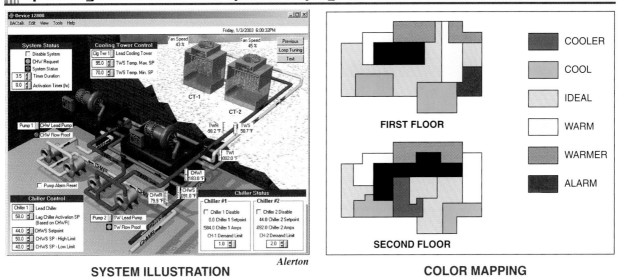

Figure 15-5. Graphical displays can show operating information in a way that is easy to understand at a glance.

Schedule Displays

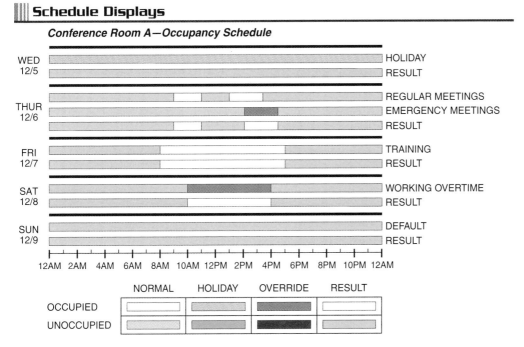

Figure 15-6. The information within a Schedule object can be displayed graphically in an easy-to-understand format. The interface may also allow schedule changes by adjusting the bars on the screen.

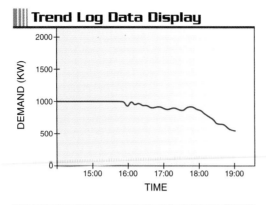

Figure 15-7. Trend Log data is often displayed as plots over time, along with historical peak information.

Troubleshooting Tools

Besides troubleshooting wiring integrity, power supplies, and configuration errors, one of the most useful technologies for finding problems in a BACnet system is a network analyzer tool. A *network analyzer tool* is a network tool that listens passively to network traffic, capturing message frames and saving them for later analysis. This tool is also known as a protocol analyzer tool or a sniffer. The software tool has access to every message that is transmitted on the network segment where the tool is connected. This is known as promiscuous sniffing. The tool keeps the messages it has been configured to capture and discards all others. For example, it may log all of the messages to or from a certain node or of a certain type.

Network analyzer tools display each message as it arrives or as soon as it can be managed. The display shows both the actual hexadecimal code of the message and an interpretation that is human-readable. **See Figure 15-8.** The meaning and structure within the various component fields of each message are viewable. This can be very helpful in diagnosing badly formed messages, watching the timing between and within messages, or finding certain types of network problems.

Messages can be inspected further. A *tracing detail* is a detailed display of the information contained within a BACnet message. For example, a network analyzer may capture a message from one node that is a ReadProperty service request, along with the answer to that request from another node. The hexadecimal code of the octets in the message is matched line-by-line with the human-readable meaning. The network layer, MAC layer, and timing information can also be viewed. Such detail is important for troubleshooting problems with vague symptoms, such as poor or intermittent communications.

INSTALLATION

The installation of a BACnet system, as with any building automation system, requires attention to detail. Careful planning, documentation, and incremental testing during installation greatly reduce the chances of error and aids troubleshooting if a problem does occur.

Planning and Preparation

When working on any BACnet system, it is vital to have a complete picture of all the nodes, network devices, and network segments involved prior to beginning installation. This is not only important for the new devices to be installed, but also for the existing hardware in a system to be expanded or interoperated with. It is recommended to create a device information table and a network information table to manage this data. This can be organized on paper or in a computer database.

A device information table defines the parameters of each BACnet node in the entire system. **See Figure 15-9.** It should include the Device instance, Device Name, Device Description, physical location, MAC address, network number, and vendor identifier code.

Network Analyzer

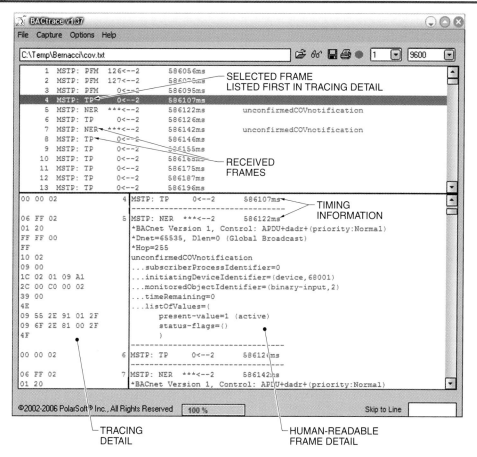

Figure 15-8. A network analyzer, also known as a sniffer, captures message frames and converts their content into human-readable form.

Device Information						
Device Instance	Device Name	Device Description	Location	MAC Address	Network Number	Vendor
1102	Router A2	2nd Floor HVAC Segment Router	2nd Floor, Cable Closet	02	3	BAS Controls
2002	VAV1	VAV Controller	2nd Floor, Mechanical Room	91	3	Climate Solutions
2003	Temp201	Room 201 Thermostat	2nd Floor, Room 201	101	3	Climate Solutions
2004	Temp202	Room 202 Thermostat	2nd Floor, Room 202	102	3	Climate Solutions

Figure 15-9. Network planning and troubleshooting relies on complete information about every device on the network, which is organized into a table.

A network information table defines the parameters for each BACnet router, or controller-router in the entire system. **See Figure 15-10.** This is typically a two-level table that lists each router, along with each port managed by that router. It should contain the router name, vendor, physical location, Device instance (if router contains a Device object), home port (if router contains a Device object), port ID, and network number and LAN type for each port.

Bench Testing

Although it is possible to physically install and wire BACnet nodes prior to configuring them, this is not recommended. Even when powered OFF, nodes can influence a network if termination, biasing, and wiring options are not properly selected, or if there are wiring faults such as shorts, opens, or improper grounding. For these reasons, the best practice is to verify the integrity of the wiring infrastructure and grounding first, and then perform a basic setup of each node individually.

Each BACnet node is connected to a testing device (usually a notebook computer with the appropriate interface) that attempts communication with the node. **See Figure 15-11.** The node can be verified as fully operational before integrating it with the rest of the installed network. By limiting communications to a one-on-one environment, any problems that occur must be caused by the node, making troubleshooting much easier.

Bench Testing

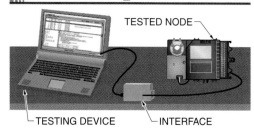

Figure 15-11. Bench testing involves testing each node individually in an isolated environment to confirm that it is working properly before connection to the installed network.

Wiring Infrastructure

When there are many nodes on a network, it can be difficult to find and isolate wiring problems. The best practice is to test each wiring segment as it is installed for shorts and opens, and then create a daisy chain. Daisy chains are formed by twisting incoming and outgoing conductor ends together and capturing the twist in the corresponding screw terminal for the network. **See Figure 15-12.** Exposed wire should be minimized so that adjacent conductors are not shorted.

In order to connect network tools for later troubleshooting, installers should include access points inside a wiring cabinet for BACnet nodes.

Network Information

Router Name	Device Instance	Vendor	Location	Home Port	Port IDs	Network Number	LAN Type
Router A1	1101	BAS Controls	1st Floor, Cable Closet	1	1	1	BACnet/IP
					2	2	MS/TP
Router A2	1102	BAS Controls	2nd Floor, Cable Closet	1	1	1	BACnet/IP
					2	3	MS/TP
					3	4	MS/TP
Router A3	1103	BAS Controls	3rd Floor, Cable Closet	1	1	1	BACnet/IP
					2	5	MS/TP

Figure 15-10. Each port of a router has a different network identity. This information should be organized in a table to facilitate planning and troubleshooting.

Daisy-Chain Wiring

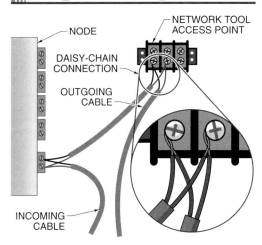

Figure 15-12. Daisy chains are formed by twisting together incoming and outgoing conductors and capturing the twist in the corresponding screw terminal. Adding separate access points for service tools can be very helpful for later servicing and troubleshooting.

Conductor specifications and wiring techniques vary depending on the LAN type. Since several LAN types are compatible with BACnet, wiring implementations are different for many systems. Detailed information is available in the BACnet standard and other documents, though the following points are especially important to reiterate:

- The shield of twisted-pair cabling must be grounded at only one point. Shields on wiring entering and leaving an enclosure should be twisted together and crimped or terminated to screw terminals that have no other connection.
- The extreme ends of MS/TP networks should include a 120 Ω termination resistance across the conductors. Active biasing should also be included at both end nodes.
- When MS/TP nodes are optically and magnetically isolated, the node-to-node wiring must include a ground conductor connecting the isolated references together.
- If the MS/TP segments span buildings, fiber optic modems should be used to provide electrical isolation between buildings.
- When planning the installation of ARCNET 156K networks, it is recommended to run a spare shielded twisted-pair cable to each wiring or controller cabinet along the same logical segment. This allows the option of replacing individual ARCNET 156K nodes with similar MS/TP nodes at a future time, without requiring additional cabling. A router might be required in order to implement this option, but it would be supported by the existing wiring infrastructure.

Service Tool Access Points

It is recommended to allow for access points to the network conductors at multiple locations throughout a network, preferably at each wiring cabinet. Access points simplify the physical connection of service tools, or notebook computers with network tool software, to the network for testing, troubleshooting, or diagnostics. There is no standard connector for MS/TP or ARCNET 156K networks, so dedicated screw terminals connected to the network conductors can be used as access points.

For BACnet/IP or Ethernet 8802-3 networks, a local hub or switch can be installed in the wiring cabinets. Although switches are preferred in most applications, a hub allows promiscuous sniffer software to see network traffic arriving at the hub without having to unplug an active controller.

CONFIGURATION

During bench testing, a node is tested to ensure proper operation, and then it is configured with the settings needed to function with other nodes on a network. Most of this configuration involves writing to certain properties of the node's Device object, which can be accomplished with any network tool that supports ReadProperty and WriteProperty services. If the network is already established, information is first needed about the installed nodes in order to determine some of the configurations.

Although many settings for specific BACnet nodes only affect that one node, sometimes settings can have system-wide effects as well. Understanding some of these scenarios enables the integrator to tune the performance of a large portion of any BACnet system.

MAC Address Assignment

Nodes in some LAN types have permanent MAC addresses. However, MAC addresses for MS/TP and ARCNET nodes are assigned when they are installed. The addresses are restricted to the range 0 to 255, except for those reserved for broadcasting (255 for MS/TP and 0 for ARCNET). Also, for MS/TP networks, slave nodes can use any of the addresses 0 to 254, but master nodes can only use 0 to 127. ARCNET nodes can use any of the non-broadcast MAC addresses equally.

Some MS/TP and ARCNET nodes are factory preset to a specific MAC address, requiring bench testing to communicate to that special MAC address in order to reset the MAC address. For example, resetting may involve writing to a property of a special object. These kinds of schemes are all proprietary because BACnet does not define a standard mechanism to set up the MAC address.

However, the MAC address for an MS/TP or ARCNET node is usually set manually by a DIP switch, rotary switch, or jumper selection. **See Figure 15-13.** Since their addresses can be set before installation, these nodes should immediately work properly when powered ON in the field. However, bench testing is still recommended.

Device Instance Assignment

Every BACnet node must be assigned a unique Device instance number as well as a unique Device Object_Name. These are entirely up to the integrator to choose, since the standard has no requirements about how they are to be assigned. Therefore, the assignment of Device instances can be organized in a number of ways. An established method should be used when adding new nodes to a system.

Device Instance Allocation. Since the Device instance number can range up to 4,194,303, available Device instances can be allocated in ways that help convey information about the node. A popular scheme is to assign blocks of numbers for each physical area, such as the floors of a building, or the buildings in a campus. The floor or building number then becomes part of the Device instance number. For example, if 1000 numbers are allotted for each building in a campus, the Device instance equals the building number times 1000, plus a node number. Therefore, building 131 contains the Device instances 131001, 131002, and 131003. This idea can be extended to also encode floor numbers or portions of the building, such as the west side of the 6th floor.

Alternatively, Device instances can be organized based on logical arrangements. Blocks of numbers can be allocated by subsystems, such as HVAC, lighting, or access control. For example, subsystem 5 may consist of all of the nodes related to control lighting, including Device instances 5001, 5002, and 5003. Similarly, Device instances can be grouped based on the building's functional departments, such as the Accounting or Maintenance department. This can be useful if the departments use the building space differently, such as having different occupancy or lighting requirements.

Device instance numbers can also be divided based on the vendor identifier code, allowing contractors to manage their own Device instances. This is a less common scheme, as it is considered less useful than organizing by location or subsystem.

Regardless of the allocation scheme, it should be relatively simple and preferably administered by one authority, such as a facility manager, that is responsible for assigning

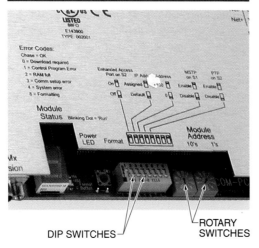

Figure 15-13. Switches are commonly used to set the addresses for MS/TP and ARCNET nodes.

Device instances. This is increasingly vital for larger BACnet systems.

Automated Device Instance Discovery. If there is no allocation convention or centralized authority for an existing system, an integrator working on a new installation or expansion may have no documentation or conflicting documentation about the current system. In those cases, it is recommended to first identify and document the Device instances and other information about all of the existing nodes. Then new nodes can be added to the system.

The easiest way to find existing nodes is with discovery tools. A *discovery tool* is software that sends dynamic device-binding messages, usually Who-Is broadcasts, and listens for the replies. The discovery software then sorts out the incoming information about the existing nodes, including their network locations, network numbers and MAC addresses, vendor identifying code, and BACnet capabilities. **See Figure 15-14.**

However, in even modest-sized networks, an inexperienced approach to discovery can generate thousands of messages that can choke routers and slow down the system for a long time. So many replies may be generated that BACnet routers between the discovery tool and the nodes being discovered are overwhelmed with replies and only some get through. Unfortunately, while some information is gained, this means that some nodes remain undiscovered.

More sophisticated discovery tool software anticipates these kinds of issues and uses a more gradual approach to restrict the number of simultaneous replies. While these measures make the discovery process more reliable, they also increase the time required to complete the discovery process.

> By using a strict numbering plan, technicians can immediately narrow down the location and type of node from the Device instance alone.

Automated Device Instance Discovery

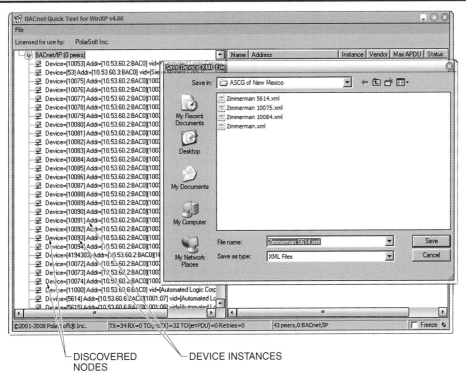

Figure 15-14. Automated discovery software tabulates the Device instances, addresses, and other information about discovered nodes. Some software can then save this information in XML or tabular text format for importing into spreadsheet and database programs.

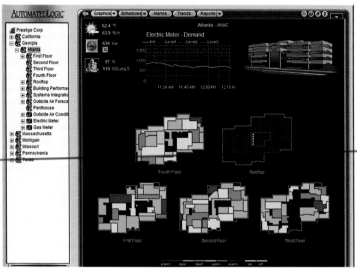

Automated Logic Corporation

Improperly configured Max_Master settings can slow down network performance, which can also affect real-time monitoring.

Discovery can also be performed in a more limited scope, such as segment by segment, in order to reduce router issues. This can be a lot more tedious because each network segment must be physically visited individually in order to attach a discovery tool, discover the nodes, and collect replies. The resulting node data is combined later to form a complete picture of the entire system.

Some discovery tools create a database of discovered node information that can be saved into text or XML form for use with an external database, spreadsheet, and other types of organizational software.

Object Discovery

If there is either no documentation or incomplete documentation on the objects in each node, an object database must be compiled that includes information based on object discovery. If an object instance or name is known, such as Analog Value 5 or OutdoorAirTemp, dynamic object binding can be used to locate the nodes that contain such an object. However, it is common during configuration to require a complete object database for the entire network.

The procedure for discovering each object in a node is simple, though tedious. The Object_List property of each node's Device object contains an array listing each object in the node. However, without knowing the size of the array, it may not be possible to read the entire property. Therefore, the first step is to read the Object_List[0] slot, which contains the size of the array. **See Figure 15-15.** For example, if Object_List[0] equals 4, then there are four more slots in the array, and therefore four objects in the node. An Object_Identifier for one of the objects is stored in each of the slots: Object_List[1], Object_List[2], Object_List[3], and Object_List[4]. ReadProperty services are then used to read each slot and compile a list of objects. Once each object is identified, its Object_Name property may also be read. Any BACnet client program that is capable of reading properties of objects in a node can use this procedure.

This procedure is effective but relatively inefficient for many nodes. There are methods for discovering the entire object database that are more sophisticated, but they use correspondingly complex procedures.

MS/TP Configuration

Several key parameters must be configured on MS/TP nodes before they can be placed in service. All MS/TP nodes on the same segment must use the same speed, which is a choice among 9600 bit/s, 19.2 kbit/s, 38.4 kbit/s, or 76.8 kbit/s. This is usually manually configured with DIP switches or jumpers on the node. **See Figure 15-16.** Some MS/TP nodes can implement a proprietary automatic speed detection scheme, which is usually enabled by a switch.

Max_Master Setting. Each MS/TP master node has a Max_Master property, which is generally configured as the highest master node's MAC address on the segment. Some MS/TP master nodes do not allow Max_Master to be changed. In those cases, the standard requires that Max_Master default to a value of 127, though this causes a performance penalty.

Any MS/TP segment that has fewer than 128 master nodes wastes a large portion of available network bandwidth constantly hunting for nodes that are not actually there.

Object Discovery

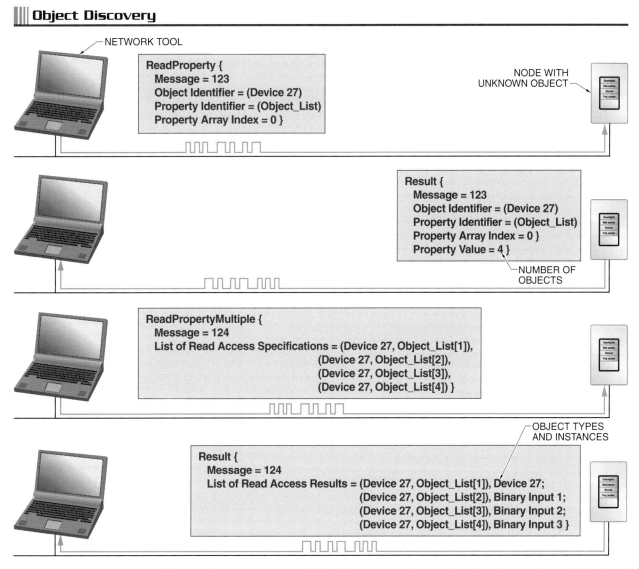

Figure 15-15. One method of object discovery requires only ReadProperty services to determine the number and identity of every object within a node.

Bitrate Configuration

Automated Logic Corporation

Figure 15-16. The bitrate of MS/TP nodes is often configured manually using small switches.

For example, an MS/TP network running at 76.8 kbit/sec has 30 master nodes, with MAC addresses 00 through 29. **See Figure 15-17.** A complete token cycle takes approximately 45 ms at this speed. For every 50 token cycles (about 2250 ms), there is a mandatory PollForMaster (PFM) cycle. If the Max_Master is set to 127, the PFM cycle looks for 98 "missing" master nodes. Since each PFM requires that the token holder wait for at least 20 ms for a reply (plus 1.5 ms for the message itself), the entire PFM cycle lasts 2752 ms, of which 2107 ms is spent looking for non-existent nodes. This is a waste of about 42% of the available network bandwidth.

Effects of Max_Master Setting

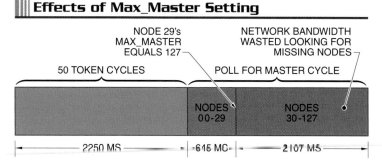

Figure 15-17. The value of the Max_Master setting determines how much time is wasted looking for non-existent nodes during the regular Poll For Master cycles.

The percentage of wasted bandwidth also varies depending on the network bandwidth, number of master nodes, the amount of normal network activity, and the waiting time for each PFM request ($T_{usage\_timeout}$). However, timing issues are a significant concern for every MS/TP segment. The following actions can be taken to minimize the impact of these factors:

- The last MAC address position in a segment should use a master node with a writable Max_Master property. By using a writable node in this position, its Max_Master property can be adjusted downward to the actual highest-used master MAC address.
- MAC addresses should be assigned without gaps in the sequences for each segment. For example, sequences such as 00, 01, 02, 27, 28, 45 should be avoided. Each gap wastes network time during the PFM cycle.
- The parameter $T_{usage\_timeout}$ should be set to the smallest possible value that is still effective for the particular MS/TP segment. Master nodes are not required to allow for adjusting the $T_{usage\_timeout}$ parameter and most fix the value at some compromise setting. If the parameter is writable, it should be as close to 20 ms as possible. Most nodes should respond within about 20 ms, but the value may need to be adjusted slightly for slower nodes.

Max_Info_Frames Setting. The Device object's Max_Info_Frames property defines the number of messages that a master node may send each time it gets the token. Most MS/TP master nodes use a setting of 1 for this property. If an MS/TP node's server application needs to initiate many replies to Reply Postponed requests, or a client application potentially initiates requests to nodes that Reply Postponed, setting Max_Info_Frames to a larger value can improve performance.

A special case is routers that route onto MS/TP network segments. Because other LAN types are typically faster than MS/TP, a router to MS/TP may accumulate many packets to be forwarded onto the MS/TP segment while waiting for the token. It is important to be able to pass these packets through to the MS/TP segment as quickly as possible. Consequently, the Max_Info_Frames property for MS/TP routers should be set to a high value, such as 50, to allow inbound traffic to be distributed without undue delay.

Router Configuration

BACnet routers vary considerably from one implementation to the next because the standard does not make specific requirements about how they should be configured. Some BACnet routers include a Device object and application entity, while others simply perform routing and are not strictly speaking BACnet devices. In general, BACnet routers provide two or more ports, each of which supports a specific type of LAN, such as MS/TP, ARCNET, PTP, Ethernet 8802-3, or BACnet/IP. It is also common to have routers with more than two ports, which are called N-way routers. There can be multiple instances of a port type, for example two MS/TP ports and two BACnet/IP ports.

BACnet routers are configured through a proprietary mechanism, proprietary set-up software running on a personal computer, or a web-based user interface. The most critical parameter to set up for each router port is a BACnet network number. **See Figure 15-18.** The same network number is assigned to all of the router ports connected to the same segment. LAN-specific configuration parameters may also need to be set up in each router.

Router Configuration

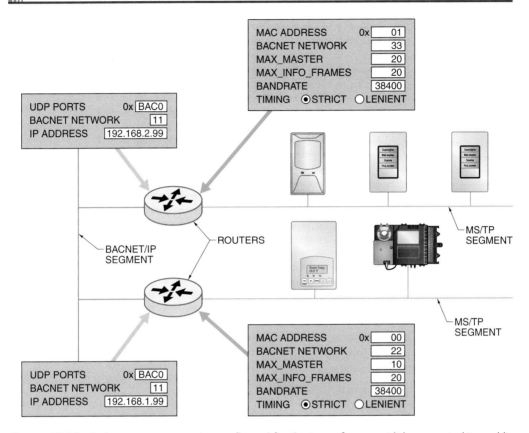

Figure 15-18. Each router port must be configured for the type of segment it is connected to and its BACnet network number.

BACnet/IP Subnet Configuration with BBMDs

If all of the BACnet/IP nodes in a BACnet system reside on the same IP subnet, the BACnet/IP configuration is fairly simple. Each node has a unique IP address and communicates using the same UDP port number. However, if the BACnet/IP nodes are on different subnets, then BACnet broadcast management devices (BBMDs) are needed to manage broadcasts between subnets. (In this context, a BBMD means any device that has BBMD functionality, which could mean a stand-alone device or a controller/router that also can perform BBMD functions.)

Generally, each IP subnet that has one or more BACnet/IP node(s) requires a BBMD. BACnet does not define a standard mechanism for BBMDs to configure each other. As a rule, BBMDs are either configured manually with the IP addresses of their peer BBMDs, or one BBMD is configured and its broadcast distribution table (BDT) is transmitted to each peer BBMD. **See Figure 15-19.** Typically, BBMDs use a proprietary or web-based interface for these definition functions.

It may be costly or impractical to include a BBMD on every subnet, especially those that include very few BACnet/IP nodes. Instead, a single BACnet/IP node on a subnet can be configured to register as a foreign device with a BBMD on another subnet. **See Figure 15-20.** Although the mechanism for foreign registration is defined in the BACnet standard, the exact procedure for its configuration is not specified in BACnet, so a proprietary setup is required in the foreign device. It may also be necessary to enable foreign device registration in the BBMD.

BBMD Address Configuration

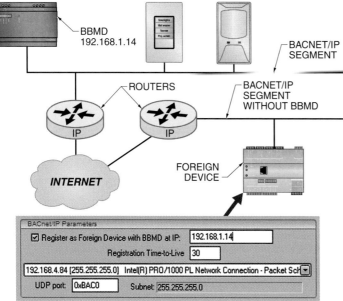

Figure 15-19. The addresses of all the BACnet broadcast management devices (BBMDs) may be configured on one BBMD and spread through the network via a broadcast distribution table.

Foreign Device Registration

Figure 15-20. A foreign device registers with a BBMD to send and receive broadcasts on its behalf. Proprietary interfaces may be necessary to configure foreign device registration in BACnet/IP nodes.

While it is possible to only use a single BBMD and configure all BACnet/IP nodes to use foreign device registration, the reduced network performance due to the additional foreign device traffic typically negates the cost savings (from fewer BBMDs) in all but the most limited situations. There are additional reasons that limit foreign device registrations. Not all BACnet/IP nodes (or BBMDs) support foreign device registration. The BBMDs that do support foreign devices may limit the number of simultaneous registrants. Traffic and local application performance requirements may necessitate fewer delays in broadcasts, implying the need for BBMDs at least on some subnets.

Client Application Binding

BACnet provides several mechanisms for client application programs to acquire real-time information from the objects and properties in other nodes. Values may be read directly using ReadProperty services and change-of-value (COV) subscriptions. Values may be changed using WriteProperty services.

The client application program must know which node/object/property is to be read or written, but they may be different from case to case. That means that every BACnet client application requires a means of being configured to use potentially different node/object/property references. *Client application binding* is the configured association between a BACnet application logical parameter and its actual source or destination object property. Although there are circumstances in the BACnet standard that call for references like this, there is not (presently) a single generic mechanism for establishing or altering the bindings in a standardized manner. As a result, there are four binding configuration scenarios.

Proprietary Binding Setup. Some implementations provide proprietary mechanisms for setting up binding references. These may require proprietary software tools or be configurable through web browser interfaces. This is the most common of the four scenarios.

Static Bindings. With a static binding, the client application does not change its device/object/property references, which are built into the code of the application program. Generally, these kinds of references can only be altered by changing the program code of the application, which requires proprietary software tools.

Standard Object Property References. Some standard object types, such as Command, Group, Loop, and Schedule objects, have standard properties that reference other object properties, sometimes including properties in other nodes. If a given implementation of these objects allows writing to the reference property, then they can be configured to reference arbitrary properties as appropriate to the application. Standard object property references may be configurable with general-purpose network tools, since only reading and writing properties are required.

Proprietary Object Binding Properties. Because of the lack of a standard binding object, some BACnet implementations allow configuration of client bindings using properties of proprietary object types. This concept is similar to standard Schedule and Loop objects, which have reference properties. This kind of configuration is usually possible with general-purpose network tools, but since the objects and properties are proprietary, detailed documentation is required from the node vendor that clearly explains the procedure.

Bursting Avoidance

All BACnet nodes have limited resources in terms of memory and CPU time. When a request message is received by a BACnet node, it may or may not be able to be processed immediately. Most nodes can save a number of incoming requests in a queue (waiting list) to be handled as time permits, along with the BACnet server's other tasks. In some BACnet nodes, this queue is relatively small. Therefore, a common source of bandwidth issues comes from client nodes that query BACnet servers too aggressively, overwhelming their ability to respond. One of the common effects is bursting, also known as a burst packet response.

For example, a BACnet node may be able to handle two requests at a time. A BACnet client has five requests and sends them all in a single burst, one right after the other. **See Figure 15-21.** The receiving node saves the first two in its queue, but the next three are discarded because the queue is full. The client, having received no response to three of the requests, resends them after a time out period, which is typically 2 sec to 4 sec. Again, the first two are queued, but the third is discarded. After the time out period expires, the last request is sent again, and is answered this time. Not only does it waste bandwidth to have to resend messages, but the client suffers a significant time penalty by waiting for a time out period before retrying the request. This penalty is incurred each time the burst rate exceeds the server's capabilities.

To address this problem, the client must be set to restrain itself to no more than two outstanding requests at a time. BACnet does not standardize a parameter for setting this behavior. If it is adjustable at all, it is handled in a proprietary way.

If excessive bursting can be avoided, however, the server node is not overwhelmed and the requests are serviced immediately without any time out delays or retries. There are two significant benefits from this technique. First, only 5 total messages are transmitted instead of 9, which is a more efficient use of the network. Second, the total time required to execute the 5 requests is considerably less. Although the capacities of BACnet nodes vary, this technique applies generally to all non-MS/TP nodes. It can be scaled accordingly for the number of requests that are appropriate for each node.

> In the field, bench testing is also known as "sandbox testing." This is because the testing device and procedures allow the new node to be safely tested within the confines of a controlled environment without affecting any outside systems.

TROUBLESHOOTING

As with any automation system, a BACnet system may have problems that require troubleshooting. The occurrence of hardware failures or flawed BACnet implementations is possible, though rare. Most problems are caused by human error in design, installation, or configuration.

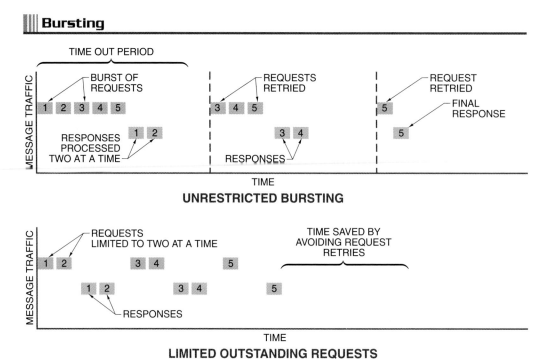

Figure 15-21. Bursts of many request messages may exceed the buffering capacity of some nodes, requiring that some messages be retransmitted after a timeout period. Limiting the number of simultaneous outstanding requests does not overwhelm the node and may require less time overall.

Regardless of the cause of a problem, troubleshooting often requires the use of a network analyzer tool. There are several third-party network analyzer tools available, some even at no cost. However, it takes a great deal of expertise to use the tools effectively and interpret their results, though manufacturer and third-party technical support can be helpful. Once the cause of a problem is determined, it can usually be remedied by using a general-purpose tool for reading and writing object properties.

The placement of a network analyzer tool into the network is also important. When working with a network that uses switches, a network analyzer tool can only see broadcasts and messages destined for a particular segment or node. All others are blocked. Therefore, the network analyzer tool may need to be connected in a special way to receive the desired traffic. For example, to see the messages going to and coming from a particular node, a small hub can be inserted between the network and node. **See Figure 15-22.** The sniffer or test device can be connected to one of the hub ports and can see the same traffic as the node.

Several problems are relatively common, though they often have somewhat vague symptoms that can complicate troubleshooting. Following are some of these common symptoms, along with some troubleshooting steps to follow to diagnose the possible causes.

Unresponsive MS/TP Nodes

One scenario requiring troubleshooting is a new MS/TP node added to a network that cannot

> The Max_Master setting, and also to some degree the Max_Info_Frames setting, can have a significant impact on overall MS/TP network performance. For this reason, it is surprising that the standard does not currently require them to be adjustable in every MS/TP master node. When they are writable, the Max_Master and Max_Info_Frames properties are usually changed through a proprietary mechanism, such as a local operator display/keypad, or by writing directly to these Device object properties.

be reached or discovered, or does not respond to Who-Is requests. If it is a master node and supports dynamic device binding response (DM-DDB-B), it should respond with an I-Am message. If it does not, then a network analyzer tool should be used to check for token-passing traffic involving the unresponsive node.

The unresponsive node's MAC address should be receiving token pass (TP) messages. If not, another master node (typically one with a MAC address 1 increment lower) should be sending PollForMaster (PFM) messages to that MAC address. For example, if MAC 05 is not responding, check that MAC 04 is sending PFM messages to 05, which should reply with RPFM, and 04 should then send TP to 05. If MAC 04 is not sending PFM or TP, then its Max_Master setting may be lower than the unresponsive node's address. All the nodes on the segment should have a Max_Master setting at least as large as the unresponsive node's MAC address.

If the unresponsive node is a slave node, it cannot be discovered using dynamic device binding unless there is an MS/TP proxy node, which must be enabled to do proxying. Troubleshooting should begin with the proxy node. For unresponsive slave nodes that are reached by static binding, the network analyzer tool is used to examine the network traffic. If there is a response, header or data CRC errors could be caused by dropping TxEnable or $T_{turnaround}$ issues. (Note that duplicate MAC addresses may exhibit similar symptoms because the two nodes reply slightly out of sync with each other.) Alternatively, the response may be taking too long, causing the sender to give up waiting for a DataExpectingReply frame.

Other possible causes for unresponsive MS/TP nodes include reversed TIA-485 polarity, mismatched MS/TP speed settings, incorrect MAC address setting, and hardware failures. Opens and shorts in the network conductors are also possible, but can be confirmed or ruled out easily since they would affect other nodes on the segment. Less common causes include PFM or TP sending nodes that drop TxEnable too quickly, causing a CRC error, or not waiting for the $T_{turnaround}$ (minimum silence) interval and sending messages too soon.

Troubleshooting Message Traffic

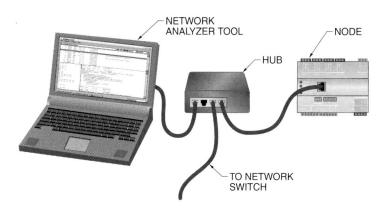

Figure 15-22. A hub can be inserted between a node and the rest of the network in order for a network analyzer tool to see all of the traffic to and from the node.

Slow Communication

Another possible problem is slow or intermittent performance of a particular MS/TP segment when compared to a similar MS/TP segment in another part of the same building.

If the Max_Master setting for the nodes on this segment is greater than the highest-used MAC address, the end node wastes time hunting for "missing" master nodes every 50 token cycles. The Max_Master setting for each node should be changed to equal the highest-used MAC address. If the Max_Master setting cannot be changed on some nodes (it is set to a default of 127), then the segment should be arranged so that none of these nodes are at the highest MAC address position. Similarly, closing any gaps in the MAC address assignments improves token-passing efficiency.

If the highest MAC address node has a lenient $T_{usage\_timeout}$ policy and there are unused master MAC addresses, then there can be significant periods of no traffic on the segment, which causes an overall slower control response. Adjusting the $T_{usage\_timeout}$ interval setting may improve performance.

There should be no more than a few message errors per day, particularly associated with the same nodes. This information can be found in a node's network statistics or with a network analyzer tool. The detection of many CRC or framing errors indicates problems like timing issues or improper network biasing.

Framing errors can also be caused by dropped tokens, which can be recognized by traffic gaps longer than 500 ms.

Communication Time Outs

Yet another problem scenario is message requests to a particular node that frequently time out (receive no response within a few seconds) and need to be retried. The cause could be timing issues, as discussed in slow communication troubleshooting, or the burst rate is being exceeded for the node.

If communication with all nodes on a segment is intermittent or frequently times out, the router to that segment may be overloaded with traffic. The client side of the router should be monitored with a network analyzer tool. If the router is generating Router-Busy-to-Network messages, excessive traffic may be caused by too many Who-Is requests or other broadcasts. If not, and the segment is MS/TP, the Max_Info_Frames for the router MS/TP port may be too low. Typically, it should be 50 messages or higher. Otherwise, there may be an issue downstream of the router (on a segment different from the client side segment to the router) such as MS/TP timing issues.

Excessive Who-Is and I-Am Traffic

In this problem, a segment with two or more routers connected to it becomes very slow. The network analyzer tool shows a lot of broadcast Who-Is and I-Am traffic continuously.

The network should be checked for circular routes. A common installation mistake is to connect two routers to the same Ethernet segment, both of which have Ethernet 8802-3 and BACnet/IP ports. Only one of the routers may have routing to both ports enabled, or a circular route is created, even if the network numbers are different. A network analyzer tool can be used to examine the Hop Count in routed Who-Is messages. The Hop Count starts at 255 and is decreased each time the message is routed to another segment. Typical networks only have two or three levels, so Hop Counts less than 252 indicate that the message has been circulating through more routers.

Broadcast storms can slow down a network. Although it is possible to send Who-Is messages without a qualifying range of Device instances, this causes a large amount of responding I-Am traffic, which can easily choke routers and MS/TP networks. The network analyzer tool can be used to look for Who-Is messages without a range qualification in order to find the sending node.

Unresponsive BACnet/IP Nodes

In this scenario, some BACnet/IP nodes can be discovered in a BACnet/IP network, but not others. This is a common problem for internetworks composed of more than one IP subnet. BACnet discovery is based on broadcast messages, which are limited by IP to nodes in the same IP subnet only. If nodes are on different IP subnets, there must be an IP route between the subnets, plus one of the following solutions:
- a BACnet BBMD on each subnet
- one BBMD that registers all of the BACnet/IP nodes on other subnets as foreign devices
- the BACnet/IP clients use static binding

Missing Alarms

In this next problem, an operator workstation is not receiving alarms. In the alarm-detecting node, the Notification Class object should be checked to confirm that it contains the following information:
- the workstation's network and MAC address (static binding) or the Device instance (using dynamic binding)
- a Process Identifier that is recognized by the workstation
- a Notification Class that is recognized by the workstation
- Valid Days, From Time, and To Time matching the period when the alarm is sent
- a destination transition that matches the transition that has occurred (for example, TO-OFFNORMAL); source object that generated the transition must have a matching Event_Enable for that transition (and Limit_Enable in the case of analog Out_Of_Range)

With a network analyzer tool, it can be verified whether the node is generating the EventNotification message. If so, further troubleshooting may be required to identify communication issues between the two nodes.

Unreadable Objects

Another scenario is a particular BACnet node that can be discovered, but not the objects within it. BACnet clients typically discover devices using Who-Is and dynamic device binding, which only returns the node's network number, MAC address, and Device instance. To discover objects in the device, the client typically uses ReadProperty or ReadPropertyMultiple to read the Device object's Object_List property. Several common problems can occur in this procedure, including the following:

- The client node tries to use ReadPropertyMultiple, but the server node does not support this service and returns an error.
- The client node tries ReadProperty to read the entire Object_List property. The Object_List is larger than 92 objects and the server node returns a segmented response. If the client node does not support segmentation, it cannot process the reply.
- The client node tries ReadProperty to read the entire Object_List property. If the Object_List is larger than 92 objects, and the server node does not support segmentation, it returns an error.

Alternatively, the client node tries ReadProperty to read Object_List[0] (containing the number of objects), then reads each Object_List[x] array slot individually. However, possible problems include the following:

- The server node improperly handles the Object_List[0] request and returns an error or times out.
- The server node improperly handles any subsequent Object_List[x] request and returns an error or times out.
- The server node takes too long to reply to an Object_List[x] request, which appears to be a time out.

These problems have no field solution because they are caused by an incorrect BACnet implementation in the client or server nodes. The network analyzer tool should be used to capture the traffic, determine which scenario it is, and get the appropriate vendor to fix that BACnet product.

If some objects and properties on a certain node can be read, but others cannot, the cause of the problem may be different. This is assuming that the node is verified to contain the desired object and/or property. A network analyzer tool is used to check the ReadProperty messages and any replies. A lack of replies indicates a problem with the BACnet implementation. However, if the server node is returning the property value, perhaps the client node does not support that property datatype. This is more common for complex datatypes such as the Schedule object's Exception_Schedule property.

Windows Network Tools

Problems may occur when using multiple BACnet/IP client tools on the same Windows computer. Most Windows-based BACnet clients use the Winsock interface. When such programs are running, they attach themselves to the standard BACnet UDP port 0xBAC0 (47808) and listen for incoming BACnet messages. However, Winsock allows only one program to listen on this port at a time. When one program is active, no other BACnet tool can also access this UDP port.

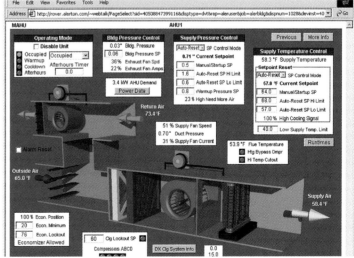

Alerton

Network information may be available through web-based interfaces, which may help troubleshoot other software problems.

The solution is to either turn off the other tool or use a second Ethernet interface with a different IP address. This gives each program its own IP address and UDP port combination. A few BACnet applications attach themselves to all available IP adapters at port 0xBAC0, so they must only be used alone.

Unchanging Commandable Properties

Output objects, such as Analog Output and Binary Output, are commandable with the BACnet prioritization mechanism. If a new Present_Value is written, but the output does not change, it usually means that some other client has previously written to the same output with a more important (numerically lower) priority. In order for a new value to take precedence, one of the following must be done:

- Command the output at a more important priority.
- Disable the control program that is blocking the new command.
- Command the active value to relinquish by writing a NULL value at its priority.

While unusual, it is possible that the output has been taken out of service (indicated by its Out_Of_Service property). In that case, writing to the Present_Value property will have no effect on the physical output. To change this state, it is necessary to write Out_Of_Service with FALSE.

Also, it is possible that the output is OVERRIDDEN, which can be determined by reading the object's Status_Flags. Nothing can be done via the network to change this state because it is overridden by some local mechanism. For example, a MAN-OFF-AUTO switch requires local and physical access to change.

Summary

- General-purpose network tools all contain certain basic functions that allow an integrator to configure at least most of a node's network and operational features.
- Automated discovery uses dynamic device binding to identify Device instances by their network numbers and MAC addresses, building a list of associations.
- Basic node configuration can often be accomplished using almost any BACnet client software tool that supports writing to object properties.
- A network analyzer tool has access to every message that is transmitted on the network segment where the tool is connected.
- A tracing detail displays the actual hexadecimal code of the octets in a message, matched line-by-line with the human-readable meaning.
- A device information table defines the parameters of each BACnet node in the entire system.
- Daisy chains are formed by twisting incoming and outgoing conductor ends together and capturing the twist in the corresponding screw terminal for the network.
- It is recommended to allow for access points to the network conductors at multiple locations throughout a network, preferably at each wiring cabinet.
- During bench testing, a node is configured by writing to certain properties of the node's Device object, which can be accomplished with any network tool that supports ReadProperty and WriteProperty services.
- MAC addresses for MS/TP and ARCNET nodes are assigned when they are installed.
- Every BACnet node must be assigned a unique Device instance number as well as a unique Device Object_Name.
- Since the Device instance number can range up to 4,194,303, available Device instances can be allocated in ways that help convey information about the node.

- The easiest way to find existing nodes is with discovery tools.
- An object database must be compiled that includes information based on object discovery.
- Several key parameters must be configured on MS/TP nodes before they can be placed in service, including speed, Max_Master, $T_{usage\_timeout}$, and Max_Info_Frames.
- The most critical parameter to set up for each router port is a BACnet network number.
- If BACnet/IP nodes are on different subnets, then BACnet broadcast management devices (BBMDs) are needed to manage broadcasts between subnets.
- Every BACnet client application must be configured with device/object/property references.
- A common source of bandwidth issues is client nodes that query BACnet servers too aggressively, overwhelming their ability to respond.
- Most problems are caused by human error in design, installation, or configuration.

Definitions

- *Automated discovery* is a method of automatically locating and identifying the node and object components of a BACnet network.
- A *network analyzer tool* is a network tool that listens passively to network traffic, capturing message frames and saving them for later analysis.
- A *tracing detail* is a detailed display of the information contained within a BACnet message.
- A *discovery tool* is software that sends dynamic device binding messages, usually Who-Is broadcasts, and listens for the replies.
- *Client application binding* is the configured association between a BACnet application logical parameter and its actual source or destination object property.

Review Questions

1. What is automated discovery?
2. How is a tracing detail used to examine a BACnet message?
3. What types of documentation are recommended to organize system information?
4. How can service tool access points be added to the network?
5. What are the common schemes for assigning Device instances?
6. What is the procedure for discovering the objects within a node?
7. Why are the Max_Master and $T_{usage\_timeout}$ settings important for network performance?
8. What factors influence the number of BACnet broadcast management devices (BBMDs) installed on a BACnet/IP network?
9. What is client application binding?
10. How does bursting waste network bandwidth?

Chapter Sixteen

System Integration

The ultimate goal of a comprehensive building automation system is the integration of multiple building systems into a common information-sharing network. Events or changes in one system can be used to trigger changes in the operation of otherwise unrelated building systems. The possible combinations of system interactions are practically infinite, though some control scenarios are relatively common. However, regardless of the required interactions, it is likely that the automation system can be designed using a variety of hardware combinations or protocol systems. It may be necessary to consider multiple possible implementations when designing a building automation system.

Chapter Objectives

- Explain how the same building automation goals can be accomplished in a variety of ways.
- Describe the example building's automation requirements and how they are addressed by the LonWorks and BACnet implementations.
- Compare the possible network infrastructures of different protocol systems.
- Evaluate the differences between LonWorks and BACnet implementations of the same control scenarios.

BUILDING AUTOMATION SYSTEM EXAMPLE

Building automation systems are increasingly popular in both new construction and existing buildings. The most common automation applications are related to HVAC, lighting, and life safety operations. Some buildings also integrate additional systems, such as security, access control, elevator control, or other special functions.

A study of an example building automation system can be used to emphasize possible applications of existing building automation technology. Common building systems are integrated together to show how various systems can be combined under a single protocol. The design focuses on implementing a robust system that limits the possibilities of a single source of failure and allows communication between various systems.

The example building belongs to the Lincoln Publishing Group (LPG). The building has 45,000 sq ft of usable space on three floors. **See Figure 16-1.** The architectural design, including the building's orientation, optimizes light penetration and shading for maximum lighting and heating benefits. LPG occupies the second and third floors. The first floor includes tenant space, plus common areas for entrances, an elevator lobby, and an exercise facility. **See Figure 16-2.** The lobby area secures access to all building spaces. The building's electrical, HVAC, fire protection, and voice-data-video (VDV) equipment is also located on the first floor.

Example Building

Figure 16-1. The integration requirements for an example automated building are useful for comparing the possible implementations of different protocol systems.

Building System Requirements

The company recognizes the importance of building automation technology in accomplishing the goals of energy efficiency, reduced maintenance, and enhanced occupant comfort and performance. A number of possible control scenarios were developed, though only some were ultimately implemented, based on a balance of benefits and costs. All building automation system designs utilize an open protocol, such as LonWorks or BACnet protocols, and all systems are interoperable with this protocol. The resulting building automation system integrates many of the building's systems together.

Electrical System. The building includes a 400 kW diesel-powered generator to serve as a back-up power source for the fire pump, elevators, computer servers, and other critical load circuits, as well as for demand shedding during peak load situations.

Lighting System. Outdoor site lighting is primarily controlled according to a schedule, which turns the lights ON from dusk until late at night, and again during a period early in the morning. Light level sensors are also used to automatically turn the lights ON if needed during the scheduled OFF times, such as during a thunderstorm.

Indoor lighting is controlled by a lighting control system that uses schedules, occupancy sensors, manual override switches, and control sequences specific to the lighting system. Light fixtures incorporate dimmable ballasts or bi-level switching to adjust lighting levels. Work areas near the perimeter utilize daylight harvesting. All life-safety-designated lighting, such as lighting in egress stairwells, is on emergency power.

HVAC System. The building is divided into HVAC zones. The HVAC system manages zone temperature with variable-volume air terminal units (ATUs), which use information from temperature sensors to control their operation. The ATUs are served by a rooftop air-handling unit (AHU) with a direct expansion (DX) cooling coil. The air conditioner (with DX evaporator coil) has an air-cooled condenser. The AHU and air conditioner are stand-alone components, not a package unit.

Setback temperature setpoints are used during unoccupied periods. The system uses an optimization strategy to provide adequate heating/cooling in anticipation of scheduled occupancy. All components of the HVAC system that are critical to life safety are on emergency power.

Plumbing System. Automated plumbing subsystems include pressure boosting and hot water circulation equipment. The critical plumbing subsystems, such as the lift station and sump pump, are on emergency power. The building automation system does not enable/disable the lift station, but monitors it for alarms.

Fire Protection System. The fire protection system must be reliable during life safety situations. Therefore, the fire protection system can share information with other building automation systems, but direct control of the fire protection system is not allowed. During an alarm condition, the fire alarm system controls critical functions such as door unlocking, HVAC shutdown, and elevator recall. The fire protection system is UL-listed and fire-marshal-approved for compatibility with BACnet and LonWorks protocols. The fire protection system is on an emergency power circuit.

Security System. The security system is separated into four partitions: one partition for each of the three floors and a fourth partition for the common areas (elevator lobbies, front entrance, employee entrances, and exercise facility). The security system monitors the closure of main entrances and windows on the first floor, and patio windows and doors on the second and third floors. Common areas on each floor are monitored with motion detectors, which are also integrated with the lighting control system to indicate occupancy.

Like the fire protection system, the security system must ensure reliability during alarm events. Therefore, critical system programming functions and critical operations, such as zone bypassing, are only allowed to be performed by authorized personnel. The security system is on emergency power.

Example Building Floor Plans

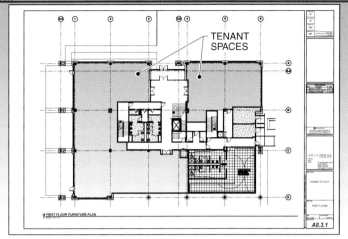
FIRST FLOOR

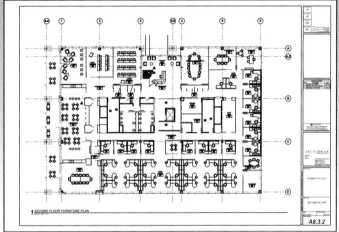
SECOND FLOOR

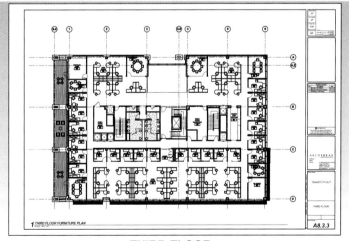
THIRD FLOOR

Figure 16-2. The office building described in the integration example consists of three floors that include open offices, individual offices, multiuse space, a training room, a lunchroom, an exercise room, and tenant spaces.

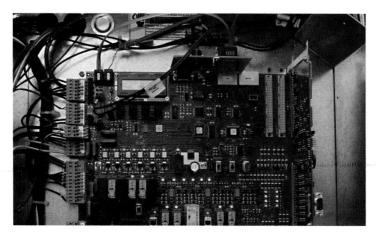

Elevator control panels may provide connection terminals for integration with a building automation system.

Access Control System. All building entrances have electric locks and card readers for employees to gain access to the building. The front doors automatically unlock and lock on a schedule programmed for normal business hours. Doors leading into tenant spaces also have electric locks and card readers that are programmed to unlock and lock according to the tenant's schedule. Stairwell doors are locked using magnetic locks and require a valid card read to enter or exit, regardless of the time of day. For safety purposes, stairwell doors utilize a touch-sensitive panic bar that releases the magnetic lock, allowing free egress from the occupied area. This event triggers an alarm condition if a valid card read was not received.

Like the security system, the access control system must be protected from unauthorized operation. The access control system can be monitored by building maintenance staff, but critical system programming and critical operations, such as remote door release, are only allowed to be controlled by authorized personnel. The access control system is on emergency power.

Voice-Data-Video (VDV) System. The VDV system operates on its own network, as opposed to the common building automation network. The voice-data-video (VDV) system is on emergency power.

Elevator System. During normal occupancy periods, the elevator controller allows unrestricted use of the elevator. During scheduled unoccupied periods, the elevator system requires access card authentication before the elevator operates. The identity of the cardholder also determines which floor the elevator provides access to, based on privileges programmed into the system. The elevator system is on emergency power.

Building Automation System Implementations

There are multiple ways to implement an effective building automation system that meets the automation and integration requirements of the various building systems. In most cases, a control scenario can be designed with any of a number of different protocol-based systems, including the two most common open protocols, LonWorks and BACnet.

The choice of protocol is typically based on cost, contractor experience, ease of programming, available compatible control devices, existing infrastructure and equipment, and other considerations. For comparison, it is particularly useful to produce preliminary implementation designs of the same control scenarios in multiple protocol systems. For the example building, the control scenarios are studied from both a LonWorks and a BACnet perspective.

LonWorks Implementation. The building automation system for the Lincoln Publishing Group building may be based on the LonWorks technology. A flat network architecture integrates the lighting, HVAC, plumbing, fire protection, security, and access control systems. Node communication is peer-to-peer without the need for programmable gateways or network supervisor nodes. A gateway is needed only to interface with the proprietary elevator control system.

The network uses twisted-pair cabling. A TP/XF-1250 channel provides a high-speed backbone, arranged in a bus topology and connecting each floor. **See Figure 16-3.** TP/FT-10 free-topology node channels are used for each building system on each floor. Multiport routers on each floor isolate network traffic for system channels. Channel terminators are installed at router ports.

LonWorks Network Infrastructure

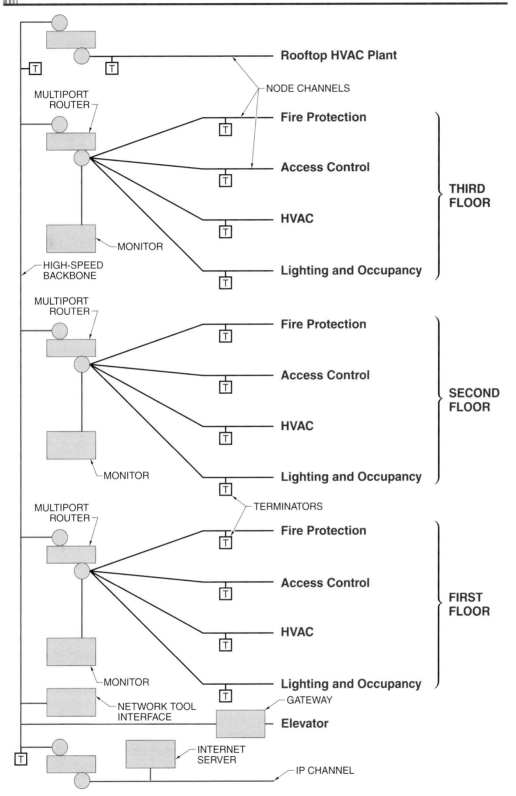

Figure 16-3. The network architecture of the proposed LonWorks implementation uses a high-speed backbone to connect to many node channels, which are divided according to building floor and system type.

> Building automation systems often allow messages to be communicated via protocols that are also common to computer networks, such as Ethernet. Some implementations may use this capability to share an existing network infrastructure between the two systems.

Nodes adhere to LonMark interoperability standards as well as local safety jurisdictions. Wherever possible, application-specific nodes utilize LonMark standard functional profiles, which provide a high degree of vendor interchangeability. The network management platform is LNS (LonWorks Network Services). Node configuration, where possible, utilizes LNS-compatible software plug-ins.

The network HMI is browser-based and provides graphical representations of equipment operating conditions and access to data logging, scheduling, and alarm-reporting functions. These monitoring and control functions are provided by an Internet server node. The access control management and database software runs on a secure computer that is accessible by authorized personnel only.

BACnet Implementation. Alternatively, the control system network for the Lincoln Publishing Group building may be based on the BACnet standard. The network architecture provides for peer-to-peer communication within each subnet. **See Figure 16-4.**

The network uses twisted-pair cabling. Integration between multiple building systems is accomplished primarily with Ethernet and ARCNET. An Ethernet network provides a high-speed connection to each floor. There, routers connect to subnets based on the building system. The lighting and HVAC systems are integrated together on ARCNET networks. Integration with proprietary systems, including the fire protection, security, access control, and elevator systems, is accomplished via a Modbus RTU protocol. The electrical demand limiting and plumbing monitoring and control are accomplished with ARCNET controllers.

Nodes adhere to BACnet interoperability standards as well as local safety jurisdictions. Wherever possible, application-specific nodes utilize BACnet standard object types, which provide a high degree of vendor interchangeability. The network management platform is an Internet server running Automated Logic Corporation's WebCTRL® software.

The network HMI is browser-based and provides graphical representations of equipment operating conditions and access to data logging, scheduling, and alarm-reporting functions. These monitoring and control functions are provided by an Internet server node. The access control management and database software runs on a secure computer that is accessible by authorized personnel only.

Twisted-pair cabling was installed throughout the building during construction for both computer networking and building automation needs.

CONTROL SCENARIO: OPENING THE BUILDING ON A REGULARLY SCHEDULED WORKDAY

A very basic control scenario concerns the actions needed by the building system when it changes to an occupied mode, such as at the beginning of a normal workday. The control scenario of opening the building on a regularly scheduled workday is initiated by an employee presenting a proximity card at either building entrance. **See Figure 16-5.** Opening scenarios on non-workdays, as determined by calendar schedule in the building automation system, may behave differently.

BACnet Network Infrastructure

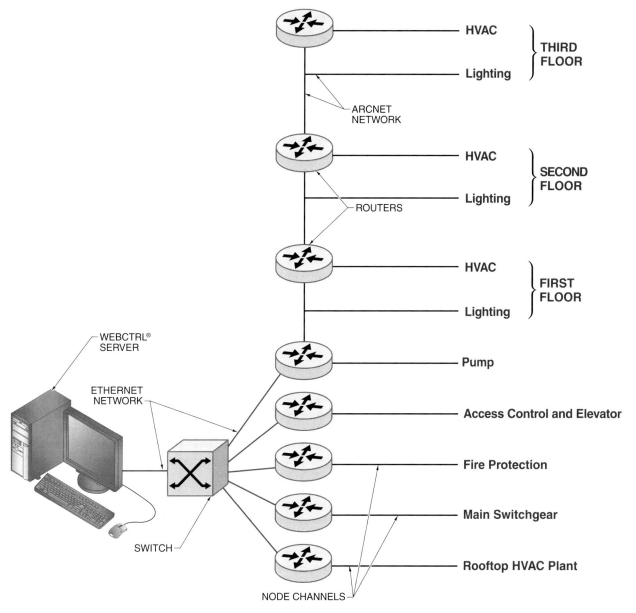

Figure 16-4. The proposed BACnet network architecture uses a fast Ethernet network to connect to system-based subnetworks, which use bus topology ARCNET networks to connect between nodes.

When the proximity card is read, the access control system detects the access request of an employee that works on the second floor. The card access system compares the credential to the access control database. Upon authorization of the request, the building automation system executes actions to make the building ready for occupancy. This scenario affects the lighting, HVAC, plumbing, access control, security, and elevator systems.

> The opening of a building at the beginning of a workday is often programmed for a certain time. However, as many occupants arrive shortly before the scheduled start time, their entry may be used to trigger the scenario early.

Opening the Building on a Regularly-Scheduled Workday

Figure 16-5. The action of a person requesting access to the building at the beginning of a regularly scheduled workday initiates changes in the building systems to prepare for occupancy.

Lighting System Response

The lighting system takes several actions to ensure adequate lighting for the occupant entering the building. This includes lighting in general work areas, plus additional areas specific to the employee's location.

- Interior lighting in the common areas of the respective employee's floor (second floor) is turned ON and will remain ON until shut OFF by the building closing scenario.
- Interior lighting in the work areas, in both enclosed office and open office areas, is controlled with occupancy sensors and turns OFF after a time delay. Alternatively, the employee can override the occupancy sensors with a local override switch.
- If the employee enters an unassigned area, such as a conference room, lighting is controlled by occupancy sensors and turns OFF again after a time delay.
- Interior work areas near the building perimeter utilize daylight harvesting and dimmable ballasts.
- Interior life safety lighting is turned ON, and stays ON until shut OFF by the building closing scenario.

All lighting actions are determined by programmed schedules of times of day and days of week. For example, if an employee enters the building at 3 AM, the lighting system actions are based on programming for unoccupied periods, and lighting in each area is controlled with occupancy sensors.

Outdoor lighting is controlled by anticipated occupancy schedules and light level sensors that may turn the site lighting ON for dark mornings before the expected arrival of the first employee.

LonWorks Lighting System Implementation. The interior lighting system includes dimmable fluorescent fixtures operated by LonWorks lighting controllers. Common area fixtures are energized from a central lighting control panel that also acts as a LonWorks node. **See Figure 16-6.** Individual offices utilize a multifunction LonWorks node that provides occupancy detection, temperature, and ambient light level measurement. A manual override switch is used to bypass the timed off function.

Nodes involved with life safety systems and demand-limiting functions also share data with lighting nodes. This is needed for performing energy-reduction strategies while maintaining safety and security.

BACnet Lighting System Implementation. The interior lighting system includes dimmable fluorescent fixtures operated by BACnet-based lighting controllers. Common area fixtures are energized from a central lighting control panel that also acts as a BACnet node. **See Figure 16-7.** Individual offices utilize a multifunction BACnet node that provides occupancy detection, temperature, and ambient light level measurement. A manual override switch is used to bypass the timed OFF function.

The first floor and second floor lighting control panels turn ON the common area lighting and emergency lighting. When the employee gains access to the second floor, the access control system sends a signal to the second floor lighting control panel. The lighting control panel turns ON the lighting in the open office area on the second floor.

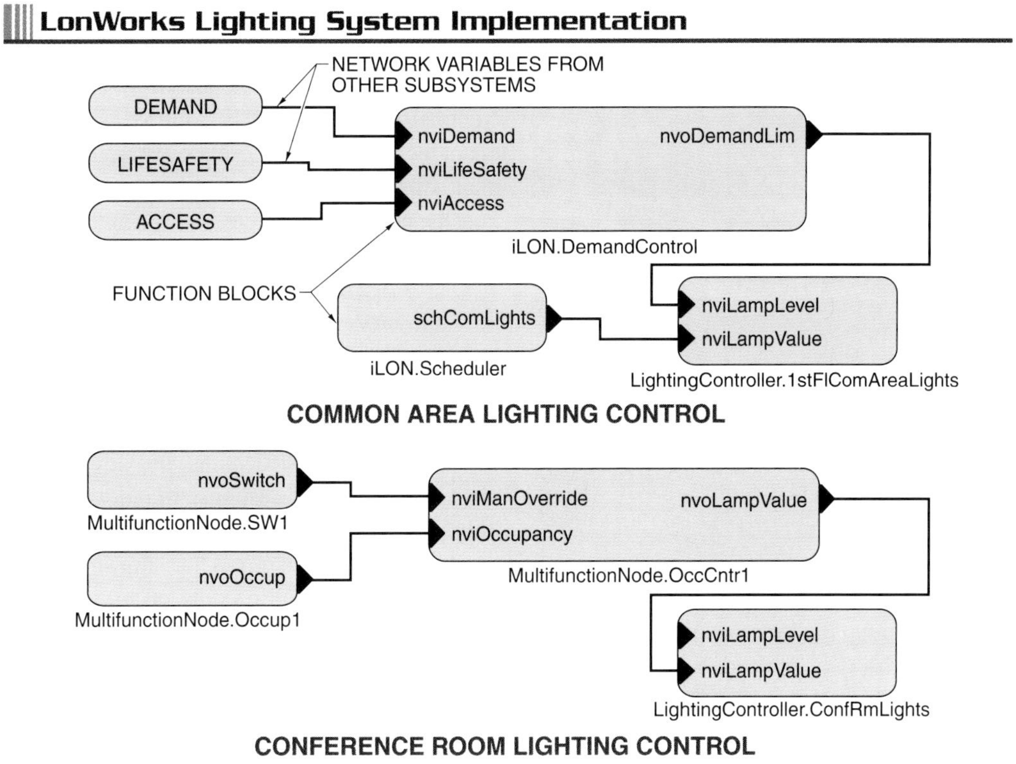

LonWorks Lighting System Implementation

Figure 16-6. A LonWorks-controlled system controls lighting according to schedules, electrical demand, access, occupancy, and life safety network data.

BACnet Lighting System Implementation

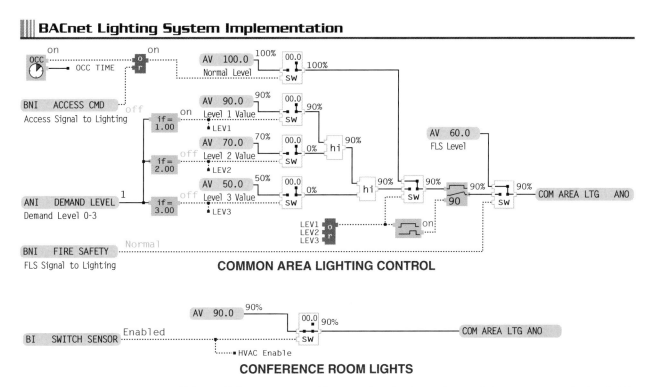

Figure 16-7. Lighting control utilizing a BACnet system uses binary and analog inputs from switches, sensors, and other building systems to determine the output of common area and individual room lighting.

HVAC units are installed in the building's ceiling to serve individual zones with the appropriate volume and temperature of conditioned air.

Nodes involved with life safety systems and demand-limiting functions also share data with lighting nodes. This is needed for performing energy-reduction strategies while maintaining safety and security.

HVAC System Response

The control of the HVAC system at the beginning of an occupancy period assumes that temperature setpoints are at setback levels during unoccupied times as an energy-conservation measure and a preheating/precooling sequence in anticipation of occupancy has not already been initiated via schedule.

Both central and employee-assigned HVAC systems are initiated and stay ON until shut OFF by the closing scenario. During scheduled unoccupied periods, occupancy sensors are used to indicate occupancy, which is used to turn ON the HVAC system equipment serving the specific occupied area.

LonWorks HVAC System Implementation. The HVAC subsystems include a rooftop mechanical plant and variable-air-volume (VAV) terminals for each floor. LonWorks nodes from nine different manufacturers share information on the network in order to perform their local applications and the specified system sequences of operations.

LonWorks VAV controllers for each HVAC zone receive space temperature and occupancy status from a ceiling-mounted multisensor node.

See Figure 16-8. Carbon dioxide levels from an indoor air quality node are also sent to VAV controllers in order to implement demand-controlled ventilation sequences. The rooftop mechanical plant includes a LonWorks programmable node that adjusts fan speed based on static pressure, mechanical cooling stages, outside air temperature, and hot water boiler operations.

The programmable plant controller is bound to each VAV zone controller's terminal load output, which indicates the level of cooling or heating required. The plant controller then resets the supply air temperature according to the zone of greatest demand and adjusts outdoor air dampers in response to indoor carbon dioxide levels.

Alarm contacts in proprietary fire and smoke detection devices in life safety systems provide system shutdown and smoke evacuation during alarm events. During peak energy conditions, a network variable from the demand-limiting subsystem initiates a load-reduction sequence that includes increasing setpoints for individual zones and reducing supply-fan speed.

BACnet HVAC System Implementation. The HVAC subsystems include a rooftop mechanical plant and VAV terminals for each floor. BACnet nodes share information on the network in order to perform their local applications and the specified system sequences of operations.

BACnet VAV controllers for each HVAC zone receive space-temperature values and occupancy status from a ceiling-mounted multisensor node. **See Figure 16-9.** Carbon dioxide levels from an indoor air quality node are also sent to VAV controllers in order to implement demand-controlled ventilation sequences. The rooftop mechanical plant includes a BACnet programmable node that adjusts fan speed based on static pressure, mechanical cooling stages, outside air temperature, and hot water boiler operations.

The programmable plant controller is bound to each VAV zone controller's terminal load output, which indicates the level of cooling or heating required. The plant controller then resets supply air temperature according to the zone of greatest demand and adjusts outdoor air dampers in response to indoor carbon dioxide levels.

Chapter 16—System Integration **339**

LonWorks HVAC System Implementation

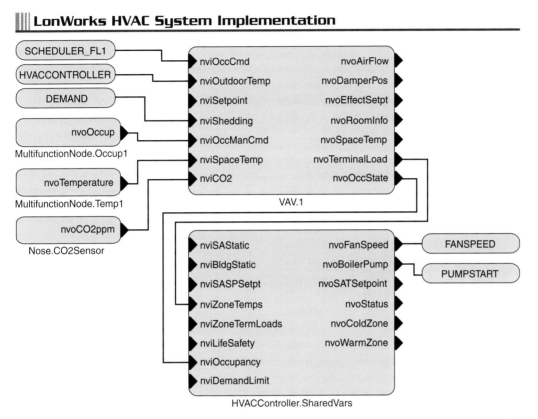

Figure 16-8. LonWorks HVAC control integration includes occupancy, temperature, indoor air quality, and demand limit variables as inputs into function blocks governing central plant controllers.

BACnet HVAC System Implementation

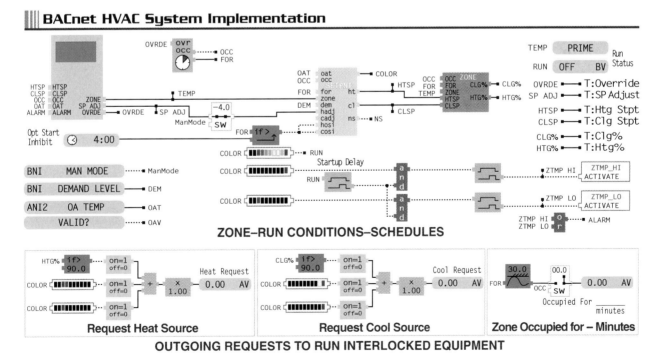

Figure 16-9. The VAV logic can be programmed with BACnet objects and their properties.

Alarm contacts in proprietary fire and smoke detection devices in life safety systems provide system shutdown and smoke evacuation during alarm events. During peak energy conditions, a signal from the demand-limiting subsystem initiates a load-reduction sequence that includes increasing setpoints for individual zones and reducing supply-fan speed.

Plumbing System Response

The plumbing system takes actions to ensure adequate water supply for the plumbing fixtures in use while the building is occupied.

- The electrical circuit serving the hot water circulation pump is switched ON and stays ON until shut OFF by the closing scenario.
- The lift station pump package is fully operational 24/7/365 with self-contained controls.
- The fire protection system provides a signal that is used to initiate an emergency mode, which shuts down gas-fired boilers in the event of a fire alarm.

LonWorks Plumbing System Implementation. The plumbing subsystems utilize general-purpose LonWorks input/output nodes as well as analog control and schedule functions provided by the network Internet server. **See Figure 16-10.** Transducers wired to analog inputs in the node measure loop water pressure and energize the booster pump when the pressure drops below a setpoint. The scheduler controls equipment operations, including pumps and domestic hot water boilers, based on expected building occupancy.

BACnet Plumbing System Implementation. The plumbing subsystems utilize general-purpose BACnet input/output nodes as well as analog control and schedule functions. **See Figure 16-11.** Transducers wired to analog inputs in the node measure loop water pressure and energize the booster pump when the pressure drops below a setpoint. The scheduler provides time of day control of equipment operations, including pumps and domestic hot water boilers.

Access Control and Security System Responses

The security system receives information from the access control system and scheduler. The access control and security systems can be a combination system covering both functions, but may have the ability to act as two individual stand-alone systems.

- The employee's credential is authenticated by the access control system and allows the door to open. The access control panel compares the access event to the programmed schedules and transmits information to the security system.
- The security system turns OFF the alarm partition of the common zone and the occupant's assigned area.
- If there is a network communication problem, the security system remains armed. The door is allowed to open, but the security system panel just inside the entrance annunciates that the security system is still armed, which requires a code to be entered during the time-delay period.

LonWorks Plumbing System Implementation

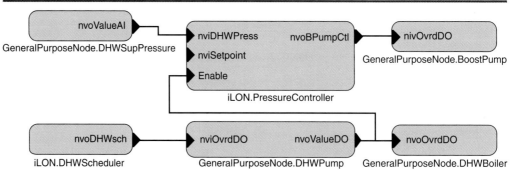

Figure 16-10. The LonWorks implementation of the plumbing system automation uses a general-purpose controller node, interacting with an Internet service, to control when pumps and the boiler are turned ON.

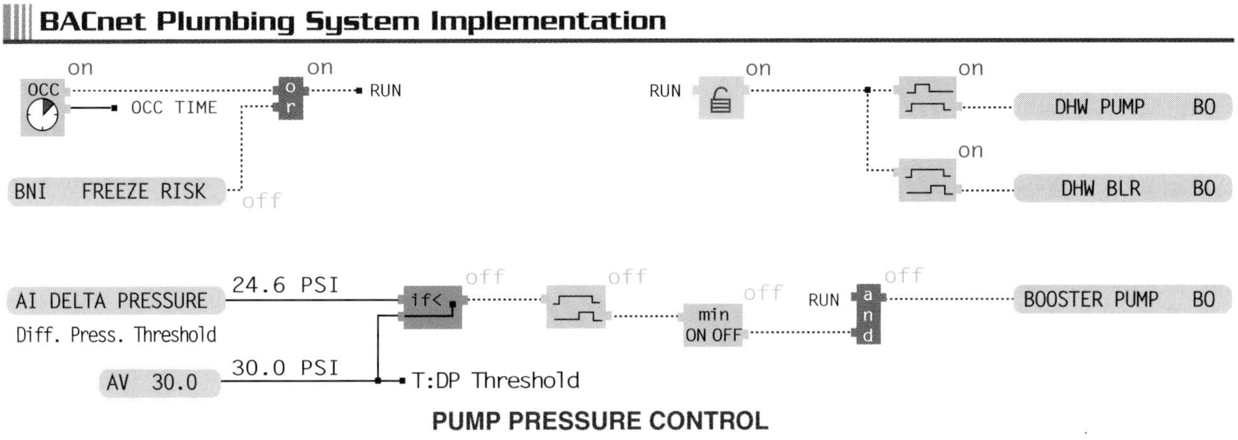

BACnet Plumbing System Implementation

PUMP PRESSURE CONTROL

Figure 16-11. Plumbing system control with a BACnet-based system determines when to activate pumps and the hot water boiler.

LonWorks Access Control and Security System Implementation. The security system uses information from LonWorks occupancy sensors, card readers, and scheduler nodes. **See Figure 16-12.** Node outputs include alarm annunciation and event notification via cell phone text messaging or email to security personnel. A LonWorks access control system controls door entries, detects unauthorized entries, and provides network data used to initiate building opening events in HVAC, plumbing, and lighting systems. Access-control software records access events and manages the users of the door access system.

BACnet Access Control and Security System Implementation. The security system uses information from BACnet-based occupancy sensors and scheduler nodes. Node outputs include alarm annunciation and event notification via cell phone text messaging or email to security personnel. A card access control system with a BACnet gateway performs several functions: it controls door entries; detects unauthorized entries; and provides network data used to initiate facility-opening events in HVAC, plumbing, and lighting systems. Access-control software records access events and manages the users of the door access system.

Elevator System Response

The elevator system receives information about the identity of the person entering the building from the access control system. The elevator controller also communicates with the scheduler to determine the appropriate action based on the programmed schedule.

- During normal occupied periods, the access control system enables unrestricted use of the elevator.
- When an employee assigned to an upper floor enters the building, the elevator is called to the first floor.
- During normally unoccupied periods, a card reader located in the elevator restricts elevator operation to authorized employees. When authenticated, the access control system then enables the elevator call button for the floor assigned to the employee.

LonWorks Elevator System Implementation. A LonWorks node integrates with the access control system and scheduler and interfaces with the proprietary elevator control system. Dry relay contacts are used to enable/disable call buttons according to access events and building schedules. Access-control software records access events and manages the users of the elevator access system.

BACnet Elevator System Implementation. A BACnet node integrates with the access control system and scheduler and interfaces with the proprietary elevator control system. Dry relay contacts are used to enable/disable call buttons according to access events and building schedules. Access-control software records access events and manages the users of the elevator access system.

LonWorks Security System Implementation

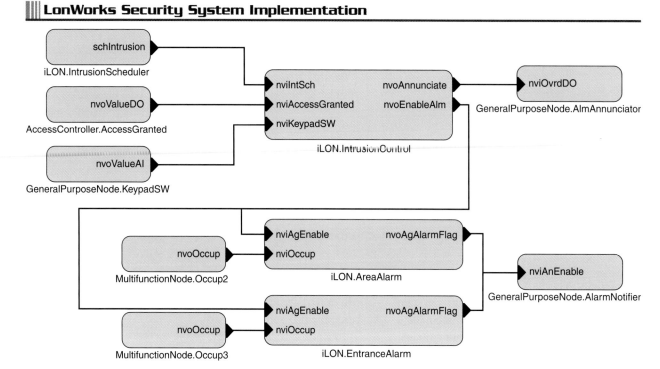

Figure 16-12. Occupancy sensors also provide information to the LonWorks security and access control systems, which initiate alarms when intrusion is detected.

Lighting in certain areas may be dimmed or turned OFF for demand-limiting purposes. However, in the event of a life safety alarm, such as a fire alarm, all lighting is restored to its full level to aid evacuation.

CONTROL SCENARIO: DEMAND LIMITING

Independent of the utility electric meter, a facility energy meter measures electrical demand and consumption for monitoring purposes. These electrical demand values are calculated over a 15-minute sliding window and evaluated according to the utility's time-of-day-demand rate schedules. During peak periods, demand-limiting strategies are initiated if electrical demand is at or above a setpoint.

When the building automation system receives the signal to initiate the demand-limiting sequence, certain noncritical loads are commanded to turn OFF or reduce their duty cycle. This consists of changes to HVAC and lighting system operations, such as dimming light fixtures, adjusting temperature setpoints, and reducing supply-fan speed, in order to reduce the overall electrical demand of the building.

LonWorks Demand Limiting Implementation

A LonWorks electrical meter node measures electrical power demand and consumption in parallel with the utility electric meter. The node also measures and reports power-quality parameters such as voltage, current, frequency, and power factor for each phase. The meter node calculates average demand over a sliding 15-minute time window and shares this information with a LonWorks-programmable node. **See Figure 16-13.** If the demand is at or above the setpoint, the node initiates a demand-limiting sequence by instructing HVAC, lighting, and plumbing control devices to turn OFF or reduce their duty cycle. Energy consumption is logged and compared with utility billing for accuracy.

LonWorks Demand Limiting Implementation

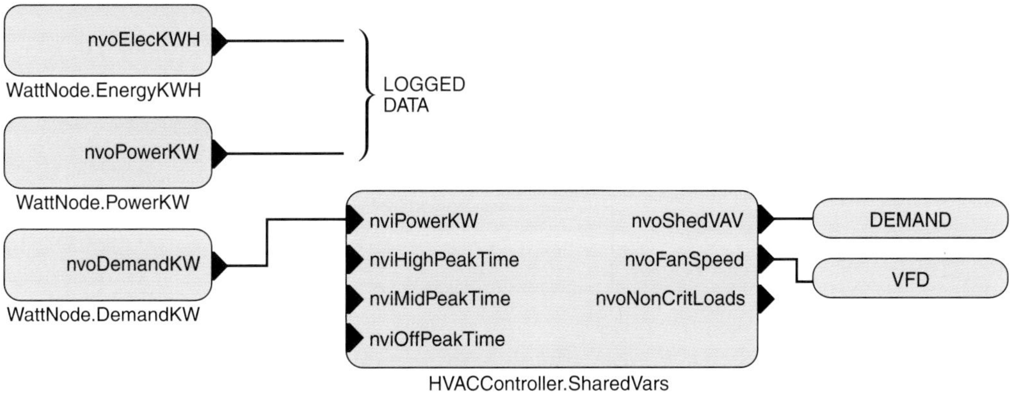

Figure 16-13. Information from a LonWorks electrical meter node is either logged or shared with a function block that determines when to initiate demand limiting.

BACnet Demand Limiting Implementation

An electrical meter node measures electrical demand and consumption in parallel with the utility electric meter. The meter also measures and reports power-quality parameters such as voltage, current, frequency, and power factor for each phase. The meter transmits this information to a BACnet node via the Modbus RTU protocol. The BACnet node calculates the average demand over a sliding 15-minute time window and compares it to the setpoint. If the demand is at or above the setpoint, the node initiates a demand-limiting sequence by instructing HVAC, lighting, and plumbing control devices to turn OFF or reduce their duty cycle. Energy consumption is logged and compared with utility billing for accuracy.

Lighting System Response

When it receives a signal to initiate demand limiting, the lighting system takes several actions to reduce the lighting levels in several areas of the building. The scheduler's programmed information may be overridden to accomplish some of the demand limiting.
- Life-safety-designated lighting remains ON.
- Lighting levels in the interior common areas are reduced, but stay ON until shut OFF by the closing scenario.
- Lighting in the work areas, in both enclosed and open office areas, is controlled with occupancy sensors and turn OFF after time delay.
- The lighting control system overrides the daylighting sequence to lower the target lighting levels.
- If lights have been enabled and demand limiting becomes active, lighting levels are reduced to one of three different presets. Abrupt changes to lighting levels are avoided via a ramping function.

LonWorks Lighting Demand Limiting Implementation. LonWorks-enabled dimmer controllers are used to reduce lighting levels. Upon receiving a signal to reduce electrical demand, LonWorks lighting nodes lower lighting energy usage while maintaining high levels of safety and security. **See Figure 16-14.**

BACnet Lighting Demand Limiting Implementation. BACnet-enabled dimmer controllers are used to reduce lighting levels. Upon receiving a signal to reduce electrical demand, BACnet lighting nodes lower lighting energy usage while maintaining high levels of safety and security.

HVAC System Response

When it receives a signal to initiate demand limiting, the HVAC system takes several actions to reduce the heating/cooling loads in several areas of the building by adjusting the zone temperature setpoints.

LonWorks Lighting Demand Limiting Implementation

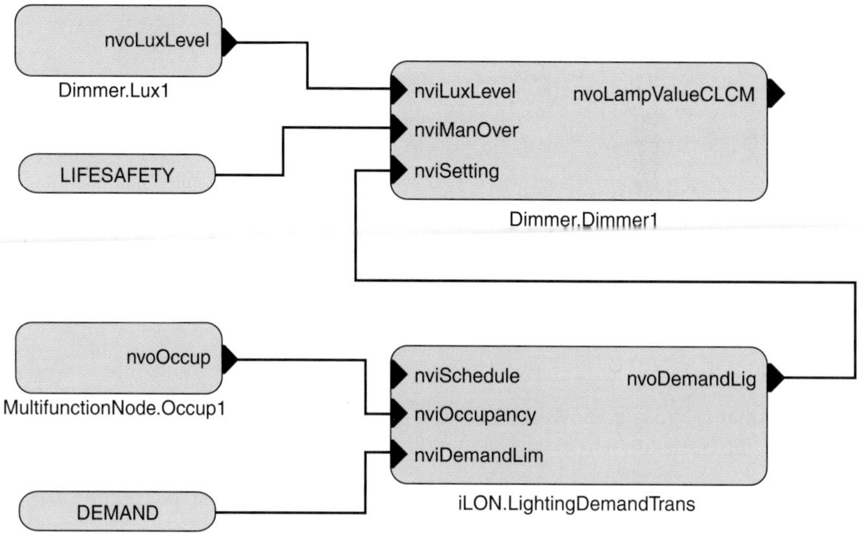

Figure 16-14. When demand limiting is initiated, the LonWorks-enabled dimmers are set to a lower light level.

- Every thermostat setpoint is raised (when in cooling mode) or lowered (when in heating mode) by 2°F.
- Occupancy sensors are used to raise or lower setpoints as much as 5°F in areas where occupancy has not been detected for a minimum period of time.

LonWorks HVAC Demand Limiting Implementation. Upon receiving a signal to reduce demand, VAV terminal devices adjust heating/cooling setpoints accordingly to reduce energy consumption. Through a browser-based user interface, individual occupants can voluntarily participate in additional demand-limiting measures by specifying an acceptable setpoint offset during periods of high energy demand.

BACnet HVAC Demand Limiting Implementation. Upon receiving a signal to reduce demand, VAV terminal devices adjust heating/cooling setpoints accordingly to reduce energy consumption. **See Figure 16-15.** Through a browser-based user interface, individual occupants can voluntarily participate in additional demand limiting measures by specifying an acceptable setpoint offset during periods of high energy demand.

BACnet HVAC Demand Limiting Settings

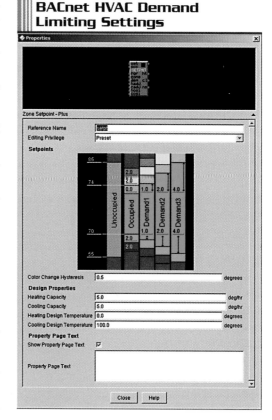

Figure 16-15. HVAC system setpoints for different demand-limiting scenarios can be set with BACnet system software.

Refer to Quick Quiz® on CD-ROM

Summary

- A study of an example building automation system can be used to emphasize possible applications of existing building automation technology.
- There are multiple ways to implement an effective building automation system that meets the automation and integration requirements of the various building systems.
- For comparison, it is particularly useful to produce preliminary implementation designs of the same control scenarios in multiple protocol systems.
- A very basic control scenario involves the actions needed by the building system when it changes to an occupied mode, such as at the beginning of a normal workday.
- When the building is opened, the lighting system takes several actions to ensure adequate lighting for the occupant entering the building.
- The control of the HVAC system at the beginning of an occupancy period assumes that temperature setpoints are at setback levels during unoccupied times and a preheating/precooling sequence in anticipation of occupancy has not already been initiated via schedule.
- The access control and security systems can be a combination system covering both functions, but may have the ability to act as two individual stand-alone systems.
- The elevator controller communicates with the scheduler to determine the appropriate action based on the scheduled occupied/unoccupied status.
- During peak periods, demand-limiting strategies are initiated if electrical demand is at or above a setpoint.
- When the building automation system receives the signal to initiate the demand-limiting sequence, certain noncritical loads are commanded to turn OFF or reduce their duty cycle.
- When it receives a signal to initiate demand limiting, the lighting system takes several actions to reduce the lighting levels in several areas of the building.
- When it receives a signal to initiate demand limiting, the HVAC system takes several actions to reduce the heating/cooling loads in several areas of the building by adjusting the zone temperature setpoints.

Review Questions

1. What types of building systems can be integrated with a building automation system?
2. Can multiple building automation solutions for a control scenario be developed from different protocol systems?
3. How is a protocol system typically chosen?
4. How is the building opening scenario initiated in the example building?
5. What is the HVAC system's response to the opening of the example building?
6. What is the elevator system's response to the opening of the example building?
7. How is the demand-limiting scenario initiated in the example building?
8. What is the HVAC system's response to demand limiting in the example building?

Chapter Seventeen

Cross-Protocol Integration

In many situations, a building automation system must be designed to integrate more than one protocol. There are three possible reasons for this: an expanded system must be built on top of an existing system, the necessary control devices are only available in certain protocols, or the building automation system must be integrated with an enterprise system. Various strategies can be used to fully integrate these systems together, including gateways and intermediary frameworks that bridge the gap between the different information structures.

Chapter Objectives

- Describe the potential issues involved in information translation.
- Compare gateways and intermediary frameworks as possible solutions for cross-protocol integration.
- Identify the considerations and advantages involved with using Extensible Markup Language (XML) for bridging between different protocols.
- Compare the possible solutions for integrating enterprise systems with building automation systems.
- Evaluate the description of an example implementation of a multi-protocol building automation system.

CROSS-PROTOCOL INTEGRATION

A building automation system based on a single protocol is usually the best solution, but is not always possible or practical. It is often necessary to design control systems using a mixture of LonWorks, BACnet, and even proprietary protocol devices. There is no one cross-protocol solution that fits every building or control situation. Often, multiple system designs can achieve the desired integration, though some may have advantages in scale, efficiency, ease of use, cost, or other aspects.

When multiple protocols are to be integrated together, the primary concern is the translation of control information from one protocol to another. **See Figure 17-1.** Designing systems that translate information accurately and reliably between two or more protocols can present significant challenges. However, since this is not an uncommon scenario, solutions are available.

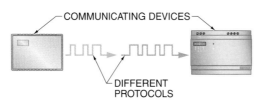

Figure 17-1. Cross-protocol integration strategies are solutions for facilitating communication between devices using different protocols.

Information Translation

The translation of information encoded in a protocol is similar to language translation, including the complexities of meaning and context. These factors must be considered when designing a cross-protocol solution for building automation systems.

For example, the English word "temperature" can easily be translated into the Italian word "temperatura" with the meaning intact. **See Figure 17-2.** However, the translation does not necessarily address every aspect of the original meaning in English, such as scale and resolution.

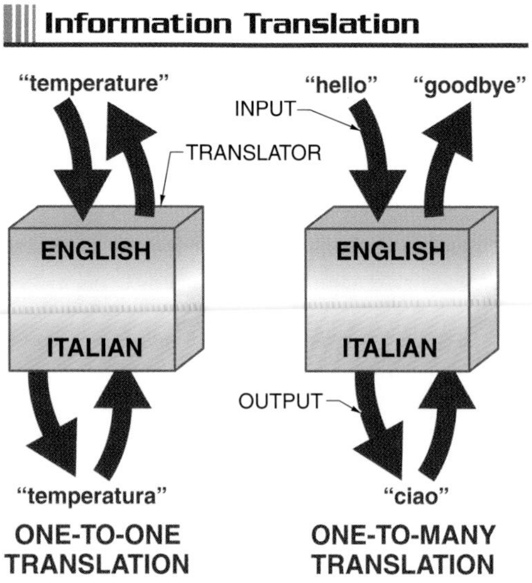

Figure 17-2. Like human languages, information translation can create problems if there is not a one-to-one relationship between words or concepts.

Translation can be further complicated by retranslation in the other direction. For example, if the English word "hello" is processed by a translator into Italian, it may be interpreted as "ciao" on the other side. However, if "ciao" is inserted on the Italian side, it may emerge on the English side as "hello," "goodbye," or "at your service" because it has multiple meanings. The idiosyncrasies and origins of different languages' words means that perfect translation is not always possible. Likewise, translating from degrees Fahrenheit to degrees Celsius, and then back into degrees Fahrenheit, could result in a loss of accuracy: 77.61000°F = 25.33888°C = 77.60998°F.

This seemingly minor change can be significant in certain situations. For example, if a chiller unit is enabled at 77.60°F, a value of 77.61°F is sent from a temperature sensor through a gateway, which maintains the value at two decimal places (disregarding the rest, instead of rounding) as it is converted to 25.33°C. Converting back to degrees Fahrenheit, the chiller receives a value of 77.59°F. Therefore, it is not enabled, though it should be.

Gateways

A gateway can be used to integrate two or more protocols together into one information communication system. A *gateway* is a network device that translates transmitted information between different application protocols. **See Figure 17-3.**

Gateways

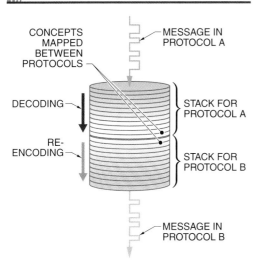

Figure 17-3. Gateways decode messages through one protocol stack and re-encode the same concept through another protocol stack to produce an equivalent message.

It is important not to confuse gateways with routers. Routers manage the forwarding of message traffic based on the destination address and may connect segments of different media types. They simply move information from one network to another without changing the content. However, gateways manipulate message traffic at an application level by translating certain information.

A gateway must have a one-to-one mapping of the information to be shared from one side to the other. *Mapping* is the process of making an association between comparable concepts in a gateway. Each concept in one protocol must have one, and only one, equivalent concept in the other protocol. If the two protocols have different features or levels of complexity, mapping can be extremely difficult. Gateways can be used for translating content or translating protocols, though often both are involved. **See Figure 17-4.**

Gateway Functions

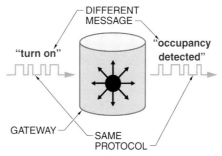

CONTENT TRANSLATION

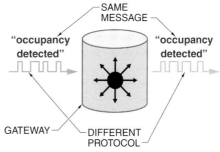

PROTOCOL TRANSLATION

Figure 17-4. Gateways provide content translation and protocol translation services, though applications often require both simultaneously.

Translating content may be required when devices communicate with the same protocol, but share incompatible information. This is often the case when the devices were not originally intended to be integrated together. For example, an occupancy sensor that is designed specifically for a lighting system cannot easily share its information with another system. If the occupancy sensor commands a set of lights to "turn ON," this output cannot be interpreted by a security system, which expects an "occupied" or "unoccupied" status. The information is different, the context is different, and the desired response of the other system may be different. A gateway can be used to translate the command into something that can be understood by the security system.

Translating protocols involves altering the structure and encoding the message. For example, the occupancy sensor may report a status, such as "occupancy detected," which can be used by both the lighting system and the security system. However, if the two systems use different languages (protocols), the meaning of the message is not conveyed. This may also involve changes to the lighting system, which must now decide if action is required for "occupancy detected," instead of being commanded directly to "turn ON." In this situation, a gateway can be used for protocol translation. This kind of translation almost always involves some content translation as well, since different protocols have different models for organizing information.

Perhaps the most common use of a gateway is in connecting a building automation network to an intranet or the Internet. The control data can then be monitored and recorded by another system for any purpose.

Intermediary Frameworks

Gateways may not be adequate for very complex systems, especially if there are multiple protocols. In response, some companies have developed intermediary frameworks. An *intermediary framework* is a complete software and/or hardware solution for integrating multiple network-based protocol systems through a centralized interface. **See Figure 17-5.**

For example, Tridium, a subsidiary of Honeywell International Inc., has developed a software platform and supporting hardware solution that integrates LonWorks, BACnet, Modbus®, and other protocols simultaneously. Their Niagara^{AX} platform transforms data from protocol formats into common software components, which bridges the gap between the control networks and enables protocol-to-protocol communications.

Intermediary frameworks provide a variety of integration options, avoiding the need for protocol-specific gateways and custom software. The disadvantage is that the entire system is tied to that framework, relying on a single solution. However, relying on a gateway may create a similar situation. An integrator must weigh the advantages and disadvantages with the end user to determine if an intermediary platform makes sense for an installation.

Extensible Markup Language (XML)

To address the issue of being locked in to specific vendors, a development of the intermediary framework idea is the standardization of the information exchange to allow other vendors to provide competing and complementary solutions for the same framework platform. Extensible Markup Language (XML) is a leading solution.

Extensible Markup Language (XML) is a general-purpose specification for annotating text with information on structure and organization. XML is used in a variety of data communication applications as a standardized way to define and structure shared data. XML is known as "extensible" because it allows the elements to be defined by the user and extended as needed.

An XML element is nothing more than a pair of tags—composed of plaintext words, values, and symbols—that surround the data being structured. **See Figure 17-6.** Each tag is enclosed by the "<" and ">" symbols. A start-tag, which is in front of the data, must be matched by a corresponding end-tag. End-tags always include the "/" character.

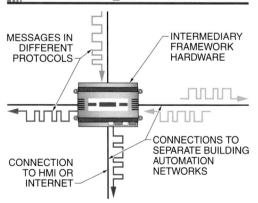

Figure 17-5. Intermediary frameworks translate all incoming messages into a common format, such as XML, which can be used for system monitoring or management, or translated into another protocol for outgoing messages.

Extensible Markup Language (XML)

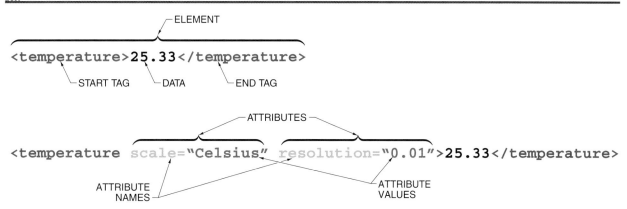

Figure 17-6. Extensible Markup Language (XML) is based on the strict formatting of data descriptions, though it allows flexibility in description choices.

For example, the temperature 25.33° could be represented in XML coding as the number "25.33" surrounded by "<temperature>" and "</temperature>" tags. The number "25.33" is the raw value that is being conveyed, and the tags provide descriptive information about the meaning of the value. The placement of the tags clearly indicates the beginning and end of the value, and the entire element.

If further descriptive information about the value is needed, attributes can be added to the start-tag. Each attribute consists of a name and value. For example, if the temperature 25.33° is in the Celsius scale, that information can be added within the start-tag as an attribute. Other attributes in this example could include the resolution, source, sensor type, or timestamp of the measurement.

Nearly any type of information can be represented in this way. With only a few basic rules, XML allows any descriptive word (or hyphenated phrase) to be used as an element. Elements can also be empty (with no value), or include other elements. Nested elements can be used to build hierarchical structures. **See Figure 17-7.**

The flexibility and simplicity of XML have made it the most widely used, interoperable information-sharing model for building automation systems. XML element structures can be used to fully describe data that must be passed between systems using different protocols. However, XML allows multiple ways to represent the same information. **See Figure 17-8.** XML elements can have exactly the same meaning with very different wording or attributes. This can be a benefit, but also a potential problem. Programmers can use the most appropriate elements and attributes for the application, but can easily introduce inconsistencies. Within a system, all senders and receivers of XML-formatted information must be programmed with the same element types, or the data could be interpreted incorrectly or not at all.

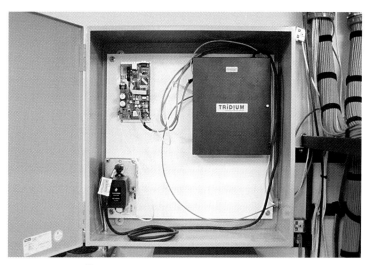

Tridium's Niagara^AX intermediary framework models the information provided by protocol-specific nodes into XML-based components, which can then be easily shared with other control devices, enterprise systems, or interfaces.

XML Structures

```xml
<lighting-system>
    <switch type="manual" location="Room 101">ON</switch>
    <switch type="occupancy" location="Room 107">OFF</switch>
    <ballast location="Room 101">
        <status type="dim-level">2</status>
        <priority type="demand-limiting">low</priority>
    </ballast>
    ...
</lighting-system>
```

Figure 17-7. XML elements that contain other XML elements form hierarchical structures that can be used to form logical relationships of building automation system devices and information.

XML Equivalent Elements

```xml
<temperature scale="Celsius" resolution="0.01"
    location="outside air">25.33</temperature>

<value type="OA temperature" units="Celsius"
    resolution="0.01">25.33</value>

<outside-air variable="temperature" units="Celsius"
    resolution="0.01">25.33</outside-air>
```

Figure 17-8. Because of its flexibility, XML can be used to create elements with very different wording that have the same meaning.

XML is used throughout the computing world, though it is often invisible to the user. For example, some computer applications use XML to organize the information stored in its files. The application program interprets the tags to determine how to display the information on screen without displaying the tags themselves.

However, custom programming adds complexity and cost, and can make a solution proprietary. Instead, standard XML formats have been developed that use the advantages of XML for gateway translation but avoid custom programming. A standard XML data-sharing model includes standardized definitions for structuring information for specific applications, such as building controls. This allows manufacturers to market off-the-shelf gateways and software interfaces that work together seamlessly, without custom programming.

oBIX. The *open building information exchange (oBIX)* is an XML-based model for conveying control information between any building automation protocols, including enterprise and proprietary protocols. This model was developed within the Continental Automated Buildings Association (CABA), and is now a responsibility of the Organization for the Advancement of Structured Information Standards (OASIS). Several companies and corporate groups, including LonMark International, have adopted oBIX as the preferred XML model for control information.

While oBIX can share information from one system to another, it does not allow one control system to control another. This is because the different protocols have different management methods and requirements.

BACnet XML. *BACnet XML* is a proposed XML-based model for conveying control information between BACnet and other protocols. Unlike oBIX, BACnet XML is specific to one underlying protocol. This model has been fully developed and proposed as an addendum to the BACnet standard, though, as of early 2009, it is not yet approved. However, some member companies of BACnet International already provide products based on this format.

The biggest benefits of BACnet XML are in the enterprise-level collection of information from one or more BACnet systems and in the sharing of XML-encoded BACnet messages between BACnet systems for control.

Both the oBIX and BACnet XML implementations require gateways or intermediary frameworks to translate information and receivers (computers, systems, and/or additional gateways) that understand the respective formats. However, a specific XML format can be self-describing by using additional files that explain, in a standard XML way, how to interpret the formats. Therefore, custom programming is usually not required if off-the-shelf tools are available.

It must be noted that some programming is still required at the gateway level in order to instruct the gateway what data to send and in what level of detail. However, this should be far less programming than what would be required in gateway-customization scenarios.

Enterprise System Integration

An *enterprise system* is a software and networking solution for managing business operations. These systems are often custom applications that include specialized services developed specifically for unique aspects of a particular company's operations. **See Figure 17-9.** For example, a large company may have an enterprise system for handling its accounting and project scheduling needs. Software applications are commonly referred to as "enterprise level" if they are part of this solution. This software is typically hosted on servers, where it simultaneously supports a large number of users over a network.

Businesses may utilize multiple enterprise systems that may need to be integrated. It may also be desirable to integrate control information from a building automation system into enterprise systems. For example, energy usage data from the building automation system can be shared directly into the accounting system for billing building tenants.

The means of sharing information with enterprise systems is similar to that for different control systems. Either a common, standardized platform is used by all systems, or an interface is required to translate and transport information between the different systems. Similar to building automation solutions for integration, gateways and intermediary frameworks are available for bridging an enterprise system and a building automation system.

Departmental Cooperation

Designing the best integration solution involves understanding the responsibilities and concerns of the building's information technology, accounting, and facility management departments.

Many information technology (IT) departments resist sharing their infrastructure with the control system. Their primary concern is the security of the facility's networks from both the outside and the inside. Adding control information traffic to the IT infrastructure may complicate the department's responsibilities, especially if they know very little about the automation system.

The accounting department can use control information for departmental/tenant billing purposes and for energy efficiency (cost savings) monitoring. However, this group may object to connecting the control system directly to the accounting enterprise system out of concern for unauthorized access to their financial information.

The facility management department uses control information to maintain building operations, and relies on the IT and accounting departments to help deliver and process the information.

All three departments, and any others involved in a particular automation solution, must work together to satisfy all the concerns while meeting the demands of integration. Control information must be communicated in a reliable and secure way.

Enterprise Systems

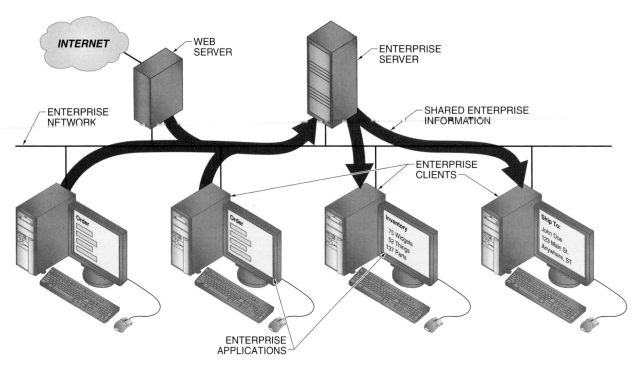

Figure 17-9. Enterprise systems are computer networks running shared software applications focused on business operations.

> Interfacing a building automation system with multiple protocols to an enterprise system can be accomplished in two different ways. Either each building automation system protocol is integrated separately, or the enterprise system is integrated with only one protocol system, which handles the through-traffic to and from the others.

Likewise, there are efforts to promote XML as an integration standard, thus avoiding the limitation of single-vendor intermediary frameworks. One such standard from OASIS is the electronic business using Extensible Markup Language (ebXML). It not only enables intra-company exchange of business information, but also inter-company exchange.

Gateway Connection. The implementation of XML for translating control information to the enterprise level may be directly specified, which can significantly affect the feasibility of integration, or it may be influenced by the capabilities of the enterprise applications. Ideally, enterprise applications must be able to accept at least one of the control system's available XML implementations. Otherwise, additional programming must convert one XML format to another that is compatible with the enterprise system. **See Figure 17-10.**

Direct Connection. If the control system includes a supervisory control device, it can be used similarly to a gateway, translating information into a format that can be read by the enterprise system. **See Figure 17-11.** The control information is shared with the central device in the native protocol. This device then interfaces directly with the enterprise system using its data exchange method, such as ebXML.

Alternatively, the enterprise system servers could be equipped with the hardware necessary to interface directly to the control system, such as an internal interface card. Software drivers translate the hardware interface card information into a data exchange format that is native to the enterprise system.

Enterprise Integration by Gateways

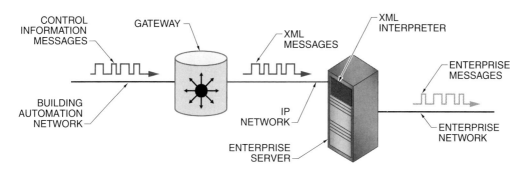

Figure 17-10. Gateways can be used to translate control information from a building automation network into XML data, which can then be processed for use in an enterprise system.

Enterprise Integration by Direct Connection

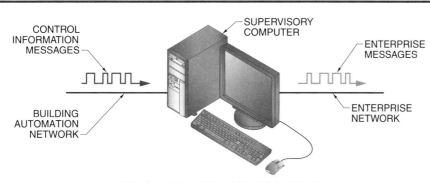

THROUGH SUPERVISORY DEVICE

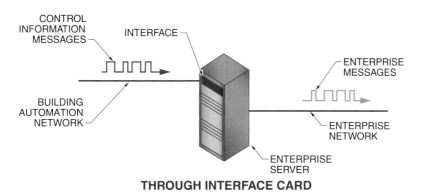

THROUGH INTERFACE CARD

Figure 17-11. Building automation networks can be directly integrated to enterprise systems through supervisory devices or simple interfaces.

Tunneled Connection. If the enterprise system is designed to understand control information formats, either natively or through additional protocol driver software, then the control information can be received directly. **See Figure 17-12.** This is not considered to be a gateway solution because there is no protocol translation. The control information is tunneled through the IP infrastructure into the enterprise system network.

Enterprise Integration by Tunneling

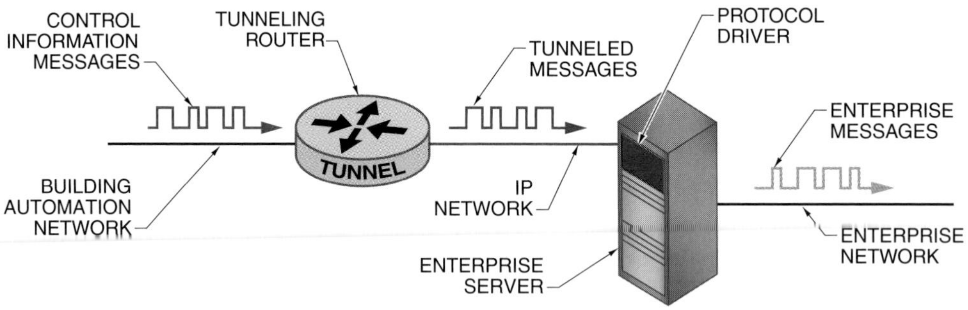

Figure 17-12. Control information messages can be tunneled through an IP network directly into an enterprise system protocol driver or interface.

CROSS-PROTOCOL IMPLEMENTATIONS

As an example of one cross-protocol integration solution, a control scenario for the Lincoln Publishing Group building with multi-protocol requirements is explained below. The resulting design is a commonly implemented solution utilizing industry-available products. This example outlines some of the strategies and implications of cross-protocol integration.

System Description

The system and building requirements are identical to those in the previous examples, except for the following:
- The lighting system uses BACnet nodes.
- The HVAC system uses LonWorks nodes.
- The plumbing system uses Modbus TCP nodes.
- The fire protection system operates on a separate proprietary network for life safety reliability reasons, but can share information onto the building automation network via a BACnet/IP integration board.
- The security system uses a Vykon Security® controller.
- The access control system uses a Vykon Security controller.
- The elevator system operates on a separate network, but can share information with the building automation network via an integration board.

Modbus®

Modbus® is another open protocol available for communicating with control devices designed for building systems and industrial process control. It was originally developed in the 1970s for communicating with programmable logic controllers (PLCs). Modbus devices often use TIA-485 signaling over twisted-pair cabling, and may be compatible with Ethernet networks. The protocol has some limitations, but is considered robust and easy to implement, so it has retained some share of the building automation industry. When implemented, it is often used for certain subsystems, which are then integrated with an overall building automation system using a more sophisticated protocol. Many cross-protocol devices, such as gateways, include Modbus as one of their compatible protocols.

Vykon is a suite of Niagara-based products designed to integrate networked control devices into a unified, Internet-enabled, web-based system. Vykon controllers integrate LonWorks, BACnet, Modbus, oBIX, the Internet, and web services protocols, and include network management tools to support the design, configuration, installation, and maintenance of interoperable networks. **See Figure 17-13.**

Example Cross-Protocol Implementation

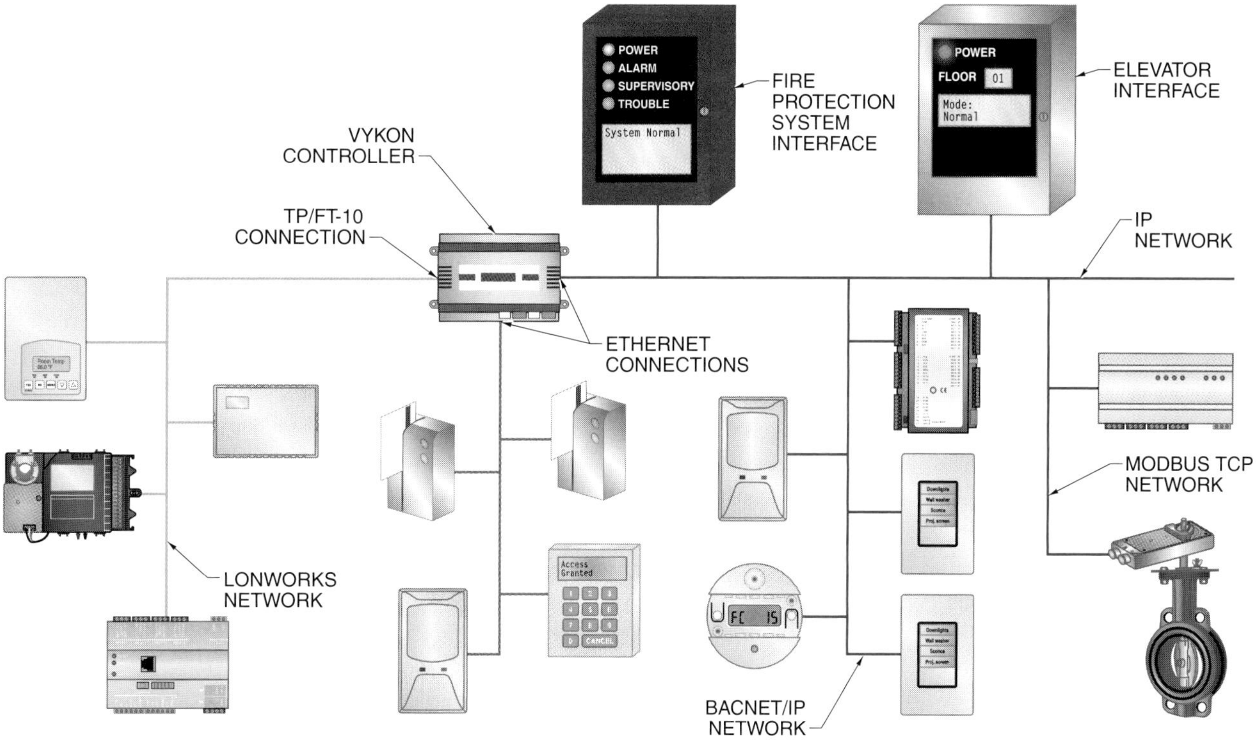

Figure 17-13. An example cross-protocol implementation may use intermediary framework controllers to connect separate networks using different building automation protocols.

The Vykon JACE® (Java Application Control Engine) is a line of controller/servers that combines control, supervision, data logging, alarming, scheduling, and network management functions. These devices can control and manage external control devices over the Internet and present real-time information to users within web-based graphical interfaces. Optional input/output modules can be added for local control applications.

Vykon Security is a security management solution that integrates with a building automation system to enable any building system to react to access events and alarm conditions. A Security JACE is used to integrate many of the building systems in the example building. The device includes several connections for integrating networks of different protocols. A TP/FT-10 network port is connected to the LonWorks network for the HVAC equipment. A primary Ethernet port is connected to operator workstations or additional JACE devices, which somewhat isolates the critical security and access control systems. A secondary Ethernet port is connected to the building's IP infrastructure that hosts BACnet/IP and Modbus communication. This integrates many of the other building systems into the overall system.

Scenario: Opening the Building on a Regularly Scheduled Workday

An employee swiping a proximity card at either building entrance initiates the control scenario of opening the building on a regularly scheduled workday. Opening scenarios on non-workdays, as determined by calendar schedule in the building automation system, may occur differently.

The access control system detects the access request of the employee and, upon authorization of the request, shares the identity of the employee with other building systems. The building systems react in various ways to this event.

Electrical System. The electrical system receives no signals from this event. The electrical system is in standard operating mode, receiving electricity from the utility's service entrance.

Lighting System. The lighting system takes several actions to ensure adequate lighting for the occupant entering the building. This includes lighting in general work areas and the employee's assigned location. The Security JACE controller sends commands to the lighting system via BACnet/IP messages. The lighting controllers interact with the scheduler to determine the appropriate actions.
- Interior life safety lighting is turned ON and stays ON until shut OFF by the building's closing scenario.
- If not already turned ON by system schedules (within the JACE controller), the interior lighting in the common areas of the employee's assigned floor is turned ON and remains ON until shut OFF by the building closing scenario.
- Interior lighting in other work areas, both enclosed offices and open office areas, is controlled with occupancy sensors and turns OFF after a time-delay period. Alternatively, the employee can override the occupancy sensors with a local override switch.
- Work areas near the building perimeter utilize daylight harvesting and dimmable ballasts.

HVAC System. The Security JACE controller utilizes schedules to automatically adjust heating and cooling setpoints during unoccupied periods. The Security JACE controller uses control algorithms to determine the optimal start time for the HVAC equipment based on current space temperatures, setpoints, scheduled event times, and dynamically adjusted recovery rates.

If the employee accesses the building prior to the optimized start event, then the Security JACE controller commands the HVAC equipment via the LonWorks network to initiate both common-area and employee-assigned HVAC systems. These systems stay ON until shut OFF by the closing scenario.

Plumbing System. The plumbing system takes several actions to ensure adequate water supply for the plumbing fixtures in use while the building is occupied. The Security JACE controller uses internal control programming and remote input/output modules for controlling the plumbing subsystem.
- The electrical circuit serving the electric water heater, heating elements, and their self-contained controls are switched ON and stay ON until shut OFF by the closing scenario.
- The electrical circuit serving the hot water circulation pump is switched ON and stays ON until shut OFF by the closing scenario.
- The pressure-boosting pump package controller is enabled and stays ON until shut OFF by the closing scenario.

Fire Protection System. The fire protection system receives no signals from this event. All fire detection, alarm notification, fire protection system monitoring, and fire safety function subsystems are fully operational 24/7/365.

Security System. The Security JACE controller closely integrates security and access control functions. When an intrusion is detected, the Security JACE controller is programmed to call the police, fire department, building security, or other authorities. Additionally, remote input/output connections are used to activate sirens and strobes. The security system receives information through an interface with the access control system, since both of these critical systems are networked separately from the rest of the building automation system.
- The alarm partitions for the common areas and the occupant's assigned floor are disabled.
- If the network connection is not available, the alarm system remains armed. The door is allowed to open, but the security panel

> Integration can also include hardware-oriented solutions with programmable protocol-to-protocol gateways, which do not require a computer to serve as an intermediary.

just inside the entrance annunciates the armed status, requiring an additional disarming code to be entered during a time-delay period.

Access Control System. During normal occupied periods, the Security JACE controller signals the credential reader module to unlock the exterior doors. If accessed during the normally unoccupied period, the Security JACE controller signals the door to be unlocked for the configured unlock period. The Security JACE controller logs the access control event based on the outcome.
- If the door is opened, an "access granted" event is logged.
- If the door is not opened, an "access granted, but not used" event is logged.
- If the door does not properly close, then the Security JACE controller initiates a door-held-open alarm.

Voice-Data-Video (VDV) System. The voice (telephony) and data portions of the system take no actions from this event. These subsystems are active 24/7/365. The video surveillance portion of the system takes no additional actions beyond those defined in the scenario controlling activation of the site when someone enters onto the property.

Elevator System. During normal occupied periods, the Security JACE controller enables unrestricted use of the elevator. When an employee assigned to an upper floor enters the building, the elevator is called to the first floor. During normally unoccupied periods, a proximity reader located in the elevator restricts elevator operation to authorized employees. When authenticated, the Security JACE controller, through distributed input/output modules, then enables the applicable elevator call buttons.

Refer to Quick Quiz® on CD-ROM

Summary

- It is often necessary to design control systems using a mixture of LonWorks, BACnet, and even proprietary protocol devices.

- When multiple protocols are to be integrated together, the primary concern is the translation of control information from one protocol to another.

- A gateway must have a one-to-one mapping of the information to be shared from one side to the other. Each concept in one protocol must have one, and only one, equivalent concept in the other protocol.

- Gateways can be used for translating content or translating protocols, though often both are involved.

- Intermediary frameworks transform data from protocol formats into common software components, which bridges the gap between the control networks.

- Extensible Markup Language (XML) is used in a variety of data communication applications as a standardized way to define and structure shared data.

- The flexibility and simplicity of XML have made it the most widely used, interoperable information-sharing model for building automation systems.

- Enterprise systems are often custom applications that include specialized services developed specifically for unique aspects of a particular company's operations.

- The means of sharing information with enterprise systems is similar to that for different control systems.

Definitions

- A *gateway* is a network device that translates transmitted information between different protocols.
- *Mapping* is the process of making an association between comparable concepts in a gateway.
- An *intermediary framework* is a complete software and/or hardware solution for integrating multiple network-based protocol systems through a centralized interface.
- *Extensible Markup Language (XML)* is a general-purpose specification for annotating text with information on structure and organization.
- The *open building information exchange (oBIX)* is an XML-based model for conveying control information between any building automation protocols, including enterprise and proprietary protocols.
- *BACnet XML* is a proposed XML-based model for conveying control information between BACnet and other protocols.
- An *enterprise system* is a software and networking solution for managing business operations.

Review Questions

1. What are the primary concerns of cross-protocol integration?
2. Why is it necessary to have a one-to-one mapping of concepts for accurate information translation?
3. What is the difference between content translation and protocol translation?
4. What are the advantages and disadvantages of an intermediary framework?
5. What role can Extensible Markup Language (XML) play in intermediary frameworks?
6. What are the advantages and disadvantages of the Extensible Markup Language (XML)?
7. How are the open building information exchange (oBIX) and BACnet XML similar, and how are they different?
8. What is an enterprise system?
9. How can enterprise systems be integrated with a building automation system?

Chapter Eighteen

Future Trends in Building Automation

The building automation industry is a relatively new participant among the building system trades, especially the portion dealing with electronic controls and networking technologies. Therefore, the industry is still maturing in a climate where building controls are becoming increasingly important, both for new construction and existing buildings. Industry watchers notice a number of trends from the past several years that are expected to influence the continuing evolution of building automation in the near future.

Chapter Objectives

- Compare the changing roles and responsibilities of key groups in the building automation industry.
- Identify the key areas in networking, protocols, and control that are expected to significantly change the building automation industry in the near future.
- Describe the factors to consider when automating existing buildings.

INDUSTRY TRENDS

The building automation industry supports a number of companies and organizations involved in choosing, specifying, designing, integrating, maintaining, and otherwise working with these systems. Changes in the building automation industry have affected the roles of these industry members and how the systems are implemented.

Integrated Building Automation Team

The traditional building automation team includes individuals and groups that take the requirements, concerns, and ideas for a new or existing building and create an automated building. Traditional teams include building owners, consulting-specifying engineers, controls contractors, and electrical contractors. As building automation systems are becoming more integrated, the roles and responsibilities of the traditional team members are changing and new members are being added.

Building Owners. Building owners decide which building automation strategies to implement based on the desired results. The concerns of building owners also include initial costs, return on investment, maintenance, impact on occupants, system flexibility, system upgradeability, and environmental impact. These factors are further complicated when retrofitting an existing building that may be partially occupied due to the impact of renovations on the operations of existing tenants.

Consulting-Specifying Engineers. The consulting-specifying engineer defines the scope of the building automation system from the owner's list of desired features, and coordinates these desires with the architectural, mechanical, and electrical systems' requirements. The engineer produces the contract documents used by the controls contractor to bid a project, and later reviews and approves the shop drawing prepared by the controls contractor to ensure that the intent of the contract documents has been met. The engineer also reviews and approves the project closeout information, such as commissioning reports.

Building Automation Roadmaps

Multiple organizations have developed technology roadmaps and other reference material regarding outlooks on the evolution of the building automation industry.

ASHRAE

The American Society of Heating, Refrigerating and Air-Conditioning Engineers (ASHRAE) focuses on the HVAC aspect of building automation. However, ASHRAE has a long history of working with other organizations, such as the Illuminating Engineering Society (IES), in the creation of building automation standards that are incorporated into the building codes. Inclusion into building codes is a major step because it makes conformance mandatory rather than at the owner's discretion.

CABA

The Continental Automated Buildings Association (CABA) has written a *Technology Roadmap for Intelligent Buildings,* which gives a good cross-disciplinary overview of building automation and the issues surrounding effective implementation. Especially useful is the companion document, *Best Practices Guide for Evaluating Intelligent Building Technologies,* which walks through the process of coming up with a plan that is applicable for the client's specific project type.

Fiatech

Fiatech is a consortium of construction industry participants. Its Capital Projects Technology Roadmap covers the entire life cycle of a building, from conceptual design through decommissioning. Fiatech addresses five areas of implementing technology in buildings, including building automation for operations and maintenance.

Contractors. Previously, in traditional installations, mechanical engineering contractors were responsible for controls involving HVAC, plumbing, and piping systems. Electrical contractors were responsible for controls involving power distribution, emergency power generation, fire protection, and lighting. There was little coordination between the two disciplines, or even among elements of the same discipline, because everything stood alone in the bid documents. Now, with integrated building controls, both engineering disciplines must work more closely together and produce a coordination document, in addition to the stand-alone components. This document details various automated operational scenarios and how each independent system participates in and responds to these scenarios.

This type of coordination document is not recognized by the Construction Specifications Institute's (CSI's) MasterFormat® standard specification format or ARCOM MasterSpec® software, which put building elements in distinct categories for the convenience of general contractors and construction managers. Therefore, contractors continue to define systems in distinct categories and also reference the requirements of the coordination document. This requires the general contractor to bundle all of the affected sections into one bid package to be assigned to the systems integrator or master systems integrator.

Systems Integrators. Controls contracting has traditionally been considered a subset of mechanical work, especially since HVAC systems are the most commonly automated building systems. Electrical contractors used to handle the remainder of the low-voltage building systems such as fire detection and lighting control. However, with systems integrated together in automated buildings, a master systems integrator is needed. Just as the mechanical and electrical design engineer work together to produce a coordination document, the master systems integrator works with both the controls contractor and the electrical contractor to create an integrated system.

Integrators are individuals and companies who engineer, implement, and maintain building automation systems, especially those using control devices from multiple vendors. They select the appropriate network architectures, choose control devices, perform installation tasks, program the network, and provide network maintenance services.

It must be noted that each project team will likely approach the coordination of their system's integration in a unique way due to the many factors involved in each scenario. For example, in retrofit installations, the owner may decide to work directly with a systems integrator and bypass the coordination provided by a consulting-specifying engineer because the other system components are already in place. The simplest solution is having an electrical contractor install both low- and medium-voltage equipment and the low-voltage control systems.

Information Technology (IT) Staff. In traditional installations, facility managers were responsible for the mechanical and electrical systems without much technical input from other areas of the organization. But with the ability to connect building automation system controllers to the building's IP network, the information technology (IT) staff is becoming involved in the decision-making process. Their input is needed to manage the integration of data communication networks and ensure reliable operation afterward.

Controls Vendors

The product lines of controls vendors vary greatly, as do the needs of building owners looking to implement building automation. The choice of controls vendors involves matching their most appropriate products with the building specifications, client expectations, budget, and other considerations. The building automation industry has responded to the needs of clients by presenting a choice of delivery options.

One-Stop Shops. A one-stop shop is a vendor that markets most of the control devices and equipment that a building owner needs. For example, some controls vendors have working agreements with fire detection manufacturers who manufacturer their own security and access control systems. Other vendors have developed a "super vendor" approach for markets such as health care. In this instance, the vendor sells everything to a hospital client, from an MRI, to switchgear, to building controls.

A variation on the one-stop shop strategy is the inclusion of factory-mounted building controls on traditional building equipment. For example, an HVAC equipment manufacturer can put controls on a package air-handling unit so that the unit is ready for service immediately after connecting power and the communication network.

Best of Breed. The best of breed strategy takes advantage of the individual strengths of various controls vendors to design an all-around robust system. One vendor may produce particularly good HVAC controls but may be weak in access control equipment. Likewise, a different vendor may have the ideal access control devices for the building owner's needs. The owner or contractor chooses each major component individually and then a systems integrator connects the individual components together to implement an operational scenario. With the interoperability of open protocols, this is a feasible strategy that can make use of the best or most appropriate devices in each building system.

Global Differences in Controls Incentives

The capabilities of building automation systems in the United States are not very different from those in other countries, especially since most controls manufacturers are multinational companies that market their products worldwide. However, building automation systems in other countries differ in the extent of implementation and integration. They are more common and typically achieve higher levels of integration by bringing together more building systems.

The biggest influence on this disparity is the desire for greater energy efficiency. This is generally not driven by the cost of energy, which is determined at the global level, but by government energy policy. For example, German utilities are required by law to buy back excess energy from renewable energy sources at prices that can be more than twice the retail rate. In this situation, it makes sense to add additional control system features that save energy because the payback profile is substantially different. The United States does not currently have any similar federal laws or incentives that encourage this trend.

NETWORKING TRENDS

Trends in networking technologies are aimed at making it easier to transmit control information quickly and reliably. Factors that influence these characteristics are the media, addressing and routing schemes, and network traffic volume. Open protocols are developed with certain network types in mind, though they are subject to addenda that expand their supported types. Improvements are also being made at migrating control data onto existing networking technologies in ways that do not degrade the performance of either system.

Utilizing Existing IP Infrastructure

The master system integrator works with a building's IT department to integrate the building automation network with the data network and IP infrastructure. This network integration is not necessary but is becoming more common.

An IP infrastructure is typically already available throughout the building and may have excess message traffic capacity. This intranet allows control data to be distributed quickly and reliably throughout a building. For sharing data between separate campus buildings, or even between remote locations, an Internet connection provides access to a global information network. This requires security measures for both obscuring the data (so that no unauthorized person can listen in) and ensuring that the data is transmitted unaltered (ensuring that the information sent is the information received). The availability of data via IP networks helps integrators fine-tune automation systems for efficient and effective control.

Many industries have a geographic structure that is ideal for adopting the Internet as a core part of their control systems. Examples include business chains, manufacturers whose parts are constructed in different areas of the country and world, and property-management or ownership firms that monitor buildings in different cities from a central location.

Wireless Networking Technologies

The growth of wireless networking technologies is one of the most significant trends in electronics. Wireless networks have become options for building automation, both as LAN choices for traditionally wired protocols and as the sole transport technology for wireless protocols. However, wireless networks have still not made strong inroads in this industry. This has always been considered a specialty technology, or at least limited in application. Several factors have kept wireless networks from capturing a large share of the building automation market, but as improvements in technologies mitigate these limitations, the future of wireless networks in this industry may also improve.

Power Sources. A large part of the problem is sourcing power for the control device nodes. The power limitations and large size of NiCad batteries were just too significant for the power-demanding signaling technology of the day. Newer battery compositions have allowed for smaller and longer-lasting power sources. The signaling methods have also improved, allowing the nodes to "sleep" (power down to a minimally operating state) when not sending and "wake" upon receiving a certain signal. Additionally, mesh networking, the ability of nodes to relay signals for other nodes, reduces the radio power output needed to deliver a message from one end of a building to the other. These advancements have greatly improved the lifetime of nodes between battery charges or replacements.

Further, there are energy-harvesting technologies that allow for battery-less nodes. *Energy harvesting* is a strategy where a device obtains power from its surrounding environment. Energy-harvesting technologies can use mechanical actions, vibrations, light, thermal gradients, or other energy sources in its area to operate its electronics or charge short-term batteries.

Wired/Wireless Technology Cycles

Shifts to wireless technologies have happened before in the communications industry, even repeatedly within the same industry. For example, television signals were originally wireless because there was no wired infrastructure to support the industry. However, the relatively low reliability of wireless television of the time (poor reception and few channel choices because of signal limitations) gradually gave way to cable programming. Later, disenchanted with the cost of cable television service, satellite (wireless) television became a practical alternative. Then, the recent standardization of high-definition television signals (requiring increased bandwidth) increased the reliability and choice of cable providers. Therefore, on-demand features have shifted television delivery back somewhat to wired services. The wired cable infrastructure is also used to bundle telephone and Internet services, which adds value to the wired platform.

However, another shift in television delivery trends is already beginning to emerge. The discontinuance of analog television in the United States encourages television stations to broadcast with greater signal clarity, even full HDTV, over the airwaves. This is causing a renewed interest in wireless television reception, especially since broadcast television is free to the consumer. High-speed cellular and Wi-Fi Internet networks have also begun negating the advantages of bundling these communication services together on wired networks.

Interference. There are many concerns about the reliability of wireless sensors and actuators. Wireless networks are vulnerable to disruptions from a variety of electromagnetic sources. Accidental disruptions are caused by stray radio waves in the same frequency, perhaps due to new equipment or transients from people and objects. Network disruption may also be caused by changes in the environment, such as the installation of foil-based wallpaper, the moving of large metallic equipment or furniture, or poorly isolated equipment power supplies. Awareness of the potential causes of wireless network interference contributes to improving the success rate of these implementations.

Security. A major concern for wireless networks is the possibility of intentional network disruptions. A person can intercept wireless signals and either attempt to jam network communications or gain control of nodes or systems. It may be possible for someone even outside the facility to infiltrate the network. With a wireless access point, a physical connection to the network is not needed. Depending on the building systems involved, a hacker could cause serious equipment damage. Encryption, authentication, and other security measures are used to minimize the potential for security breaches.

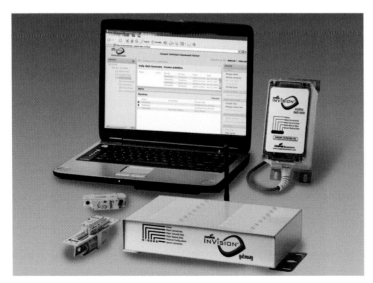

Wireless systems may be particularly suited for certain applications, such as monitoring of maintenance components that cannot be connected via physical media.

Infrared. Most wireless nodes use radio frequency (RF) as the media, though this has disadvantages. Infrared (IR) is another, perhaps more secure, wireless platform. IR requires line-of-sight arrangement of communicating nodes, which limits its application to within certain walls. While this may be a disadvantage in most applications, it can be an advantage from an information security point of view. IR communication channels can be used to complement wired and RF-wireless channels for use in particular applications within small areas.

OPEN PROTOCOL TRENDS

Today, building automation installations with open protocols are still a small minority. The majority of installations are still using proprietary technologies, either completely closed or with available application programming interfaces (APIs). Increasingly, however, more controls specifications are demanding open protocols and specifying the limits of exceptions to that rule. The various open protocols today will continue to expand while proprietary protocols become less popular. As each open protocol grapples for a larger customer base, their larger battle is more against proprietary systems than against other open protocols.

However, proprietary protocols will continue to exist and be installed in projects where specifications do not dictate otherwise. The often lower up-front costs may be more attractive to the building owner or construction contractor, despite the long-term maintenance contracts, especially if the company constructing a facility is doing so for a different company.

Perceptions of Open Protocols

What makes a protocol "open" can be very subjective. For manufacturers, an open protocol allows them to focus on their core competency, which is developing products, rather than having to define the platform. The types of data structures used to carry information and the mechanisms by which that information gets from point A to point B are already established.

Specifications for open protocols also allow the manufacturer to bid for jobs from which they might otherwise be excluded.

Ironically, it is the specification for an open protocol that discourages a manufacturer from dominating the installation and maintenance of a facility's automation system. This counters the goal of large manufacturers, which is to involve their product lines in as much of the facility as possible. However, given an increased penetration of only semi-open products in a facility, which allow for specialized additions under the standardized framework, a manufacturer may still be able to lock a facility into a maintenance contract.

For integrators, an open protocol allows the integrator to bring together nodes from one or more manufacturers to create an interoperable system. The integrator is concerned with having the nodes share the same network and not only coexist, but interact intelligibly without additional programming. This keeps system complexity and integration costs low. Also, the increasing implementation of open protocols means that integrators must learn only a few open protocols, rather than many proprietary protocols. This requires fewer employees and less training, and allows integrators to focus on actual job installations instead of the details of different product lines from multiple manufacturers.

A building owner's definition of an open protocol focuses on interchangeability of nodes on the network without the requirement for custom solutions by the manufacturer or the original systems integrator. This allows a variety of products from different manufacturers to be easily installed and integrated by a wide selection of integrators, and supported by many maintenance organizations. A building owner typically views an automation system as a long-term, capital asset. The potentially higher short-term costs and complexity of an open system are balanced with the long-term benefits: interoperability, choice of maintenance contractors, and future expandability without additional gateways, programming, or original-manufacturer interaction. However, this group is rarely concerned with which protocol is used for the facility, as long as it is an open protocol.

> Internet protocol (IP) versions 0 through 3 were development versions of what became version 4 (IPv4) in 1981. Version 5 was an experimental stream protocol. Versions 6 through 9 were proposals to replace IPv4. Version 6 was eventually chosen as the official successor Internet protocol.

IP Addressing Changes

When the present-day IP addressing scheme was invented, 4,228,250,625 (2^{32}) unique addresses seemed adequate for all future needs. However, large groups of IP addresses were assigned to organizations or companies in the early 1990s, making many addresses unavailable, even if unused. This also leaves large gaps in the numbering that prevents unique IP addresses from being doled out sequentially. For example, Apple, Inc. was granted the entire address space under 17.xxx.xxx.xxx, a total of 16,777,216 addresses. Many other companies, institutions, and organizations were granted similar blocks, known as "Class A" or "/8" addresses. Even if the numbers were sequentially assigned, there are not enough addresses for everyone in the world today.

This current IP addressing scheme, known as IP version 4 (IPv4), will eventually be replaced by IPv6, which has a far more equitable address distribution scheme and allows for 2^{128} unique addresses. This allows for more than 48,000,000,000,000,000,000,000,000,000 unique addresses per person (assuming a world population of about 7 billion). Just as this upgrade will significantly change computer-based networks, it will improve addressing opportunities for building automation nodes that also use IP.

However, the migration is complicated and will require years to complete. Support for IPv6 is being added to updates for computer operating systems, updates for new and existing home network routers, Internet service provider equipment, and major IP traffic routing systems. Upgraded equipment and systems are designed to support both IPv4 and IPv6 to facilitate a smooth transition sometime in the future.

LonWorks System Trends

Nodes in a LonWorks network are designed to interoperate and communicate in a peer-to-peer fashion without the need for a supervisory controller. Peer-to-peer communication is the strength of a LonWorks network and makes it a good choice for both small networks and commercial building integrated networks.

Standardization of data structures and LonMark functional profiles ensures a certain level of interoperability between products from multiple manufacturers. A node created today can communicate and interoperate with a node installed in the network a decade prior, without the need for additional hardware or glue logic.

The LonTalk protocol continues to be enhanced by increased addressing options, improvements in IP tunneling standards, and the addition of the Interoperable Self-Installation protocol, which standardizes a method for installing nodes in a network without the use of a network management tool. Further, LonMark International continues to develop specifications to profile the functions of existing and emerging building systems, allowing LonWorks systems to be a platform for integrating systems together in a facility or campus.

BACnet System Trends

While BACnet nodes also have the ability to communicate peer-to-peer, their strength is in their programming flexibility, allowing manufacturers to assemble different data points in a node to achieve a particular functionality. That assembly is represented in a machine-readable format that can allow a supervisory controller to integrate different nodes into the same network.

> BACnet International expands and improves the BACnet standard by maintaining an addendum-making process. The organization accepts suggested changes, evaluates the need for change through public remarks, prepares proposed addenda, collects review commentary, and votes on the final addenda.

Through addenda, BACnet International continues to enhance the protocol by integrating new data structures and control features, including security and authentication. BACnet networking is also being expanded to enhance BACnet/IP accommodation of remote operator access through network address translation (NAT) firewalls and includes the ZigBee protocol as a BACnet wireless data link layer.

Wireless Networking Protocol Trends

Wireless networks are becoming more reliable. This trend will continue since batteries last longer and transceiver designs are getting smaller and more efficient. Implementation of wireless networks for building automation is expected to increase considerably. However, specific wireless networking protocols will not entirely replace the need for other transport media, including other wireless media. For example, a specific wireless networking protocol may be appropriate for integration and data sharing within a building, but a wired protocol with a longer range is typically needed for transmission between buildings in a campus setting.

Wireless networking protocols are also expected to continue growing in proximity service applications. Applications for radio-frequency identification (RFID) include asset tracking (including salable product, movable equipment, and even personnel tracking), proximity activation (the presence of something or someone causes another action to automatically occur), and proximity security (the allowance or restriction of persons or things based on their proximity). In this regard, wireless protocols are a connection between the physical world and the control network, which may use a very different building automation protocol.

Therefore, the future of wireless networking protocols is as a complement to other building automation protocols that use various media. The success of wireless protocols will be to their benefit as a media channel option, but not as a complete building automation solution due to their media-dependent design.

Features Added to Existing Open Protocols

In many cases, technologies being developed to enhance the existing set of protocols are beyond the known desires of the user community. These include features that provide flexibility and control beyond what is contained in any buildings today, but have not yet been identified as a desirable feature from the user community. Following the development of these protocol features is training and education supporting the need for such features. The burden is on integration and design firms to educate building owners about new features because it is ultimately the owners' knowledge about an operation and its control that drives a feature to become one of their requirements. It is the application of the available toolsets rather than the tools themselves that drives their implementation.

Media Availability. With the exception of strictly wireless platforms, open building automation protocols are beginning to offer a variety of media choices. The choice of media allows the physical extension of the network into building areas that might not otherwise be reachable. In addition to the typical copper conductor and wireless media, some of the popular protocols can utilize fiber optic media, coaxial cable, powerline mains, and infrared. Built-in support for multiple media types allows data transmission from one medium to another through routers, without gateways or translators.

Also, the popular building automation protocols now add support for IP-based networks. This allows control data to leave dedicated control networks and be transmitted on IP-based networks, which are shared by personal computer and enterprise systems. Nearly all of the associations that maintain these protocols have embraced both the carrying of their messages over an IP-based technology (though perhaps limited to Ethernet in some cases) and the translation of their data structures to XML.

Integration of Multiple Building Systems. Building automation protocols with origins in HVAC applications are branching into lighting, security, elevator control, and energy management, the last of which is the latest important field in building automation. In many existing control devices, these systems are still proprietary and require low-level input/output connections, high-level gateway interfaces, or IP-based application programming interfaces to be integrated with other building systems. The integration of these applications will encourage development of new control products "native" to these protocols and bring together other distinct building systems. These protocols will be the most desired pathway for cross-system integration projects.

Advanced Cross-Protocol Integration. Both governmental and nongovernmental organizations are trying to meet mandated restrictions on carbon-output limits and energy consumption, which is creating the forum for supporters of different protocols to come together to reach these common goals. Much of the effort in cross-protocol integration is taking place outside the protocol-supporting associations. The competitive spirit between the protocol advocates and their continued desire for market penetration and dominance speeds the respective companies and organizations toward solving the cross-protocol issues as quickly as possible. Innovations that were once being driven by a push from inside the organizations are now being driven by mandates and parties outside those organizations.

Because it is part of a life safety system, a fire alarm control panel must be integrated carefully with other systems, typically through only one-way network communications.

The combined effects of organizational efforts and the energy goals will fuel cross-protocol integration at the application level, which will then cause the organizations to work more closely together at defining the interfaces for these new initiatives.

CONTROL STRATEGY TRENDS

Control strategies are used to maintain the desired indoor environment while operating the building systems in an appropriate and efficient manner. There are a number of capable control strategies that are chosen based on the particular requirements and conditions of a building and its occupants. There are also opportunities to improve control strategies in both application program algorithms and system equipment.

Individualized Controls

Individualized controls allow the occupants of the building to adjust setpoints themselves, usually within an allowable range of values, to create their own optimal indoor environment. Individualized control is typically used for HVAC systems, but sometimes also lighting systems. As occupants may have different perceptions of comfort, this strategy requires much finer control than most systems. For example, an occupant of one office may prefer a warmer temperature, while occupants at adjacent workstations may prefer cooler temperatures. Controls must also be user-friendly so that any occupant can use them effectively.

The turning point that really created an awareness of individualized control was the creation of a credit within the LEED® rating system that recognized the value of individual control for temperature and ventilation. The credit has two requirements: providing controls for at least half of the occupants to adjust their personal environment for their own comfort, and providing controls for all shared multi-occupant spaces to enable adjustments to suit group needs and preferences.

At the time, this was considered an aggressive mandate, and engineers and manufacturers tried to meet the requirements of this credit in a variety of ways. Simply increasing the number of air terminal units and thermostats proved to be cost prohibitive. A cost-effective variation of this technique is to place a temperature sensor in every space and then use an averaging sequence to provide the input to the controller.

Underfloor Air Distribution Systems. An underfloor air distribution system utilizes a raised floor as a plenum space to distribute conditioned air throughout a building space. **See Figure 18-1.** This system has the distinct advantage of placing cooling where it needs to be. Airflow into the occupied space is controlled by the placement and adjustment of floor diffusers. The use of floor diffusers makes it possible for occupants to manually control the amount of airflow into their personal space. However, the conditioning of the air is still controlled centrally with automated controls. The downside to such a system is that, because part of the system is manually controlled, it has a limited ability to set back when unoccupied.

Integrated Furniture Comfort Control Systems. An integrated furniture comfort control system is a variation of the underfloor air distribution system concept. It uses ducts integrated into the occupant's furniture to channel, mix, and filter conditioned air from an underfloor (or wall or column) plenum into the occupant's personal space. **See Figure 18-2.** These systems often also control lighting and background noise. This system allows for a great deal of individualized control but requires significant up-front costs for the furniture-integrated equipment.

Software as a Service

Supervisory software has cycled through different strategies. Early digital control systems were dependent on centralized intelligence and robust networks. After that, intelligence was distributed to the local controllers. It now appears that the next major step in the evolution of automation intelligence will be supplementing existing distributed systems with Internet-scale computing and advanced data collection.

Underfloor Air Distribution Systems

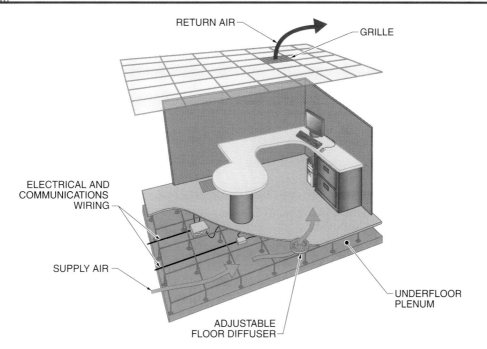

Figure 18-1. Underfloor air distribution systems allow individuals to adjust the flow of conditioned air into their personal space with the accessible floor diffusers.

Integrated Furniture Comfort Control System

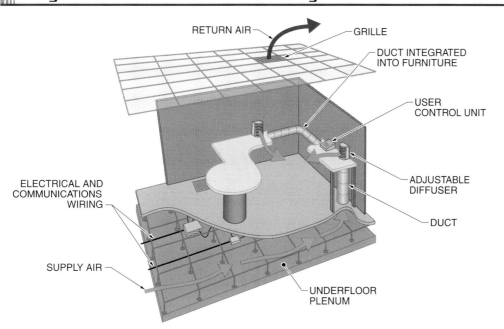

Figure 18-2. By integrating ducts, diffusers, and controls into the furniture, conditioned air from underfloor, wall, or column plenums can be directed into the occupant's workspace.

To create a data format that enables systems to work together, a gateway collects the data from the various automation and control systems in the building and pushes it out over the Internet for storage on a server in a data warehouse. **See Figure 18-3.** This data can then be used by analysis systems.

Data Warehouses. A *data warehouse* is a dedicated storage area for electronic data. The structure of data within the storage area is also organized to facilitate accessing and filtering the data at a later time for analysis, trending, and reports. Storage of large quantities of data is important for trending analyses. Control points sampled at short intervals over a long period can reveal important historical trends. Data warehouses offer several advantages over trend logging from within a building automation system.

• Data storage is not limited by the capacity of the building automation system. The initial storage capacity of a data warehouse system is typically much higher than a local automation system, and it can be much more easily upgraded as needed. A data warehouse can store the value of every control point for every minute potentially for the life of the building, and at a cost-effective price point.
• Off-site storage offers security for the stored data, which can become an extremely valuable resource when used for analyzing ways to reduce building energy consumption and other operational costs.
• The standardized cataloging and organization of the data from multiple sources by a single system facilitates data retrieval and analysis.

Fault Detection and Diagnosis. Since symptoms often precede many types of failures, automated fault detection and diagnostics software is able to predict many failures by analyzing data trends. Analysis software sorts through data warehouse archives and applies fault detection and diagnosis algorithms to uncover building operational problems. This automated system looks for trends in the data to determine if equipment is beginning to fail because failure is often preceded by measurable symptoms. **See Figure 18-4.** The availability of extensive stores of data allows the operator to determine the financial impact of the system failure.

Data Warehouses

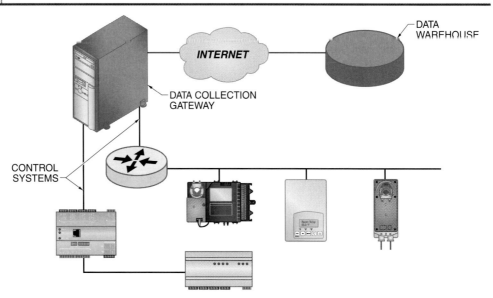

Figure 18-3. Data can be collected from building control systems, even separate or proprietary systems, and sent for storage in a data warehouse.

System Optimization. The data warehouse can also be used to calibrate an energy model. When a building is designed, the mechanical engineer typically generates a building model that simulates energy consumption of various building configurations. The system optimization software model can be extended to post-construction analysis. The model's input and output files are modified by current data from the building control system and the data warehouse, rather than being based on preconstruction assumptions. Then, operational scenarios are simulated for various desirable outcomes. For example, what is the least disruptive method of reducing demand? Since the new simulations use real control data from actual building operation, the optimization accuracy is significantly higher, and increases as the available input data grows over time.

Enterprise Integration

A building automation system can be integrated with enterprise systems to improve productivity or sales. In a retail environment, the system can use building and environment information to control ambient music, lighting, electronic advertisements, and product displays. This kind of integrated system could be sensitive to the weather, time of day, customers, sales goals, or other factors. For example, a coffee shop could display hot beverages more prominently on cooler days and frozen drinks on warmer days. Music and lighting can automatically change to create a different mood, which may influence buying habits. The result not only benefits the retailer but also gives the consumer an enhanced buying experience.

AUTOMATING EXISTING BUILDINGS

The automation of existing buildings is far more complicated than designing building automation systems for new construction. However, the increasing focus on energy use and environmental impact encourages building owners to seriously consider upgrading their existing system infrastructure, including equipment, building materials, and electronic control.

Fault Detection

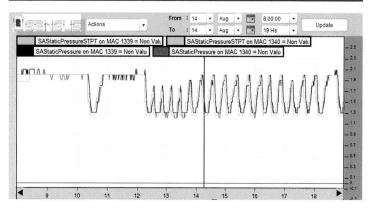

UNSTABLE SETPOINT

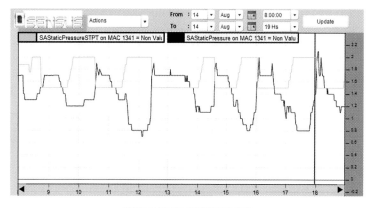

OPPOSITE RESPONSE

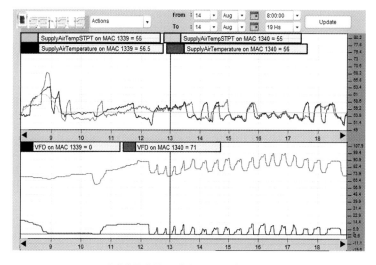

POOR SETPOINT TRACKING

Figure 18-4. Archives of control data can be analyzed by software to find problems in the automation system algorithms.

The controls contractors must consider a number of trade areas when planning a project, including the sheet metal, plumbing, electrical, and piping drawings, as well as project details such as schedule and scope. This applies to both new and retrofit projects.

Upgrading the Existing Infrastructure

The difficulty in retrofitting existing buildings requires prioritization of the possible measures for improving energy efficiency. Materials and equipment that are relatively accessible are typically replaced, while it is rarely feasible to modify the building design. Adding building automation to new systems can also be challenging. For example, it is relatively simple to replace old lighting ballasts and HVAC equipment with modern, efficient equivalents, but their controls must also be replaced with versions operable with other building systems.

Integrating those new controls with daylight-harvesting sun blinds and added solar and wind energy solutions can help reach desired, lower energy-consumption patterns. With the emerging trend of adjustable electrical rate tiers from utility companies, energy contracts can be made that benefit building owners if peak electricity demand can be reduced through integrated solar and wind power systems. If the building's electrical system can communicate expected energy requirements with the utility, better rates can be negotiated based on predictability measures.

Refer to Quick Quiz® on CD-ROM

Merging New and Existing Systems

The first decision in upgrading an existing facility is determining whether existing systems should be merged with new systems or completely removed. This is a decision that must be made while considering the life-cycle stage of the facility, the depreciation value of the existing equipment, and the cost to migrate versus the cost to remove systems. Costs are measured in terms of labor, down time, disruption, and capital expenditures.

If the decision is made to migrate from old to new systems over the course of time, finding the right players to support the migration path may be difficult. The question of "whose work is it?" arises not only in terms of installation but also in terms of merging the physical networks, merging the databases and user interfaces, and merging (or rather separating) the maintenance agreements and areas of responsibility—especially where responsibility overlaps within vertical disciplines.

Summary

- As building automation systems are becoming more integrated, the roles and responsibilities of the traditional team members are changing, and new members are being added.
- The consulting-specifying engineer produces the contract documents used by the controls contractor to bid a project, and later reviews and approves the shop drawing prepared by the controls contractor to ensure that the intent of the contract documents has been met.
- A coordination document details various automated operational scenarios and how each independent system participates in and responds to these scenarios.
- The master system integrator works with both the controls contractor and the electrical contractor to create an integrated system.
- Input from information technology (IT) staff is needed to manage the integration of data communication networks and ensure reliable operation afterward.

- A one-stop shop is a vendor that markets most of the control devices and equipment that a building owner needs.
- The best of breed strategy takes advantage of the individual strengths of various controls vendors to design an all-around robust system.
- The biggest influence on the disparity of global automation implementations is the desire for greater energy efficiency, which is driven primarily by government energy policy.
- An IP infrastructure is typically already available throughout the building and may have excess message traffic capacity. This intranet allows control data to be distributed quickly and reliably throughout a building.
- Wireless networks have become options for building automation, both as LAN choices for traditionally wired protocols and as the sole transport technology for wireless protocols.
- Wireless networks have still not made strong inroads in this industry.
- Building automation installations with open protocols are still a small minority.
- What makes a protocol "open" can be very subjective, depending on one's point of view.
- The current IP addressing scheme, known as IP version 4 (IPv4), will eventually be replaced by IPv6, which has a far more equitable address distribution scheme and allows for 2^{128} unique addresses.
- The LonTalk protocol continues to be enhanced by increased addressing options, improvements in IP tunneling standards, and the addition of the Interoperable Self-Installation protocol, which standardizes a method for installing nodes in a network without the use of a network management tool.
- Through addenda, the BACnet standard continues to enhance the protocol by integrating new data structures and control features, including security and authentication.
- The future of wireless networking protocols is as a complement to other building automation protocols that use various media.
- With the exception of strictly wireless platforms, open building automation protocols are beginning to offer a variety of media choices.
- Building automation protocols with origins in HVAC applications are branching into lighting, security, elevator control, and energy management, the last of which is the latest important field in building automation.
- Individualized controls allow the occupants of the building to adjust setpoints themselves, usually within an allowable range of values, to create their own optimal indoor environment.
- The next major step in the evolution of automation intelligence will supplement existing distributed systems with Internet-scale computing and advanced data collection.
- Data warehouses offer several advantages over trend logging from within a building automation system.
- The difficulty in retrofitting existing buildings requires prioritization of the possible measures for improving energy efficiency.

Definitions

- *Energy harvesting* is a strategy where a device obtains power from its surrounding environment.
- A *data warehouse* is a dedicated storage area for electronic data.

Review Questions

1. How does a coordination document help with integrating building systems together?

2. Why is it now important for information technology (IT) staff to be involved in planning building automation systems?

3. How have controls vendors developed ways to differentiate themselves?

4. What are the advantages of integrating a building automation system with an IP infrastructure?

5. Why have wireless networks not yet been significantly implemented in building automation applications?

6. How do perceptions of open protocols differ between industry participants?

7. How will the migration to IPv6 help the utilization of IP infrastructures for building automation applications?

8. What are the common features being added to existing open protocols?

9. How does an individualized controls strategy improve occupant comfort?

10. How is data warehouse information used to analyze building operations?

Appendix

Table of Contents

Numbering Systems ... 380
    Decimal Numbers ... 380
    Binary Numbers ... 381
    Octal Numbers ... 382
    Hexadecimal Numbers ... 382
    Numbering System Designations ... 382
    Numbering System Conversions ... 383

LonWorks Standard Network Variable Types (SNVTs) ... 385

LonWorks Standard Configuration Property Types (SCPTs) ... 386

BACnet Standard Objects ... 389

BACnet Standard Properties ... 389

NUMBERING SYSTEMS

A numbering system is a unique way to represent numerical quantities with a certain range of symbols. The same number can be represented in different ways by different numbering systems. **See Figure A-1.** Every numbering system is defined by a base that determines the total number of symbols used to represent quantities.

Numbering Systems

Decimal	Binary	Binary Coded Decimal	Octal	Hexadecimal
1	1	0001	1	1
2	10	0010	2	2
3	11	0011	3	3
4	100	0100	4	4
5	101	0101	5	5
6	110	0110	6	6
7	111	0111	7	7
8	1000	1000	10	8
9	1001	1001	11	9
10	1010	0001 0000	12	A
11	1011	0001 0001	13	B
12	1100	0001 0010	14	C
13	1101	0001 0011	15	D
14	1110	0001 0100	16	E
15	1111	0001 0101	17	F
16	10000	0001 0110	20	10
17	10001	0001 0111	21	11
18	10010	0001 1000	22	12
19	10011	0001 1001	23	13
20	10100	0010 0000	24	14
30	11110	0011 0000	36	1E
45	101101	0100 0101	55	2D
68	1000100	0110 1000	104	44
99	1100011	1001 1001	143	63
671	1010011111	0110 0111 0001	1237	29F
1876	11101010100	0001 1000 0111 0110	3524	754

Figure A-1

Decimal Numbers

The most common numbering system is decimal. A decimal number is a number expressed in a base of 10. The symbols used in this system are the digits 0, 1, 2, 3, 4, 5, 6, 7, 8, and 9.

A place value, or weight, is assigned to each position that a number greater than 9 holds. **See Figure A-2.** The weighted value of each position can be expressed as the base (10 in this case) raised to the power of n, the position. For example,

10 to the power of 2, or 10^2, is 100, and 10 to the power of 3, or 10^3, is 1000. For the decimal system, then, the position weights from right to left are 1, 10, 100, 1000, etc. Multiplying each digit by the weighted value of its position and then summing the results gives the value of the equivalent decimal number. For example, the number 9876 can be expressed as $(9 \times 1000) + (8 \times 100) + (7 \times 10) + (6 \times 1)$. This method is used with all bases to convert numbers to the decimal numbering system.

Decimal numbers are familiar from everyday use. However, in computing applications, the functions of processors and memory lend themselves better to arranging numbers in other numbering systems. Numbering systems common in computing applications include binary, octal, and hexadecimal.

Binary Numbers

A binary number is a number expressed in a base of 2. The symbols used in this system are 0 and 1. A bit is a binary digit, consisting of a 0 or a 1. Just as in the decimal system, the weighted value of each position is expressed as the base raised to the power of *n*, the position. **See Figure A-3.** For the binary numbering system, the weighted values of each position, from right to left, are 1, 2, 4, 8, 16, 32, 64, etc.

A binary number can be signed or unsigned. A signed binary number can be positive (no sign) or negative (–). The sign uses the left-most position of a binary number. The use of a signed binary number reduces the size of the largest decimal number represented. A 16-bit unsigned binary number is always positive and can represent a maximum decimal number of 65,535. A 16-bit signed binary number can represent a decimal range of –32,767 to +32,767.

Communications using binary numbers typically organize the information in groups of 4, 8, or 16 bits. **See Figure A-4.** Smaller numbers may not require all of the bits in a logical group to represent its value, but since a certain number of bits may be expected by the message receiver, leading zeroes are sometimes added. For example, the decimal number 18 can be expressed in binary as 10010, but if 8 bits were reserved in a communication protocol to represent this number, the transmitted signal would be 00010010.

Decimal Numbering System

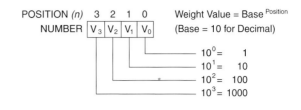

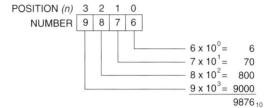

Figure A-2

Binary Numbering System

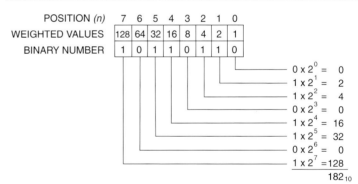

Figure A-3

Binary Numbers

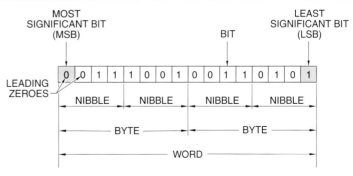

Figure A-4

A variation of the binary numbering system is binary coded decimal (BCD). This system uses groups of four bits to represent each decimal digit. For example, the decimal number 87 is represented as binary 1000 0111. **See Figure A-5.** The first group of four binary digits (1000) represents decimal 8 and the second group (0111) represents 7. Spaces are typically added between the 4-bit groups to indicate each digit. Because the 4-bit groups for each decimal digit must be preserved in this scheme, leading zeroes are included as needed to each group.

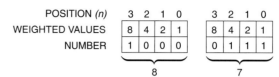

Figure A-5

It is interesting to compare a number in BCD to its representation in pure decimal. The binary number 87 is 1010111. BCD encoding requires more digits than traditional binary to represent the same number, but it is typically easier for people to convert to and from decimal digits.

While nearly all raw communication between computer-based devices is binary, the sequences for representing numbers can be very long and difficult for a person to interpret. This is not a common necessity, but sometimes troubleshooting communication problems can involve inspecting the raw message codes. In these cases, the information is often represented in octal or hexadecimal format, which shortens the sequences and makes the codes more recognizable.

Octal Numbers

An octal number is a number expressed in a base of 8. The symbols used in this system are the digits 0, 1, 2, 3, 4, 5, 6, and 7. The octal numbering system is often used to represent binary numbers using fewer digits, as each octal digit represents three bits in a binary system. According to the same position weighting scheme as other numbering systems, the weighted values of each position, from right to left, are 1, 8, 64, 512, etc. **See Figure A-6.**

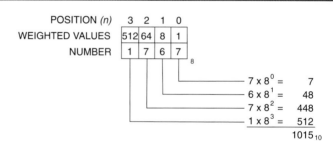

Figure A-6

Hexadecimal Numbers

A hexadecimal number is a number expressed in a base of 16. Since sixteen different symbols are required for this system, the digits 0 to 9 are insufficient. The letters A, B, C, D, E, and F are added for the numbers 10 through 15. The hexadecimal numbering system uses one digit to represent four bits in the binary numbering system. Just as in the decimal system, the weighted value of each position is expressed as the base raised to the power of n. **See Figure A-7.** The weighted values of each position, from right to left, are 1, 16, 256, 4096, etc.

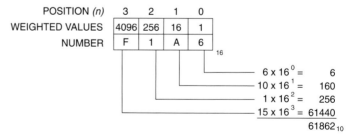

Figure A-7

Numbering System Designations

When there is the possibility of confusion between different numbering systems, subscripts or symbols are added to distinguish the base. These are not standardized, so there are several different designations that are commonly used.

See Figure A-8. The most common may depend on the industry or field using the designations.

Decimal is often assumed to be the default numbering system, unless the context specifically explains otherwise. When it is necessary to designate decimal numbers, such as when they are used in combination with numbers in other systems, decimal numbers often use "dec" as a prefix or suffix. They may instead use a subscript "10" after the number.

Binary numbers are commonly indicated with "bin" or "b" as a prefix or suffix, subscript "2" after the number, or the symbol "%." The "%" has become a common convention because the symbol resembles the numeral one with two zeroes.

Octal numbers are commonly indicated with an "oct" subscript or an "o" prefix, such as "47_{oct}" or "o47." However, the "o" can be confused with 0 (zero).

Hexadecimal numbers are commonly written with a "0x" or "h" prefix, "hex" subscript, or "#" or "h" suffix. The "0x" prefix is the most common.

| Common Numbering System Designations ||||
Decimal	Binary	Octal	Hexidecimal
37	0b100101	0o47	0x56A
37_{10}	100101_2	47_8	$56A_{16}$
37_{dec}	bin 100101	47_{oct}	$56A_{hex}$
	100101_b	o47	h56A
	%100101		56A#
			56Ah

Figure A-8

Numbering System Conversions

It is often necessary to convert numbers between decimal and other bases. Computer programs or calculators can be used to automatically perform conversions, but the processes are actually very simple and involve only multiplication and addition. It is important to understand the concepts of numbering system conversions so that they can be done manually if necessary.

Converting Numbers in Other Bases to Decimal. The same method can be used with all numbering systems to convert numbers to the decimal numbering system. The weighted value of each position is expressed as the base raised to the power of *n*, the position. The weighted value is multiplied by the digit in that position, and all the products are added together, according to the following formula:

$$N_{10} = \ldots + (d_{3,b} \times b^3) + (d_{2,b} \times b^2) + (d_{1,b} \times b^1) + (d_{0,b} \times b^0)$$

where:

N_{10} = number represented in decimal

$d_{n,b}$ = nth-position digit of number to be converted to decimal

b = base of number to be converted to decimal

The decimal equivalent of a binary number can be calculated by multiplying each bit by the position weighting. For example, the decimal equivalent of the binary number 10110110 is calculated with the following formula:

$$N_{10} = (1 \times 2^7) + (0 \times 2^6) + (1 \times 2^5) + (1 \times 2^4) + (0 \times 2^3) + (1 \times 2^2) + (1 \times 2^1) + (0 \times 2^0)$$

$$N_{10} = (1 \times 128) + (0 \times 64) + (1 \times 32) + (1 \times 16) + (0 \times 8) + (1 \times 4) + (1 \times 2) + (0 \times 1)$$

$$N_{10} = 128 + 32 + 16 + 4 + 2$$

$$N_{10} = \mathbf{182}$$

The decimal equivalent of an octal number can be calculated by multiplying each octal digit by the position weighting. For example, the decimal equivalent of the octal number 1767 is calculated with the following formula:

$$N_{10} = (1 \times 8^3) + (7 \times 8^2) + (6 \times 8^1) + (7 \times 8^0)$$

$$N_{10} = (1 \times 512) + (7 \times 64) + (6 \times 8) + (7 \times 1)$$

$$N_{10} = 512 + 448 + 48 + 7$$

$$N_{10} = \mathbf{1015}$$

The decimal equivalent of a hexadecimal number can be calculated by multiplying each hexadecimal digit by the position weighting. For example, the decimal equivalent of the hexadecimal number F1A6 is calculated with the following formula:

$N_{10} = (15 \times 16^3) + (1 \times 16^2) + (10 \times 16^1) + (6 \times 16^0)$

$N_{10} = (15 \times 4096) + (1 \times 256) + (10 \times 16) + (6 \times 1)$

$N_{10} = 61{,}440 + 256 + 160 + 6$

$N_{10} = \mathbf{61{,}862}$

Converting Numbers in Decimal to Other Bases. To convert a decimal number to its equivalent in any base, a series of divisions is performed. The conversion process starts by dividing the decimal number by the base. **See Figure A-9.** If there is a remainder, it is placed in the least significant digit (LSD) right-most position of the new base number. If there is no remainder, a 0 is placed in the LSD position. The result of the division is then brought down, and the process is repeated until the final result of the successive divisions is 0. This methodology is a bit cumbersome, but it is the easiest conversion method to understand and use.

For example, for the binary equivalent of the decimal number 35, the first step is to divide the 35 by 2. The result is 17 with a remainder of 1, which becomes the LSD. The next step is to divide the 17 by 2, resulting in 8 with a remainder of 1. This 1 is the next LSD. Then the 8 is divided by 2, getting 4 with a remainder of 0. This 0 is the next LSD. Next, divide the 4 by 2, getting 2 with a remainder of 0. This 0 is the next LSD. The next step is to divide the 2 by 2, getting 1 with a remainder of 0. This 0 is the next LSD. Lastly, the 1 is divided by 2, getting 0 with a remainder of 1. This is the final digit. Therefore, the number 35_{10} is equivalent to 100011_2.

The same method can be used to convert decimal numbers to any other base, except that the decimal number is divided by that base instead of by 2. For example, for the hexadecimal equivalent of 1355_{10}, it is divided by 16, getting 84 with a remainder of 11. Decimal 11 is equivalent to hexadecimal B. Therefore, 0xB is the LSD. The 84 is divided by 16, getting 5 with a remainder of 4. This is the next LSD. The next step is to divide 5 by 16, getting 0 with a remainder of 5. This is the last digit. Therefore, the number 1355_{10} is equivalent to 0x54B.

Decimal Conversion

Division	Remainder
35 ÷ 2 = 17	1
17 ÷ 2 = 8	1
8 ÷ 2 = 4	0
4 ÷ 2 = 2	0
2 ÷ 2 = 1	0
1 ÷ 2 = 0	1

$35_{10} = 100011_2$

Division	Remainder
1355 ÷ 16 = 84	11
84 ÷ 16 = 5	4
5 ÷ 16 = 0	5

$1355_{10} = 54B_{hex}$

Figure A-9

LonWorks Standard Network Variable Types (SNVTs)

SNVT	Index	SNVT	Index	SNVT	Index	SNVT	Index
SNVT_abs_humid	160	SNVT_elapsed_tm	87	SNVT_log_status	191	SNVT_sec_state	178
SNVT_address	114	SNVT_elec_kwh	13	SNVT_lux	79	SNVT_sec_status	179
SNVT_alarm	88	SNVT_elec_kwh_l	146	SNVT_magcard	86	SNVT_setting	117
SNVT_alarm_2	164	SNVT_elec_whr	14	SNVT_mass	23	SNVT_smo_obscur	129
SNVT_amp	1	SNVT_elec_whr_f	68	SNVT_mass_f	56	SNVT_sound_db	33
SNVT_amp_ac	139	SNVT_ent_opmode	168	SNVT_mass_kilo	24	SNVT_sound_db_f	61
SNVT_amp_f	48	SNVT_ent_state	169	SNVT_mass_mega	25	SNVT_speed	34
SNVT_amp_mil	2	SNVT_ent_status	170	SNVT_mass_mil	26	SNVT_speed_f	62
SNVT_angle	3	SNVT_enthalpy	153	SNVT_motor_state	155	SNVT_speed_mil	35
SNVT_angle_deg	104	SNVT_evap_state	118	SNVT_muldiv	91	SNVT_state	83
SNVT_angle_f	49	SNVT_ex_control	157	SNVT_multiplier	82	SNVT_state_64	165
SNVT_angle_vel	4	SNVT_file_pos	90	SNVT_multiplier_s	188	SNVT_str_asc	36
SNVT_angle_vel_f	50	SNVT_file_req	73	SNVT_nv_type	166	SNVT_str_int	37
SNVT_area	110	SNVT_file_status	74	SNVT_obj_request	92	SNVT_switch	95
SNVT_btu_f	67	SNVT_fire_indcte	133	SNVT_obj_status	93	SNVT_switch_2	189
SNVT_btu_kilo	5	SNVT_fire_init	132	SNVT_occupancy	109	SNVT_telcom	38
SNVT_btu_mega	6	SNVT_fire_test	130	SNVT_override	97	SNVT_temp	39
SNVT_char_ascii	7	SNVT_flow	15	SNVT_ph	125	SNVT_temp_diff_p	147
SNVT_chlr_status	127	SNVT_flow_dir	171	SNVT_ph_f	126	SNVT_temp_f	63
SNVT_clothes_w_a	187	SNVT_flow_f	53	SNVT_pos_ctrl	152	SNVT_temp_p	105
SNVT_clothes_w_c	184	SNVT_flow_mil	16	SNVT_power	27	SNVT_temp_ror	131
SNVT_clothes_w_m	185	SNVT_flow_p	161	SNVT_power_f	57	SNVT_temp_setpt	106
SNVT_clothes_w_s	186	SNVT_freq_f	75	SNVT_power_kilo	28	SNVT_therm_mode	119
SNVT_color	70	SNVT_freq_hz	76	SNVT_ppm	29	SNVT_time_f	64
SNVT_color_2	190	SNVT_freq_kilohz	77	SNVT_ppm_f	58	SNVT_time_hour	124
SNVT_config_src	69	SNVT_freq_milhz	78	SNVT_preset	94	SNVT_time_min	123
SNVT_count	8	SNVT_gfci_status	154	SNVT_press	30	SNVT_time_passed	40
SNVT_count_32	183	SNVT_grammage	71	SNVT_press_f	59	SNVT_time_sec	107
SNVT_count_f	51	SNVT_grammage_f	72	SNVT_press_p	113	SNVT_time_stamp	84
SNVT_count_inc	9	SNVT_hvac_emerg	103	SNVT_privacyzone	151	SNVT_time_stamp_p	192
SNVT_count_inc_f	52	SNVT_hvac_mode	108	SNVT_ptz	150	SNVT_time_zone	134
SNVT_ctrl_req	148	SNVT_hvac_overid	111	SNVT_pump_sensor	159	SNVT_tod_event	128
SNVT_ctrl_resp	149	SNVT_hvac_satsts	172	SNVT_pumpset_mn	156	SNVT_trans_table	96
SNVT_currency	89	SNVT_hvac_status	112	SNVT_pumpset_sn	158	SNVT_turbidity	143
SNVT_date_cal	10	SNVT_hvac_type	145	SNVT_pwr_fact	98	SNVT_turbidity_f	144
SNVT_date_day	11	SNVT_ISO_7811	80	SNVT_pwr_fact_f	99	SNVT_valve_mode	163
SNVT_date_event	176	SNVT_length	17	SNVT_rac_ctrl	181	SNVT_vol	41
SNVT_date_time	12	SNVT_length_f	54	SNVT_rac_req	182	SNVT_vol_f	65
SNVT_defr_mode	120	SNVT_length_kilo	18	SNVT_reg_val	136	SNVT_vol_kilo	42
SNVT_defr_state	122	SNVT_length_micr	19	SNVT_reg_val_ts	137	SNVT_vol_mil	43
SNVT_defr_term	121	SNVT_length_mil	20	SNVT_res	31	SNVT_volt	44
SNVT_density	100	SNVT_lev_cont	21	SNVT_res_f	60	SNVT_volt_ac	138
SNVT_density_f	101	SNVT_lev_cont_f	55	SNVT_res_kilo	32	SNVT_volt_dbmv	45
SNVT_dev_c_mode	162	SNVT_lev_disc	22	SNVT_rpm	102	SNVT_volt_f	66
SNVT_dev_fault	174	SNVT_lev_percent	81	SNVT_sblnd_state	180	SNVT_volt_kilo	46
SNVT_dev_maint	175	SNVT_log_fx_request	193	SNVT_scene	115	SNVT_volt_mil	47
SNVT_dev_status	173	SNVT_log_fx_status	194	SNVT_scene_cfg	116	SNVT_zerospan	85
SNVT_earth_pos	135	SNVT_log_request	195	SNVT_sched_val	177		

LonWorks Standard Configuration Property Types (SCPTs)...

SCPT	Index	SCPT	Index
SCPTactFbDly	1	SCPTdeltaNight	134
SCPTactuatorCharacteristic	284	SCPTdeviceControlMode	238
SCPTactuatorType	41	SCPTdeviceGroupID	172
SCPTahamApplianceModel	304	SCPTdevListDesc	322
SCPTairTemp1Alrm	132	SCPTdevListEntry	323
SCPTairTemp1Day	126	SCPTdevMajVer	165
SCPTairTemp1Night	131	SCPTdevMinVer	166
SCPTalrmClrT1	2	SCPTdial3tring	178
SCPTalrmClrT2	3	SCPTdiffNight	122
SCPTalrmIhbT	4	SCPTdiffTempSetpoint	201
SCPTalrmSetT1	5	SCPTdiffValue	130
SCPTalrmSetT2	6	SCPTdirection	44
SCPTareaDuctHeat	266	SCPTdischargeAirCoolingSetpoint	183
SCPTaudibleLevel	228	SCPTdischargeAirDewpointSetpoint	204
SCPTaudOutput	144	SCPTdischargeAirHeatingSetpoint	184
SCPTautoAnswer	177	SCPTdrainDelay	108
SCPTbaseValue	164	SCPTdriveT	8
SCPTblockProtectionTime	251	SCPTdriveTime	45
SCPTbrightness	230	SCPTductArea	46
SCPTbuildingStaticPressureSetpoint	193	SCPTductStaticPressureLimit	192
SCPTbuttonColor	312	SCPTductStaticPressureSetpoint	189
SCPTbuttonHoldAction	314	SCPTeffectivePeriod	272
SCPTbuttonPressAction	311	SCPTemergCnfg	258
SCPTbuttonRepeatInterval	313	SCPTemergencyPosition	250
SCPTbypassTime	34	SCPTenergyCntInit	137
SCPTclockCalibration	300	SCPTexhaustEnablePosition	202
SCPTclOffDelay	85	SCPTfadeTime	95
SCPTclOnDelay	86	SCPTfanDifferentialSetpoint	195
SCPTcombFlowCharacteristic	287	SCPTfanInEnable	328
SCPTcontrolPriority	171	SCPTfanOperation	260
SCPTcontrolSignal	245	SCPTfieldCalib	90
SCPTcontrolTemperatureWeighting	215	SCPTfireIndicate	153
SCPTcoolingLockout	209	SCPTfireInitType	38
SCPTcoolingResetEnable	211	SCPTfireTxt1	149
SCPTcoolLowerSP	76	SCPTfireTxt2	150
SCPTcoolSetpt	75	SCPTfireTxt3	151
SCPTcoolUpperSP	77	SCPTflashFreq	145
SCPTcutOutValue	125	SCPTfreeCoolPosition	247
SCPTdayDateIndex	103	SCPTgain	31
SCPTdayNightCntrl	121	SCPTgainVAV	66
SCPTdebounce	139	SCPTgainVAVHeat	268
SCPTdefaultAutoPanSpeed	176	SCPTheatingLockout	210
SCPTdefaultPanTiltZoomSpeeds	175	SCPTheatingResetEnable	212
SCPTdefaultSetting	297	SCPTheatLowerSP	79
SCPTdefaultState	295	SCPTheatSetpt	78
SCPTdefInput	305	SCPTheatUpperSP	80
SCPTdefltBehave	71	SCPThighLimDefrDly	133
SCPTdefOutput	7	SCPThighLimDly	124
SCPTdefrostCycles	219	SCPThighLimit1	9
SCPTdefrostDetect	225	SCPThighLimit1Enable	302
SCPTdefrostFanDelay	222	SCPThighLimit2	10
SCPTdefrostHold	224	SCPThighLimit2Enable	303
SCPTdefrostInternalSchedule	217	SCPThighLimTemp	123
SCPTdefrostMode	106	SCPTholdTime	91
SCPTdefrostRecoveryTime	223	SCPThumSetpt	36
SCPTdefrostStart	218	SCPThvacMode	74
SCPTdefScale	162	SCPThvacType	169
SCPTdefWeekMask	102	SCPThystHigh1	11
SCPTdelayTime	96	SCPThystHigh2	12

...LonWorks Standard Configuration Property Types (SCPTs)...

SCPT	Index	SCPT	Index
SCPThystLow1	13	SCPTmaxRemotePressureSetpoint	240
SCPThystLow2	14	SCPTmaxRemoteTempSetpoint	244
SCPTidentity	294	SCPTmaxReturnExhaustFanCapacity	187
SCPTifaceDesc	318	SCPTmaxRnge	20
SCPTinFbDly	15	SCPTmaxSendTime	49
SCPTinjDelay	109	SCPTmaxSetpoint	50
SCPTinstallDate	146	SCPTmaxSndT	22
SCPTinstalledLevel	232	SCPTmaxStroke	253
SCPTinvrtOut	16	SCPTmaxSupplyFanCapacity	185
SCPTlightingGroupEnable	342	SCPTminDefrostTime	220
SCPTlimitChlrCap	81	SCPTminDeltaAngl	43
SCPTlimitCO2	42	SCPTminDeltaCO2	63
SCPTlinkPowerDetectEnable	320	SCPTminDeltaFlow	47
SCPTlocation	17	SCPTminDeltaLevel	88
SCPTlogAlarmThreshold	339	SCPTminDeltaRH	62
SCPTlogCapacity	324	SCPTminDeltaTemp	64
SCPTlogFileHeader	338	SCPTminDischargeAirCoolingSetpoint	206
SCPTlogHighLimit	330	SCPTminDischargeAirHeatingSetpoint	208
SCPTlogLowLimit	331	SCPTminDuctStaticPressureSetpoint	191
SCPTlogMinDeltaTime	333	SCPTminFlow	54
SCPTlogMinDeltaValue	334	SCPTminFlowHeat	55
SCPTlogNotificationThreshold	325	SCPTminFlowHeatStby	263
SCPTlogRecord	337	SCPTminFlowSetpoint	236
SCPTlogRequest	340	SCPTminFlowStby	56
SCPTlogResponse	341	SCPTminFlowUnit	261
SCPTlogSize	326	SCPTminFlowUnitHeat	270
SCPTlogTimestampEnable	329	SCPTminFlowUnitStby	264
SCPTlogType	327	SCPTminOutdoorAirFlowSetpoint	198
SCPTlowLimDly	129	SCPTminPressureSetpoint	234
SCPTlowLimit1	18	SCPTminRemoteFlowSetpoint	241
SCPTlowLimit1Enable	298	SCPTminRemotePressureSetpoint	239
SCPTlowLimit2	19	SCPTminRemoteTempSetpoint	243
SCPTlowLimit2Enable	299	SCPTminReturnExhaustFanCapacity	188
SCPTlowLimTemp	128	SCPTminRnge	23
SCPTluxSetpoint	82	SCPTminSendTime	52
SCPTmaintDate	147	SCPTminSetpoint	53
SCPTmanfDate	148	SCPTminSndT	24
SCPTmanOvrTime	35	SCPTminStroke	252
SCPTmanualAllowed	101	SCPTminSupplyFanCapacity	186
SCPTmasterSlave	97	SCPTmixedAirLowLimitSetpoint	196
SCPTmaxCameraPrepositions	174	SCPTmixedAirTempSetpoint	197
SCPTmaxDefrostTime	221	SCPTmodeHrtBt	105
SCPTmaxDefrstTemp	110	SCPTmonInterval	319
SCPTmaxDefrstTime	107	SCPTname1	306
SCPTmaxDischargeAirCoolingSetpoint	205	SCPTname2	309
SCPTmaxDischargeAirHeatingSetpoint	207	SCPTname3	310
SCPTmaxDuctStaticPressureSetpoint	190	SCPTneuronId	301
SCPTmaxFanIn	332	SCPTnightPurgePosition	246
SCPTmaxFlow	51	SCPTnomAirFlow	57
SCPTmaxFlowHeat	37	SCPTnomAirFlowHeat	267
SCPTmaxFlowSetpoint	237	SCPTnomAngle	58
SCPTmaxFlowUnit	262	SCPTnomFreq	159
SCPTmaxNVLength	255	SCPTnomRPM	158
SCPTmaxOut	93	SCPTnormalRotationalSpeed	180
SCPTmaxPower	317	SCPTnumDampers	269
SCPTmaxPressureSetpoint	235	SCPTnumDigits	293
SCPTmaxPrivacyZones	173	SCPTnumValves	59
SCPTmaxRcvT	21	SCPTnvDynamicAssignment	256
SCPTmaxRcvTime	48	SCPTnvPriority	296
SCPTmaxRemoteFlowSetpoint	242	SCPTnvType	254

...LonWorks Standard Configuration Property Types (SCPTs)

SCPT	Index	SCPT	Index
SCPTnwrkCnfg	25	SCPTsensConstTmp	65
SCPTobjMajVer	167	SCPTsensConstVAV	67
SCPTobjMinVer	168	SCPTserialNumber	179
SCPToemType	61	SCPTsetPnts	60
SCPToffDely	30	SCPTsetpoint	213
SCPToffset	26	SCPTsluiceCnfg	259
SCPToffsetCO2	68	SCPTsmokeDayAlrmLim	40
SCPToffsetFlow	265	SCPTsmokeDayPreAlrmLim	138
SCPToffsetRH	69	SCPTsmokeNightAlrmLim	127
SCPToffsetTemp	70	SCPTsmokeNightPreAlrmLim	140
SCPTonOffHysteresis	84	SCPTsmokeNomSens	39
SCPTorientation	231	SCPTsndDelta	27
SCPToutdoorAirEnthalpySetpoint	200	SCPTsourceAddress	336
SCPToutdoorAirTempSetpoint	199	SCPTspaceHumSetpoint	203
SCPTovrBehave	32	SCPTstandbyRotationalSpeed	181
SCPTovrValue	33	SCPTstep	83
SCPTpartNumber	182	SCPTstepValue	92
SCPTpollRate	335	SCPTstrtupDelay	111
SCPTpowerupState	87	SCPTstrtupOpen	115
SCPTprimeVal	155	SCPTsummerTime	99
SCPTpulseValue	292	SCPTsuperHtRefInit	114
SCPTpumpCharacteristic	233	SCPTsuperHtRefMax	118
SCPTpumpDownDelay	113	SCPTsuperHtRefMin	116
SCPTpwmPeriod	216	SCPTtemperatureHysteresis	214
SCPTpwrSendOnDelta	315	SCPTtempOffset	227
SCPTpwrUpDelay	72	SCPTtermTimeTemp	112
SCPTpwrUpState	73	SCPTthermAlrmROR	142
SCPTrampDownTm	161	SCPTthermMode	120
SCPTrampUpTm	160	SCPTthermThreshold	152
SCPTreflection	89	SCPTtimeEvent	104
SCPTrefrigGlide	117	SCPTtimeout	170
SCPTrefrigType	119	SCPTtimePeriod	291
SCPTregName	163	SCPTtimeZone	154
SCPTreturnFanStaticPressureSetpoint	194	SCPTtrnsTblX	28
SCPTrunHrAlarm	136	SCPTtrnsTblX2	285
SCPTrunHrInit	135	SCPTtrnsTblX3	288
SCPTrunTimeAlarm	290	SCPTtrnsTblY	29
SCPTsafExtCnfg	257	SCPTtrnsTblY2	286
SCPTsaturationDelay	271	SCPTtrnsTblY3	289
SCPTscanTime	321	SCPTupdateRate	98
SCPTscene	307	SCPTvalueDefinition	276
SCPTsceneColor	343	SCPTvalueName	277
SCPTsceneName	316	SCPTvalveFlowCharacteristic	248
SCPTsceneNmbr	94	SCPTvalveKvs	282
SCPTsceneOffset	157	SCPTvalveNominalSize	281
SCPTsceneTiming	308	SCPTvalveOperatingMode	249
SCPTschedule	274	SCPTvalveStroke	280
SCPTscheduleDates	273	SCPTvalveType	283
SCPTscheduleInternal	226	SCPTvisOutput	143
SCPTscheduleName	279	SCPTweeklySchedule	278
SCPTscheduleTimeValue	275	SCPTwinterTime	100
SCPTscrollSpeed	229	SCPTzoneNum	141
SCPTsecondVal	156		

BACnet Standard Objects

Object	Identifier	Object	Identifier	Object	Identifier	Object	Identifier
Access Credential*	32	Averaging	18	File	10	Multi-State Value	19
Access Door	30	Binary Input	3	Global Group	26	Notification Class	15
Access Point*	33	Binary Output	4	Group	11	Program	16
Access Rights*	34	Binary Value	5	Life Safety Point	21	Pulse Converter	24
Access User*	35	Calendar	6	Life Safety Zone	22	Schedule	17
Access Zone*	36	Command	7	Lighting Output*	31	Structured View	29
Accumulator	23	Credential Data Input*	37	Load Control	28	Trend Log	20
Analog Input	0	Device	8	Loop	12	Trend Log Multiple	27
Analog Output	1	Event Enrollment	9	Multi-State Input	13		
Analog Value	2	Event Log	25	Multi-State Output	14		

* standard objects proposed but not yet approved, as of December 2008

BACnet Standard Properties . . .

Property	Identifier	Property	Identifier	Property	Identifier
accepted-modes	175	derivative-constant-units	27	log-enable	133
acked-transitions	0	description	28	log-interval	134
ack-required	1	description-of-halt	29	logging-object	183
action	2	device-address-binding	30	logging-record	184
action-text	3	device-type	31	low-limit	59
active-text	4	direct-reading	156	maintenance-required	158
active-vt-sessions	5	effective-period	32	manipulated-variable-reference	60
active-cov-subscriptions	152	elapsed-active-time	33	manual-slave-address-binding	170
adjust-value	176	error-limit	34	maximum-output	61
alarm-value	6	event-enable	35	maximum-value	135
alarm-values	7	event-state	36	maximum-value-timestamp	149
all	8	event-time-stamps	130	max-apdu-length-accepted	62
all-writes-successful	9	event-type	37	max-info-frames	63
apdu-segment-timeout	10	event-parameters	83	max-master	64
apdu-timeout	11	exception-schedule	38	max-pres-value	65
application-software-version	12	fault-values	39	max-segments-accepted	167
archive	13	feedback-value	40	member-of	159
attempted-samples	124	file-access-method	41	minimum-off-time	66
auto-slave-discovery	169	file-size	42	minimum-on-time	67
average-value	125	file-type	43	minimum-output	68
backup-failure-timeout	153	firmware-revision	44	minimum-value	136
bias	14	high-limit	45	minimum-value-timestamp	150
buffer-size	126	inactive-text	46	min-pres-value	69
change-of-state-count	15	in-process	47	mode	160
change-of-state-time	16	input-reference	181	model-name	70
client-cov-increment	127	instance-of	48	modification-date	71
configuration-files	154	integral-constant	49	notification-class	17
controlled-variable-reference	19	integral-constant-units	50	notification-threshold	137
controlled-variable-units	20	last-notify-record	173	notify-type	72
controlled-variable-value	21	last-restore-time	157	number-of-APDU-retries	73
count	177	life-safety-alarm-values	166	number-of-states	74
count-before-change	178	limit-enable	52	object-identifier	75
count-change-time	179	limit-monitoring-interval	182	object-list	76
cov-increment	22	list-of-group-members	53	object-name	77
cov-period	180	list-of-object-property-eferences	54	object-property-reference	78
cov-resubscription-interval	128	list-of-session-keys	55	object-type	79
database-revision	155	local-date	56	operation-expected	161
date-list	23	local-time	57	optional	80
daylight-savings-status	24	location	58	out-of-service	81
deadband	25	log-buffer	131	output-units	82
derivative-constant	26	log-device-object-property	132	polarity	84

... BACnet Standard Properties

Property	Identifier	Property	Identifier	Property	Identifier
prescale	185	record-count	141	time-of-active-time-reset	114
present-value	85	reliability	103	time-of-state-count-reset	115
priority	86	relinquish-default	104	time-synchronization-recipients	116
pulse-rate	186	required	105	total-record-count	145
priority-array	87	resolution	106	tracking-value	164
priority-for-writing	88	scale	187	units	117
process-identifier	89	scale-factor	188	update-interval	118
profile-name	168	schedule-default	174	update-time	189
program-change	90	segmentation-supported	107	utc-offset	119
program-location	91	setpoint	108	valid-samples	146
program-state	92	setpoint-reference	109	value-before-change	190
proportional-constant	93	slave-address-binding	171	value-set	191
proportional-constant-units	94	setting	162	value-change-time	192
protocol-object-types-supported	96	silenced	163	variance-value	151
protocol-revision	139	start-time	142	vendor-identifier	120
protocol-services-supported	97	state-text	110	vendor-name	121
protocol-version	98	status-flags	111	vt-classes-supported	122
read-only	99	stop-time	143	weekly-schedule	123
reason-for-halt	100	stop-when-full	144	window-interval	147
recipient-list	102	system-status	112	window-samples	148
records-since-notification	140	time-delay	113	zone-members	165

Glossary

A

acknowledged (ACK) message service: A LonWorks message service type where a sending node expects a response from the receiving node, confirming receipt of the message packet.

adaptive control algorithm: A control algorithm that automatically adjusts its response time based on environmental conditions.

adaptive start time control: A process that adjusts the actual start time for HVAC equipment based on the building temperature responses from previous days.

address space: The logical collection of all possible LAN addresses for a given MAC layer type.

ad hoc program design: A method of creating a new LonWorks network design while the network management tool is attached to the network.

alarm differential: The amount of change required in a variable for an alarm to return to normal after it has been in alarm status.

alarming: The detection and notification of abnormal building conditions.

alarm setpoint: The control point value that should trigger an alarm.

algorithm: A sequence of instructions for producing the optimal result to a problem.

algorithmic change reporting: A BACnet alarming feature that uses Event Enrollment objects to configure the alarm/event behavior of some other object property.

alias network variable: A software duplicate of a network variable that allows a different Selector ID to be used for the same network variable data.

alternate scheduling: The programming of more than one unique time schedule per year.

application-generic node: A control device that converts traditional control signals into network variable data.

application layer: The OSI Model layer that provides communication services between application programs.

application program: The software that implements the node's functionality.

application-specific node: A control device that performs dedicated control functions defined by the application program loaded by the manufacturer.

authenticated message service: A LonWorks message service type where the receiving node determines if the sending node is authorized to communicate with it.

automated discovery: A method of automatically locating and identifying the node and object components of a BACnet network.

averaging control: A control strategy that calculates an average value from multiple inputs, which is then used in control decisions.

B

BACnet broadcast management device (BBMD): A BACnet device that embeds broadcast messages in unicast messages (which can pass through IP routers) and send them to a peer device for further distribution.

BACnet interoperability building block (BIBB): A standardized name that is associated with some BACnet feature or capability.

BACnet/IP-to-BACnet/IP (BIP/BIP) router: A BACnet/IP router that routes between two or more BACnet/IP UDP ports.

BACnet stack: The portion of the application software that manages BACnet protocol communication over the LAN.

BACnet XML: A proposed XML-based model for conveying control information between BACnet and other protocols.

bandwidth: The maximum rate at which bits can be conveyed by a signaling method over a certain media type.

biasing: At a transceiver, the connection of network conductors to the reference voltage and ground through pull-up and pull-down resistors.

binding: A connection between a network variable input and a network variable output for the one-way sharing of dynamic control data.

bridge: A network device that joins two LANs at the data link and physical layers.

bridging router: A router that bridges non-routed messages at the data link layer.

broadcast: The transmission of a message intended for all nodes on the network.

broadcast addressing: A LonWorks addressing mode where all nodes on a subnet (for a subnet broadcast) or domain (for a domain broadcast) are identified as receivers of a network variable update.

bus topology: A linear arrangement of networked nodes with two specific endpoints.

C

calibration: The adjustment of control algorithm parameters to the optimal values for the desired control response.

channel: One or more contiguous network segments that do not span routers.

clamping: The restriction of lighting level between minimum and maximum levels, as long as the light is ON.

client application binding: The configured association between a BACnet application logical parameter and its actual source or destination object property.

closed-loop control system: A control system in which the result of an output is fed back into a controller as an input.

coaxial cable: A two-conductor cable in which one conductor runs along the central axis of the cable and the second conductor is formed by a braided wrap.

collapsed architecture: A protocol stack that does not include layers that are not needed for the application.

collision: The interaction of two messages on the same network media, which can cause data corruption and errors.

commandable property: A property that may be written with a priority on a scale from 1 (most important) to 16 (least important).

configuration property: An adjustable value that affects node, function block, or network variable behavior.

connection bleed: A situation where a node that is not a member of a broadcast group erroneously responds to a network variable update.

control logic: The portion of controller software that produces the necessary outputs based on the inputs.

control loop: The continuous repetition of the control logic decisions.

control point: A variable in a control system.

control strategy: A method for optimizing the control of building system equipment.

D

daily multiple time period scheduling: The programming of time-based control functions for atypical periods of building occupancy.

daisy chain: A wiring implementation of bus topology that connects each node to its neighbor on either side.

data link layer: The OSI Model layer that provides the rules for accessing the communication medium, uniquely identifying (addressing) each node, and detecting errors produced by electrical noise or other problems.

data trending: The recording of past building equipment operating information.

datatype: A specification for the type and format of information.

data warehouse: A dedicated storage area for electronic data.

derivative control algorithm: A control algorithm in which the output is determined by the instantaneous rate of change of a variable.

device commissioning: The process of assigning logical addresses and loading network program information onto installed nodes.

Device instance: The unique 22-bit code number of the Device object.

device management: See *node management*.

device profile: A collection of BIBBs that are required for the interoperable functionality of a general type of node.

duty cycling control: A supervisory control strategy that reduces electrical demand by turning OFF certain HVAC loads temporarily.

dynamic device binding: The automatic discovery of network numbers and MAC address bindings for corresponding BACnet Device instances.

dynamic object binding: The identification of nodes containing certain objects with the Who-Has and I-Have services.

E

electrical demand control: A supervisory control strategy designed to reduce a building's overall electrical demand.

energy harvesting: A strategy where a device obtains power from its surrounding environment.

engineered program design: A method of creating a new LonWorks program without the network management tool being attached to the network.

enterprise system: A software and networking solution for managing business operations.

estimation start time control: A process that calculates the actual start time for HVAC equipment based on building temperature data and a thermal recovery coefficient.

Extensible Markup Language (XML): A general-purpose specification for annotating text with information on structure and organization.

external interface file (XIF): A file on a LonWorks node that documents its Program ID, network variables, configuration properties, function blocks, and number of address table entries.

F

fading: The changing from one lighting level to another over a fixed period.

fan-in binding: A group of bindings where two or more function blocks send network variable updates to one receiving function block.

fan-out binding: A group of bindings where one function block sends network variable updates to two or more receiving function blocks.

feedback: The measurement of the results of a control action by a sensor.

fiber optics: A form of signaling based on light pulses to convey signals.

firewall: A router-type device that allows or blocks the passage of packets depending on a set of rules for restricting access.

flat network architecture: A network configuration where control devices are arranged in a peer-to-peer way.

foreign device registration: A technique that allows a BACnet/IP node to register with a central BBMD for the purpose of sending and receiving broadcast messages.

frame: A packet surrounded by additional data to facilitate its successful transmission and reception by delineating the start and end (or length) of the packet.

freely programmable node: A control device that is loaded with a custom control application.

free topology: An arrangement of nodes that does not require a specific structure and may include any combination of buses, stars, rings, and meshes.

full client: A non-local LNS network management tool that communicates with the LNS Server and database computer through the LonWorks network.

full-duplex communication: A system where data signals can flow in both directions simultaneously.

functional profile: A specification of a particular control function.

function block: A software object within a node that represents a certain control task.

G

gateway: A network device that translates transmitted information between different protocols.

gateway network architecture: A network configuration where a gateway is used to integrate separate control systems based on different protocols.

group addressing: A LonWorks addressing mode where a sending node uses a Group ID to send message packets to a certain set of nodes.

Group ID: A unique number that identifies the set of nodes that all must receive a certain network variable update.

H

half-duplex communication: A system where data signals can flow in both directions, but only one direction at a time.

heartbeat: A LonWorks network variable update that continuously repeats at a configured time interval.

hierarchical network architecture: A network configuration where control devices are arranged in a tiered network and have limited interaction with other control devices.

high-limit control: A control strategy that makes system adjustments necessary to maintain a control point below a certain value.

high/low signal select: A control strategy in which the building automation system selects the highest or lowest values from among multiple inputs for use in the control decisions.

high-priority load: A load that is important to the operation of a building and is shed last when demand goes up.

holiday and vacation scheduling: The programming of time-based control functions during holidays and vacations.

Hop Count: The number of segments that the message is allowed to traverse before reaching its destination.

host-based node: A LonWorks node that employs the Microprocessor Interface Program (MIP) to use a supplemental microprocessor.

hub: A network device that repeats messages from one port onto all of its other ports.

hunting: An oscillation of output resulting from feedback that changes from positive to negative.

I

independent client: An LNS network application that accesses the network only to perform monitoring and control of network variables.

integral control algorithm: A control algorithm in which the output is determined by the sum of the offset over time.

intermediary framework: A complete software and/or hardware solution for integrating multiple network-based protocol systems through a centralized interface.

internetwork: A network that involves the interaction between LANs through routers at the network layer.

interoperability: The capability of network devices from different manufacturers and systems to interact using a common communication network and language framework.

intrinsic reporting: An alarming feature of certain BACnet objects where the alarm or event originates from the object.

IP-only router: A router that only handles IP packets and discards or ignores any others.

L

latency: The time delay involved in the transmission of data on a network.

lead/lag control: A control strategy that alternates the operation of two or more similar pieces of equipment in the same system.

life safety control: A supervisory control strategy for life safety issues such as fire detection and suppression.

lightweight client: An LNS network management tool that communicates with the LNS Server computer directly over a TCP/IP network.

local area network (LAN): The infrastructure for data communication within a limited geographic region, such as a building or a portion of a building.

local client: An LNS network management tool that resides on the same computer as the LNS Server and database.

local operating network (LON): A network of intelligent devices sharing information using the standard communication protocol LonTalk®.

logical segment: A combination of multiple segments that are joined together with network devices that do not change the fundamental behavior of the LAN.

LonTalk®: The open protocol standard used in LonWorks control networks.

LonWorks technology: A platform developed by Echelon Corporation designed for networked control applications.

low-limit control: A control strategy that makes system adjustments necessary to maintain a control point above a certain value.

low-priority load: A load that is shed first for electrical demand control.

M

MAC address: A node's address that is based on the addressing scheme of the associated data link layer protocol.

MAC layer: A sublayer of the OSI Model that combines functions of the physical and data link layers to provide a complete interface to the communications medium.

mapping: The process of making an association between comparable concepts in a gateway.

master node: An MS/TP node that shares equal responsibility for administering the network by taking turns to use the network and control the access to the network by other nodes.

media type: The specification of the characteristics and/or arrangement of the physical conductors or electromagnetic frequencies used for digital communication.

mesh topology: An interconnected arrangement of networked nodes.

Microprocessor Interface Program (MIP): Software that transforms the Neuron chip into a co-processor and moves the upper two layers of the LonTalk® protocol, the presentation and application layers, off the Neuron chip and onto a more sophisticated external processor.

multicast: The transmission of a message intended for multiple nodes, which are all assigned to the same multicast group.

multicast binding: A group of two or more bindings involving a common function block.

N

network address translation (NAT): A technique that changes source and destination IP addresses, as well as TCP and UDP port numbers, of incoming and outgoing IP packets as they pass through a NAT-capable router.

network analyzer tool: A network tool that listens passively to network traffic, capturing message frames and saving them for later analysis.

network interface: A hardware device and/or software driver installed in the network management tool computer for communicating on the network media.

network layer: The OSI Model layer that provides for the interconnection of multiple LAN types (MAC layers) into a single internetwork.

network management platform: A software operating system that performs network management services and connectivity tasks utilized by network management tools.

network program: A collection of all node configurations and binding definitions within the control network.

network recovery: An LNS service used by network management tools to create an LNS network database from information recovered from nodes.

Network Service Device (NSD): A software object used by LNS to communicate with nodes and routers during network management tasks.

network tool: A software application that runs on a computer connected to a network and is used to make changes to the operation of the nodes on a network.

network variable: A basic unit of shared control information that conforms to a certain data type.

network variable input (nvi): The representation of data that a node can receive for use within its application.

network variable output (nvo): The representation of data that a node can send to one or more nodes on the network.

Neuron ID: A 48-bit serial number that is unique to a single Neuron chip.

node: A computer-based device that communicates with other similar devices on a shared network.

node management: The manual forcing of a node's function blocks into specific operating modes in order to send control data onto the network. Also known as device management.

O

object: An abstract container that organizes related information and makes it accessible to other nodes in a standard way.

object identifier: A 32-bit number that is unique to each BACnet object within a node.

object instance: A 22-bit code number in the object identifier that assigns a number to each individual object of a certain type within a node.

object-oriented modeling: The concept of organizing many different types of information into defined and structured units.

object type: A 10-bit code number in the object identifier that represents the kind of object.

octet: A sequence of 8 bits.

OFFNET: A LonWorks commissioning mode in which any network configuration changes are applied only to the LNS database.

offset: The difference between the value of a control point and its corresponding setpoint.

ONNET: A LonWorks commissioning mode in which network configuration changes are loaded onto the nodes at the time of commissioning.

open building information exchange (oBIX): An XML-based model for conveying control information between any building automation protocols, including enterprise and proprietary protocols.

open-loop control system: A control system in which decisions are made based only on the current state of the system and a model of how it should work.

open protocol: A standardized communications and network protocol that is published for use by any device manufacturer.

Open Systems Interconnection (OSI) Model: A standard description of the various layers of data communication commonly used in computer-based networks.

optimum start control: A supervisory control strategy in which the HVAC load is turned ON as late as possible to achieve the indoor environment setpoints by the beginning of building occupancy.

optimum stop control: A supervisory control strategy in which the HVAC load is turned OFF as early as possible to maintain the proper building space temperature until the end of building occupancy.

P

packet: A collection of data message information to be conveyed.

permanent holiday: A holiday that remains on the same date each year.

physical layer: The OSI Model layer that provides for signaling (the transmission of a stream of bits) over a communication channel.

plug-in: A third-party software add-on that works within a network management tool that is based on LNS to provide a user-friendly interface for adjusting configuration properties.

port: A virtual data connection used by nodes to exchange data directly with certain application programs on other nodes.

powerline signaling: A communications technology that encodes data onto the alternating current signals in existing power wiring.

predictive maintenance: The monitoring of wear conditions and equipment characteristics in comparison to a predetermined tolerance to predict possible malfunctions or failures.

presentation layer: The OSI Model layer that provides transformation of the syntax of the data exchanged between application layer entities.

preventive maintenance: Scheduled inspection and work that is required to maintain equipment in peak operating condition.

Program ID: A 64-bit number that provides specific details and unique identification for a LonWorks node.

property: One item of information about an object.

property identifier: A 32-bit code number that identifies a property in an object.

proportional control algorithm: A control algorithm in which the output is in direct response to the amount of offset in the system.

proprietary protocol: A communications and network protocol that is developed and used by only one device manufacturer.

protocol: A set of codes, message structures, signals, and procedures implemented in hardware and software that permits the exchange of information between nodes.

protocol analyzer: Software that provides detailed channel traffic statistics, packet logs, individual packet analysis, and other information about network traffic.

protocol data unit: The portion of the frame containing fields belonging to a certain OSI Model layer and the layers above it.

protocol stack: A combination of OSI layers and the specific protocols that perform the functions in each layer.

R

radio frequency signaling: A communications technology that encodes data onto carrier waves in the radio frequency range.

ramping: The changing from one lighting level to another at a fixed rate of change of level per second.

repeated (UNACK_RPT) message service: A LonWorks message service type where a sending node transmits a series of identical message packets.

repeater: A network device that amplifies and repeats the electrical signals, providing a simple way to extend the length of a segment.

request/response (REQ-RESP) message service: A LonWorks message service type where a sending node requests control data from a receiving node and requires a response message packet with the requested information.

reset control: A control strategy in which a primary setpoint is adjusted automatically as another value (the reset variable) changes.

restored load: A shed load that has been turned ON after electrical demand control.

resynchronization: An LNS network management function used to update the network program, node, configurations, and the network database so that they all reflect the same design.

ring topology: A closed-loop arrangement of networked nodes.

rotating priority load shedding: An electrical demand control strategy in which the order of loads to be shed is changed with each high electrical demand condition.

router: A network device that joins two or more LANs together at the network layer and manages the transmission of messages between them.

routing: The process of determining the path between LANs that is required to deliver a message.

S

scheduled control: A supervisory control strategy in which the date and time are used to determine the desired operation of a load or system.

schedule linking: The association of loads within the building automation system that are always used during the same time.

segment: A portion of a network where all of the nodes share common wiring.

segmentation: A protocol mechanism that controls the orderly transmission of large data in small pieces.

Selector ID: A unique identifying number that is shared by the network variables involved in a binding.

service: A formal procedure for BACnet nodes to make requests of other nodes.

service LED: An LED that indicates the node's operating status.

service pin: A node pushbutton that causes the node to transmit an identifying message onto the network.

session layer: The OSI Model layer that provides mechanisms to manage a long series of messages that constitute a dialog.

setback: The unoccupied heating or cooling setpoint.

setpoint: The desired value to be maintained by a system.

setpoint control: A control strategy that maintains a setpoint in the system.

setpoint schedule: A description of the amount a reset variable resets the primary setpoint.

seven-day scheduling: The programming of time-based control functions that are unique for each day of the week.

shed load: An electric load that has been turned OFF for electrical demand control.

shed table: A table that prioritizes the order in which electrical loads are turned OFF.

signal: The conveyance of information.

signaling: The use of electrical, optical, and radio frequency changes in order to convey data between two or more nodes.

simplex communication: A system where data signals can flow in only one direction.

slave node: An MS/TP node that cannot participate in token management or recovery.

slave proxy: An MS/TP master node that acts on behalf of slave nodes on its segment for dynamic device binding.

standard network variable type (SNVT): A LonWorks network variable with a format, structure, and intended use that is defined and publicly documented for use by any node manufacturer.

standard object type: An object type defined in the BACnet standard.

star topology: A radial arrangement of networked nodes.

static device binding: The manual association of network numbers and MAC addresses with BACnet Device instances.

stepping: The incremental increasing or decreasing of lighting level in fixed steps.

subnet/node addressing: A LonWorks addressing mode where a sending node identifies a single receiving node by its logical subnet/node address.

supervisory control strategy: A method for controlling certain overall functions of a building automation system.

switch: A network device that can forward messages selectively to one of its other ports based on the destination address.

system-wide controller: A control device that provides general-purpose network functions for multiple nodes.

T

temporary scheduling: The programming of time-based control functions for a one-time temporary schedule.

terminal: A human-machine interface (HMI) that serves as an input and display device for text-based communication with another computer-based device.

terminator: A resistor-capacitor circuit connected at one or more points on a communication network to absorb signals, avoiding signal reflections.

thermal recovery coefficient: The ratio of a temperature change to the length of time it takes to obtain that change.

throughput: The actual rate at which bits are transmitted over a certain media at a specific time.

timed override: A control function in which occupants temporarily change a zone from an UNOCCUPIED to OCCUPIED state.

topology: The shape of the wiring structure of a communications network.

tracing detail: A detailed display of the information contained within a BACnet message.

transceiver: A hardware component that provides the means for nodes to send and receive messages over a network.

transient holiday: A holiday that changes its date each year.

transport layer: The OSI Model layer that manages the end-to-end delivery of messages across multiple LAN types.

tunnel router: A type of BACnet router that bridges segments across IP boundaries by encapsulating the BACnet messages into UDP/IP packets.

turn-around binding: A binding where a function block sends network variable updates to itself.

twisted-pair cable: A multiple-conductor cable in which pairs of individually insulated conductors are twisted together.

U

unacknowledged (UNACK) message service: A LonWorks message service type where a sending node transmits message packets without expecting any subsequent action by the receiving node.

unicast binding: A single binding where one sending function block transmits network variable updates to one receiving function block.

user-defined network variable type (UNVT): A LonWorks network variable that is not standardized by LonMark and may be unique to individual node manufacturers.

X

XML: See *Extensible Markup Language (XML)*.

Index

Page numbers in italic refer to figures.

A

Access Control object, 292, *293*
access control system response, 340–341
AcknowledgeAlarm service, 277, *278*
acknowledged (ACK) message service, *148*, 148–149
adaptive control algorithm, *35*, 35–36
adaptive start time control, *39*, 39
address space, 61
ad hoc program design, 141–142, *142*
alarm differential, *43*, 43
alarm/event summary service, 280, *281*
alarming, 273–281, *274*
   AcknowledgeAlarm service, 277, *278*
   alarm/event summary service, 280, *281*
   algorithmic change reporting, *275*, 275–276
   Event Enrollment object, 275–276, *276*, 277
   Event Log object, 279–280, *280*
   EventNotification service, 277, *278*
   intrinsic reporting, 274, *275*
   Notification Class object, 278–279, *279*
alarm monitoring, 42–43, *43*
alarm setpoint, 42
algorithm, 30
   control algorithm, 31–36
      adaptive, *35*, 35–36
      derivative, *33*, 33–34
      integral, *33*, 33
      proportional, 32, *33*
      two-position, 31–32, *32*
algorithmic change reporting, *275*, 275–276
   Event Enrollment object, 275, *276*, 277
alias network variable, 145, *146*
alternate scheduling, 38
application-generic node, *134*, 134
application layer, 63
application program, 207
application-specific node, *133*, 133
architectures, 10–12. *See also* network architectures
   BACnet, 204–206, *206*, *218*, 218–219
      backbone, 204–206, *206*

flat, *12*, *13*
gateway, *13*
hierarchical, *11*
information architecture for LonWorks, 88–93
authenticated message service, *150*, 150
automated building systems, 12–25
   access control, 20–21, *21*, 357
   alarm monitoring, 42–43, *43*
   closed-loop control, *31*, 31
   control scenarios, 334–344
      access control system response, 340–341
      demand limiting, 342–344, *343*, *344*
      elevator system response, 341
      HVAC system response, 338–340, *339*, 343–344
      lighting system response, 336–338, *337*, 343, *344*
      plumbing system response, *340*, 340, *341*
      security system response, 340–341
   data trending, *43*, 43–44
   electrical systems, 13–15, 358
      demand limiting, *14*
      on-demand distributed generation, *15*
   elevators, 22, *23*, 357
   example, *330*, 330–334, *331*
      implementations, 332–334, *333*
      requirements, 330–332
   fire protection, 19, *20*, 358
   graphical interfaces, 44, *45*
   HVAC, 16–17, 358
      control devices, *18*
      smoke control modes, *18*
   lighting, 15–16, *17*, 358
      daylighting, *16*
   management, 42–44
   open-loop control, *30*, 30–31
   plumbing, 18–19, 358
      water heating, *19*
   preventive maintenance, 44, *45*
   roadmaps, 364
   security, 19–20, *21*, 358–359
   voice-data-video (VDV), 21–22, *23*, 359

automated discovery, 306, *307, 315*
AutomicReadFile service, 299–300, *300*
AutomicWriteFile service, 300, *301*
averaging control, *30*, 30

B

BACnet broadcast management device (BBMD), 240–241, *241*
BACnet interoperability building block (BIBB), 209
BACnet/IP-to-BACnet/IP (BIP/BIP) router, *243*, 243
BACnet objects, *248*, 248–258, *249*
BACnet services, 258–267
  CreateObject, 263
  DeleteObject, 263
  ListElement, 262–263, *263*
  ReadProperty, 259–261, *259, 260*
  ReadRange, 263, *264*
  remote device management, 264–267
    DeviceCommunicationControl, 264
    I-Am service, 266–267, *267*
    I-Have service, *267*, 267
    PrivateTransfer, *266*, 266
    ReinitializeDevice, *264*, 264
    TextMessage, *265*, 265
    TimeSynchronization, 264–265, *265*
    Who-Has service, *267*, 267
    Who-Is service, 266–267, *267*
    WriteProperty, *261*, 261–262
BACnet stack, 207–208, *208*
  basic, 248–251
    Analog object, 249–250, *250*
    Binary object, 250, *251*
    Device object, 249, *250*
    Multi-state object, 251, *252*
  scheduling, 282–284
    Calendar object, *282*, 282–283
    Schedule object, *283*, 283–284
  special function, 251–258
    Accumulator object, 251–252, *252, 253*
    Averaging object, *253*, 253, *254*
    Command object, 253–254, *254*
    Global Group object, *255*, 255
    Group object, *255*, 255
    Loop object, *256*, 256
    Program object, 256–257, *257, 258*
    Pulse Converter object, *252*, 252, *253*
    Structured View object, 257–258, *258*
  trending, 284–286, *285, 286*
BACnet systems, 8–9, *192*, 192–211
  alarming, 273–280, *274*
    AcknowledgeAlarm service, 277, *278*
    alarm/event summary service, 280, *281*
    algorithmic change reporting, *275*, 275–276
    Event Enrollment object, 275–276, *276*, 277
    Event Log object, 279–280, *280*
    EventNotification service, 277, *278*
    intrinsic reporting, 274, *275*
    Notification Class object, 278–279, *279*
  architecture, 204–208
    BACnet nodes, 206–208, *207*
    network infrastructure, 204–206, *205, 206*
    network tools, 208, *209*
  change-of-value notification, *272*, 272–273
    COVNotification service, 272–273, *273*
    SubscribeCOV service, 272, *273*
  configuration, 313–321
    with BBMDs, 319–320, *320*
    bursting avoidance, 321, *322*
    client application binding, 320–321
    device instance assignment, 314–316, *315*
    MAC address assignment, *314*, 314
    MS/TP, 316–318, *317, 318*
    object discovery, 316, *317*
    router, 318, *319*
  development, 192–193, *193*
  features, 194–196
  information architecture, 197–203
    objects, 197–202, *198, 199, 200*
    services, 202–203, *203*
  installation, 310–313
    bench testing, *312*, 312
    planning and preparation, 310–312, *311, 312*
    service tool access point, 313
    wiring infrastructure, 312–313, *313*
  LAN types, *219*, 219–227
    ARCNET, *225*, 225–226, *226*
    BACnet/IP, 226–227, *227*
    Ethernet, 223–225, *224, 225*
    LonTalk foreign frame, 227
    Master-Slave/Token-Passing (MS/TP), 219–223, *220, 221, 222*
    point-to-point (PTP), 226
  limitations, 196
  methodology, 193–194
    application language, *194*
    message transportation, *195*
  misconceptions, 197
  MS/TP nodes and token passing, 227–232, *228*
    slave node, 232, *233*
    slave proxy, 232, *233*
    timing, 230–232, *231, 232*
    token passing, 230
    token recovery, 229–230, *231*
  network layer, 233–237, *234*
    messages, *237*, 237

network headers in transit, 236–237, *237*
network protocol header, 233–236, *234, 235, 236*
network tools, *306,* 306–310, *307*
installation and configuration, 306–307, *308*
operation, 307–309, *308, 309, 310*
troubleshooting, 310, *311*
over IP infrastructures, 238–243
broadcasting considerations, 240–242, *241*
network address translation (NAT), 242–243, *243*
routing considerations, *238,* 238–240, *239, 240*
physical architecture, *218,* 218–219
isolation, *224*
scheduling, 282–284
Calendar object, *282,* 282–283
Schedule object, *283,* 283–284
special application, 292–302
Access Control object, *292, 293*
AutomicReadFile service, 299–300, *300*
AutomicWriteFile service, 300, *301*
File object, *299,* 299–300
Life Safety object, 297–299, *298*
Lighting Output object, 292–297, *294, 295, 296*
Load Control object, *297,* 297
virtual terminal service, 300–302, *301, 302*
testing and certification, 208–211
BTL mark, *209*
device profile, 211, *212*
interoperability criteria, 209–211, *210, 211*
trending, 284–286
data sampling, 284–285, *285*
Trend Log object, 285–286, *286*
troubleshooting, 321–326, *323*
communication time outs, 324
excessive Who-Is and I-Am traffic, 324
missing alarms, 324–325
slow communication, 323–324
unchanging commandable properties, 326
unreadable objects, 325
unresponsive BACnet/IP nodes, 324
unresponsive MS/TP nodes, 322–323
window network tools, 325–326
BACnet Testing Laboratory (BTL), 208–209
BTL mark, *209*
BACnet XML, 353
bandwidth, 53, 54
effects, *53*
Basic Reference Model. *See* Open Systems Interconnection (OSI) Model
biasing, *221,* 221–222, *222, 223*
BIBB, 209
binding, 91
BIP/BIP router, *243,* 243
bridge, *67,* 67

bridging router, *238,* 238
broadcast, 56
broadcast addressing, 147, *148*
building automation systems. *See* automated building systems
building automation trends, 364–375
automating existing buildings, 375–376
control strategy, 372–375
industry, 364–366
controls vendors, 365–366
global differences in control incentives, 366
integrated building teams, 364–365
networking, 366–368
utilizing IP infrastructure, 366–367
wireless networking technologies, 367–368
open protocol, 368–372
BACnet system, 370
features added to, 371–372
IP addressing changes, 369
LonWorks system, 370
perceptions, 368–369
wireless networking protocol, 370
building system management, 42–44
alarm monitoring, 42–43, *43*
data trending, *43,* 43–44
graphical interfaces, 44, *45*
preventive maintenance, 44, *45*
building systems. *See* automated building systems
bus topology, *70,* 70–71

C

Calendar object, *282,* 282–283
calibration, 34–35, *35*
channel, 93
capacities, *174*
channel types (for LonWorks), 108–112, *109*
codes, *126*
FO-20 channels, 112
IP-852, 111
PL-20x, 111
TP/FT-10, 109–110, *111*
TP-RS485-39, 111
TP/XF-78, 111
TP/XF-1250, 110
clamping, *293,* 293
client application binding, 320–321
closed-loop control system, *31,* 31
coaxial cable, 73
collapsed architectures, 64–65, *65*
collision, 57, *58*
commandable property, 261–262, *262*
communication, *4,* 4–12. *See also* data communications

interoperability, 9–10
network architectures, 10–12, *11, 12, 13*
network messages, *5*
open protocol, 5–9, *6, 7, 8, 9*
proprietary protocol, 5, *6*
configuration property, 91, *151,* 151–154
plug-in, *131, 152,* 152
connection bleed, 147
control algorithm, 31–36
adaptive, *35,* 35–36
derivative, *33,* 33–34
integral, *33,* 33
proportional, 32, *33*
two-position, 31–32, *32*
control logic, 30–36
adaptive control algorithms, *35,* 35–36
calibration, 34–35, *35*
closed-loop control, *31,* 31
hunting, *35,* 35
open-loop control, *30,* 30–31
tuning, 34–35, *35*
control loop, 30
control point, 28
control scenarios, 333–344
access control system response, 340–341
demand limiting, 342–344, *343, 344*
elevator system response, 341
HVAC system response, 338–340, *339,* 343–344, *344*
lighting system response, 336–338, *337,* 343, *344*
plumbing system response, *340,* 340, *341*
security system response, 340–341
control strategy, 28–30
averaging control, *30,* 30
data warehouse, *374,* 374
high/low signal select, 29–30
integrated furniture comfort control system, *373*
lead/lag control, *29,* 29
low-limit/high-limit control, 28–29, *29*
reset control, 28
setpoint control, *28,* 28
setup/setback control, 28, *29*
underfloor air distribution system, *373*
COVNotification service, 272–273, *273*
cross-protocol implementation, 356–359, *357*
opening building—scenario, 357–359
system description, 356–357
cross-protocol integration, *348,* 348–355
enterprise system, 353–355, *354, 355, 356*
extensible markup language (XML), 350–353, *351,* 352
gateways, *349,* 349–350, *354, 355*
information translation, *348,* 348
intermediary framework, *350,* 350

D

daily multiple time period scheduling, 37
daisy chain, 70, *313*
data communications, *52,* 52–58
media access, 57–58, *58*
token passing, *59*
message frames, *55,* 55–57, *58*
protocols, 52
signaling, 52–55, *53*
data link layer, 60–61
local area network (LAN), 61, *62*
MAC address, 61
MAC layer, *61,* 61
data trending, *43,* 43–44
datatype, 199
data warehouse, *374,* 374
daylighting, *16*
deadband, *32,* 32, 35
demand limiting, *14,* 14, 342–344, *343, 344*
derivative control algorithms, *33,* 33–34
PID control, *34*
device commissioning, 154–157
commissioning modes, 156
Neuron ID identification, 154–155, *155*
node configuration, 156
router commissioning, 156, *157*
subnet/node address assignment, 155–156
wink command, *155,* 155
DeviceCommunicationControl service, *264,* 264
device instance, 201
device management, 164, *165*
services, 264–267
DeviceCommunicationControl, *264,* 264
I-Am, 266–267, *267*
I-Have, *267,* 267
PrivateTransfer, *266,* 266
ReinitializeDevice, *264,* 264
TextMessage, *265,* 265
TimeSynchronization, 264–265, *265*
Who-Has, *267,* 267
Who-Is, 266–267, *267*
WriteProperty, *261,* 261–262
device profile, 211, *212*
duty cycling control, 39–40, *40*
dynamic device binding, 266

E

electrical demand control, *40,* 40–42
electrical demand targets, 42
rotating priority load shedding, *41,* 41
shedding strategies, 41–42
shed table, 40

targets, *42*
electrical systems, 13–15
elevator system response, 341
endianness, 57
energy harvesting, 367
engineered program design, 142, *143*
enterprise systems, 353–355, *354, 355*
   integration, 375
estimation start time control, 39
Event Enrollment object, 275, *276, 277*
Event Log object, 279–280, *280*
EventNotification service, 277, *278*
Extensible Markup Language (XML), 350–353, *351, 352*
external interface files (XIF), 131, *132*

F

fading, *295*, 295–297
fan-in binding, 144, *145*
fan-out binding, 144, *145*
feedback, 30
fiber optics, *75*, 75
File object, *299*, 299–300
firewall, 69
flat network architecture, *12*, 12
foreign device registration, 241, *242*
frame, *55*, 55
   fields, 56
freely programmable node, 134, *135*
free topology, 72, *73*
full client, 105, *106*, 107
full-duplex communication, *54*, 55
functional profile, 90, *127*
function block, 90–91, *91, 128*

G

gateway, *69*, 69–70, *349*, 349–350, 354, *355*
gateway network architecture, 12, *13*
group addressing, 147
Group ID, 147

H

half-duplex communication, *54*, 54
heartbeat, 152–153, *153, 167*
hierarchical network architecture, *11*, 11–12
high-limit control, *29*, 29
high/low signal select, 29–30
high-priority load, 40
holiday scheduling, 37
Hop Count, 236
host-based node, 121

hub, *67*, 67
human machine interface (HMI) traffic, 174
hunting, *35*, 35
HVAC system response, 338–340, *339*, 343–344, *344*

I

I-Am service, 266–267, *267*
I-Have service, *267*, 267
independent client, 107, *108*
individualized controls, 372, *373*
integral control algorithm, *33*, 33
intermediary framework, *350*, 350
International Organization for Standardization (ISO), 58, 192
internetwork, *68*, 68
interoperability, 9–10
   building system integration, 10, *11*
   manufacturer and vendor integration, 10
intrinsic reporting, 274, *275*
IP-only router, *238*, 238
ISO, 58, 192
isolation, 223, *224*, 313

L

LAN, 61, *62*
   and routers, 68–69
   types, *204, 205, 219*, 219–227
latency, 54
lead/lag control, *29*, 29
life safety control, 36
Life Safety object, 297–299, *298*
Lighting Output object, 292–297, *294, 295, 296*
lighting system response, 336–338, *337*, 343, *344*
lightweight client, *107*, 107
LNS, 102–104, *103*
Load Control object, *297*, 297
load shedding, *14*, 14
local area network (LAN), 61, *62*
   and routers, 68–69
   types, 204, *205, 219*, 219–227
local client, 105, *106*
local operating network (LON), 86
logical segment, 66
LON®, 86
LonMark certification, *120*, 120
LonTalk®, 86, 112
LonWorks, 86–96, *337, 339, 340, 342, 343, 344*
   development, 86–87
     LonMark® International, 86, *87*
     standards, 86–87
   device commissioning, 154–157

commissioning modes, 156
Neuron ID identification, 154–155, *155*
node configuration, 156
router commissioning, 156, *157*
subnet/node address assignment, 155–156
wink command, *155*, 155
infrastructure planning, 112–115, *113, 333*
node installation, 114–115
node power supplies, *115*, 115
twisted-pair cabling installation, 112–114, *114*
network architectures, 102–108
client functionality, *108*, 108
client/server, 104–107
full client, *106*
independent client, *108*
lightweight client, *107*
local client, *106*
LonWorks Network Services (LNS), 102–104, *103*
network interface, *104*, 105
Object Server, *103*
network documentation, 186–187, *187*
node health statistics, 187, *188*
traffic reports, 187
network infrastructure, 108–112
channel types, 108–112, *109*
large networks, *109*
LonTalk repeaters, 112
LonTalk routers, 112
network maintenance, 180–186
network design backups, *183*, 183
network management tool relocation, *183*, 183
network recovery, *186*, 186
network resynchronization, *185*, 185
node additions, 180–181
node application updates, *182*, 182–183
node relocation, *181*, 181–182, *182*
node replacement, *180*, 180
remote client connections, *184*, 184–185
network performance optimization, 173–176, *174*
binding properties, 176
high-speed backbone channels, 176
human machine interface (HMI) traffic, 174
network variable update traffic, 174–176, *175*
network programming, *140*, 140–142
management tools, *140*, 140–141
program design, 141–142, *142, 143*
network testing, 164–173
control sequence operation, 164–167, *165, 166, 167, 168*
network infrastructure integrity, *170*, 169–170, *171, 172*
node error statistics, 168–169, *169*
protocol analyzer, 171–173, *172, 173*
network variable bindings, 143–154, *144*
addressing modes, *146*, 146–147, *148*
arrangements, 144

channel priority, 150–151, *151*
configuration properties, *151*, 151–154, *152, 153*
message service types, 147–150, *148, 149, 150*
Selector ID, 144–145, *145*
Neuron chip, *86, 121*
node hardware components, *120*, 120–124
external memory, 122
input/output connections, 123
microprocessors, 120–122
power supplies, 123
service LED, *123*, 123–124, *124*
service pin, *123*, 123
transceivers, *122*, 122
node software components, 124–133, *125*
application programs, 124–126, *125, 126*
configuration properties, 130, *131*
external interface files (XIF), 131, *132*
function blocks, 126–127, *127, 128*
network management tool resource files, 131–133
network variables, 127–129, *129, 130*
node types, 133–134
application-generic, *134*, 134
application-specific, *133*
freely programmable, 134, *135*
system-wide controller, *134*, 134
LonWorks Network Services (LNS), 102–104, *103*
LonWorks systems, 7–8
communication, 7
Neuron chip, 7, 120–121
LonWorks technology, 86, 87–96, *88*
control decisions, *89*
information architecture, 88–93
binding, 91
configuration property, 91, *93*
function block, 90–91, *91, 92*
network program, 92–93, *93*
network variable, 88–90, *89*
network addressing, 96, *97*
network management platform, 95–96, *96*
system architecture, 93–95
network infrastructure, 93–94, *94*
network tools, 94–95, *95*
nodes, *92*, 94
low-limit control, 28–29, *29*
low-priority load, 40

M

MAC address, 61
MAC layer, *61*, 61
mapping, 70, 347
master node, 227, *228*
media types, 53, 73–77
copper conductors, 73–75, *74, 75*

fiber optics, *75,* 75
powerline, *77,* 77
radio frequency signaling, *76,* 76
mesh topology, 71–72, *72*
microprocessor interface program (MIP), 121
microprocessors, 120–122
multicast, 56
multicast binding, 144, *145*

N

network address translation (NAT), 242–243, *243*
network analyzer tool, 310, *311,* 322
network architectures, 10–12, 66–72, 102–108
   devices, 66–70
   flat, *13*
   gateway, *12*
   hierarchical, *11*
   for LonWorks, 102–108
   network tool, 72
   network topology, 70–72
   termination, 72
network interface, *104,* 104, *105*
network layer, 61–62
network management platform, 95–96
network program, 92–93, *93*
network recovery, *186,* 186
network service device (NSD), 103
network tool, 72
network variable, 88–90, *89,* 127–129, *129, 130*
   bindings, 143–154, *145*
network variable input (nvi), 143
network variable output (nvo), 143
Neuron chip, *86, 121*
Neuron ID, 122
   identification, 154–155, *155*
node, *52,* 52, *91, 92,* 94
   additions, 180–181
   application updates, *182,* 182–183
   BACnet, 206–208, *207,* 227, *228*
      BIP, *240*
   error statistics, 168–169, *169*
   health statistics, 187, *188*
   installation, 114–115
   LonMark certified, *120*
   MS/TP, 227–232
   power supplies, *115,* 115
   relocation, *181,* 181–182, *182*
   replacement, *180,* 180
   types, 133–134
      application-generic, *134,* 134
      application-specific, *133,* 133
      freely programmable, 134, *135*
      slave node, 232, *233*
      system-wide controller, *134,* 134
node management, 164, *165*
Node Object, 164, *165*
Notification Class object, 278–279, *279*
NSD, 103

O

oBIX, 350
object, 194, *198, 199. See also* BACnet objects
object identifier, 200–201, *201*
object instance, 201
object-oriented modeling, 197
object type, 200, *201*
octet, 55, *56*
OFFNET, 156–157
offset, 28
ONNET, 157
open building information exchange (oBIX), 352
open-loop control, *30,* 30–31
open protocol, 5–9, *6*
   BACnet systems, *8,* 8–9
   LonWorks systems, 7–8
   trends, 366–370
   wireless systems, 9
Open Systems Interconnection (OSI) Model, 58–65, *60,* 120
   analogy, 64
   application layer, 63
   collapsed architectures, 64–65, *65*
   data link layer, 60–61
   network layer, 61–62
   physical layer, 60
   presentation layer, 63
   protocol data unit, *65,* 65
   session layer, *63,* 63
   transport layer, 62
optimum start/stop control
   optimum start control, *38,* 38–39
   optimum stop control, 39
   thermal recovery coefficients, 39
OSI Model, 58–65, *60,* 120
   analogy, 64
   application layer, 63
   collapsed architectures, 64–65, *65*
   data link layer, 60–61
   network layer, 61–62
   physical layer, 60
   presentation layer, 63
   protocol data unit, *65,* 65
   session layer, *63,* 63
   transport layer, 62

P

packet, 55
permanent holiday, 37
physical layer, 60
plug-in, 130, *131*, *152*, 152
plumbing system response, *340*, 340, *341*
port, 56
powerline signaling, *77*, 77
predictive maintenance, 44
presentation layer, 63
preventive maintenance, 44
   software, *45*
PrivateTransfer service, *266*, 266
Program ID, *125*, 125–126
   channel type codes, *126*
   standard usage codes, *126*
property, 197, *198*, *199*
property identifier, 202
proportional control algorithm, 32, *33*
proprietary protocol, 5
   separate system interfaces, *6*
protocol, 4, 52
   terminology, 59
   translation, *70*
protocol analyzer, 171–173, *172*
   packet statistics and details, 172, *173*
protocol data units, *65*, 65
protocol stack, 59, *60*

R

radio frequency signaling, *76*, 76
ramping, *295*, 295–297
ReinitializeDevice service, *264*, 264
remote device management, 264–267
   DeviceCommunicationControl, *264*, 264
   I-Am, 266–267, *267*
   I-Have, *267*, 267
   PrivateTransfer, *266*, 266
   ReinitializeDevice, *264*, 264
   TextMessage, *265*, 265
   TimeSynchronization, 264–265, *265*
   Who-Has, *267*, 267
   Who-Is, 266–267, *267*
   WriteProperty, *261*, 261–262
repeated (UNACK_RPT) message service, 149
repeater, *66*, 66
request/response (REQ-RESP) message service, 149
reset control, 28
restored load, 40
resynchronization, *185*, 185
ring topology, *71*, 71
roadmaps, 362

rotating priority load shedding, *41*, 41
router, 68–69, *69*, *236*, *239*
   BIP/BIP, *243*, 243
   commissioning, 156, *157*
   configuration, 318, *319*
   LonTalk, 112
routing, 62, *63*

S

scheduled control, 36–38
   alternate scheduling, 38
   daily multiple time period scheduling, 37
   holiday scheduling, 37
   schedule linking, 38
   seven-day scheduling, *36*, 36
   temporary scheduling, 38
   timed override, *37*, 37
   vacation scheduling, 37
schedule linking, 38
Schedule object, *283*, 283–284
segment, *66*, 66
segmentation, *57*, *58*
Selector ID, 144–145, *145*
service, 192
service LED, *123*, 123–124, *124*
service pin, *123*, 123
session layer, *63*, 63
setpoint control, *28*, 28
setpoint schedule, 28
setup, 28
seven-day scheduling, *36*, 36
shed load, 40
shed table, 40
short cycling, 32
signal, 52
signaling, 52–55, *53*
   bandwidth, 53
   latency, 54
   throughput, 53
simplex communication, *54*, 54
slave node, 230
slave proxy, 230
SNVT, 89
software as service, 372–375, *374*, *375*
sound-masking technology, 22
standard network variable type (SNVT), 89
standard object type, 196
star topology, *71*, 71
static device binding, 266
stepping, *295*, 295–297
subnet/node addressing, 147
SubscribeCOV service, *272*, *273*
supervisory control, 36–42

adaptive start time, *39*
duty cycling control, 39–40, *40*
electrical demand control, *40*, 40–42, *42*
life safety control, 36
optimum start/stop control, *38*, 38–39
scheduled control, 36–38, *37*
supervisory control strategy, 36
switch, 67–68, *68*
systems. *See* automated building systems
system-wide controller, *134*, 134

T

temporary scheduling, 38
terminal, 300
terminator, 72
TextMessage service, *265*, 265
throughput, 53
timed override, *37*, 37
TimeSynchronization service, 264–265, *265*
tools
   network tools for BACnet, 306–310. *See also*
      BACNET: network tools
   troubleshooting, 310, *311*
topology, 70–72
   bus, *70*, 70–71
   free, 72, *73*
   mesh, 71–72, *72*
   ring, *71*, 71
   star, *71*, 71
tracing detail, 310
transceivers, *122*, 122
transient holiday, 37
transport layer, 62
trends
   automating existing buildings, 375–376
   control strategy, 372–375
   industry, 364–366
      controls vendors, 365–366
      global differences in control incentives, 366
      integrated building teams, 364–365
   networking, 366–368
      utilizing IP infrastructure, 366–367
      wireless networking technologies, 367–368
   open protocol, 368–372
      BACnet system, 370
      features added to, 371–372
      IP addressing changes, 369
      LonWorks system, 370
      perceptions, 368–369
      wireless networking protocol, 370
troubleshooting
   BACnet, 321–326, *323*
      communication time outs, 324
      excessive Who-Is and I-Am traffic, 324
      missing alarms, 324–325
      slow communication, 323–324
      tools, *306*, 306–310
      unchanging commandable properties, 326
      unreadable objects, 325
      unresponsive BACnet/IP nodes, 324
      unresponsive MS/TP nodes, 322–323
      window network tools, 325–326
   LonWorks network testing, 164–173
      control sequence operation, 164–167, *165, 166, 167, 168*
      network infrastructure integrity, 169–170, *170, 171*
      node error statistics, 168–169, *169*
      protocol analyzer, 171–173, *172, 173*
tuning, 34–35
   effects of increasing PID parameters, *35*
tunnel router, *239*, 239
turn-around binding, 144
twisted-pair cable, *73*, 73
   installation, 112–114
      lightning protection, *114*
      segment terminators, *114*
      shielded-cable grounding, *114*
   link-power networks, *111*
   TP/FT-10 channel, 109
two-position control algorithms, 31–32, *32*

U

unacknowledged (UNACK) message service, *149*, 149–150
unicast binding, *144*, 144
user-defined network variable type (UNVT), 90

V

vacation scheduling, 37
virtual terminal service, 300–302, *301, 302*

W

Who-Has service, *267*, 267
Who-Is service, 266–267, *267*
wink command, *155*, 155
wireless networking protocol, 368
wireless systems, *9*, 9
WriteProperty service, *261*, 261–262

X

XML, 348–351, *349, 350*

USING THE *BUILDING AUTOMATION: SYSTEM INTEGRATION WITH OPEN PROTOCOLS* CD-ROM

Before removing the CD-ROM from the protective sleeve, please note that the book cannot be returned for a refund or credit if the CD-ROM sleeve seal is broken.

System Requirements

To use this Windows®-compatible CD-ROM, your computer must meet the following minimum system requirements:
- Microsoft® Windows Vista™, Windows XP®, Windows 2000®, or Windows NT® operating system
- Intel® Pentium® III (or equivalent) processor
- 256 MB of available RAM
- 90 MB of available hard-disk space
- 800 × 600 monitor resolution
- CD-ROM drive
- Sound output capability and speakers
- Microsoft® Internet Explorer 5.5, Firefox 1.0, or Netscape® 7.1 web browser and Internet connection required for Internet links

Opening Files

Insert the CD-ROM into the computer CD-ROM drive. Within a few seconds, the home screen will be displayed allowing access to all features of the CD-ROM. Information about the usage of the CD-ROM can be accessed by clicking on USING THIS INTERACTIVE CD-ROM. The Quick Quizzes®, Illustrated Glossary, Flash Cards, Media Clips, and ATPeResources.com can be accessed by clicking on the appropriate button on the home screen. Clicking on the American Tech web site button (www.go2atp.com) accesses information on related educational products. Unauthorized reproduction of the material on this CD-ROM is strictly prohibited.

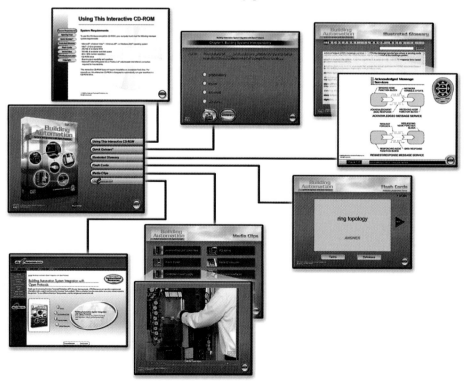

Intel and Pentium are registered trademarks of Intel Corporation or its subsidiaries in the United States and other countries. Microsoft, Windows Vista, Windows XP, Windows 2000, Windows NT, and Internet Explorer are either registered trademarks or trademarks of Microsoft Corporation in the United States and/or other countries. Adobe, Acrobat, and Reader are either registered trademarks or trademarks of Adobe Systems Incorporated in the United States and/or other countries. Netscape is a registered trademark of Netscape Communications Corporation in the United States and other countries. Quick Quiz and Quick Quizzes are registered trademarks of American Technical Publishers, Inc. All other trademarks are the properties of their respective owners.